Low Temperature Preservation in Medicine and Biology

'O, call back yesterday, bid time return'

William Shakespeare

from Richard II (Act III, Scene 2)

Low Temperature Preservation in Medicine and Biology

Edited by

M J Ashwood-Smith, BSc, MSc, PhD

Department of Biology, University of Victoria
British Columbia, Canada

J Farrant, B Pharm, PhD

Clinical Research Centre, Harrow,
Middlesex, United Kingdom

PITMAN MEDICAL

First published 1980

Catalogue Number 21.0021.81

Pitman Medical Limited
57 High Street, Tunbridge Wells, Kent

Associated Companies

Pitman Publishing Pty Ltd, Melbourne

Pitman Publishing NZ Ltd, Wellington

British Library Cataloguing in Publication Data

Low temperature preservation in medicine and biology.
1. Natural history
2. Cryobiology
I. Ashwood-Smith, M J
II. Farrant, J
574'.0724 QH61

ISBN: 0-272-79557 7

Set in 11 on 12 pt IBM Press Roman by
Gatehouse Wood Limited, Cowden
Printed by offset-lithography and bound
in Great Britain at The Pitman Press, Bath

Contents

Foreword

During the past 30 years there have been remarkable advances in low temperature biology. In particular, it has become possible to store a wide variety of living cells and tissues at very low temperatures in a state of suspended animation. After intervals varying from a few days to a quarter of a century they have been thawed with little loss of viability. Subsequently, they have resumed normal functions either *in vitro* or after restoration to their normal environment in the animal body. Several books have been written and many conferences held in which different techniques, results and interpretations have been recorded. Many of these techniques have involved the use of glycerol, dimethylsulphoxide, sugars, alcohols and other protective agents in the media, and various rates of cooling and rewarming.

The fundamental causes of damage during freezing and thawing and the principles of avoiding them have been hotly debated. As a result, the whole subject has become confused, so that newcomers to the field of banking living material at low temperatures feel confused and intimidated; often they reject the entire mass of apparently conflicting information and ideas and embark on a solitary voyage, sometimes doomed to disappointment. This new book should illuminate the situation. It has a strongly practical bias. The scientific principles which best fit observed facts and on which success or failure depend are clearly and simply stated.

The editors are valued ex-colleagues; both have approximately 20 years experience in the theory and practice of preserving living cells at low temperatures; their contributions are significant and up-to-date. The authors of successive chapters are all leaders in their chosen branches of biology and my old friend and ex-collaborator, Dr. C. Polge, was and is a pioneer. The preservation of mammalian embryos at low temperatures is fully described; it is one of the most exciting advances of recent years with important applications to animal husbandry, genetics and embryology. Haematologists and immunologists will be specially interested in the chapters on banking red blood cells and platelets, leucocytes and bone marrow stem cells. The sections on protozoa and helminth parasites and on insects and their cells will rivet the attention of medical and veterinary pathologists and agricultural scientists everywhere. The chapter on plant cells made me long to get back to a laboratory and make a fresh start.

I welcome this book and wish it a wide circulation. In particular, I hope that it will encourage a new generation of young biologists to bring their training, talents and imagination to bear on outstanding problems of life and death at low temperatures.

Audrey U. Smith
Mill Hill, August 1979

Editors' Introduction

The aim of this book is to provide practical help for those working in medicine and biology who need to preserve cells in the living state. It is particularly directed at those who are not working primarily in low temperature biology but who want to use preservation efficiently for application within their own different fields. We have asked workers with experience in the preservation of different biological systems each to contribute a chapter presenting concise information on both theoretical and practical aspects relevant to their system. The idea has been to illustrate the basis for practical preservation with examples that underscore the theoretical foundation upon which the practical techniques are based.

Some of the observed phenomena that are common to most biological systems are covered in Chapter 1. This book does not deal with the physical chemistry of low temperatures, hibernation and hypothermia or the destructive effects of freezing as used in cryosurgery. For physical chemistry we refer the reader to the recent and comprehensive book by Douzou and for cryosurgery to the companion volume to the present book by H. Holden (see *Book List*).

Both of the editors wish to express their gratitude to Dr Audrey Smith for writing the Foreword as she is one of the first and perhaps the greatest of low temperature biologists. Both editors worked in her laboratory at the National Institute for Medical Research, Mill Hill, London at the formative stage of their careers. The important publication in 1949 of the cryoprotective properties of glycerol (Chapter 1, reference 1) was written by Dr Audrey Smith together with Sir Alan Parkes, FRS and Dr C. Polge. We are indebted to Dr C Polge for the chapter on the preservation of spermatozoa (Chapter 3).

We hope that the topics covered in this book will be of general as well as specific interest. Blood and bone marrow are covered by Chapters 5, 6 and 7, spermatozoa and embryos by Chapters 3 and 4, bacteria and viruses by Chapter 10, insects by Chapter 9 and multicellular systems by Chapter 2. We feel that the two chapters on parasites (Chapter 8) and plant cells (Chapter 11) represent some of the first comprehensive and critical accounts in these fields of growing interest.

In conjunction with the individual authors Chapter 12 has been prepared which contains some useful information for practical procedures in low temperature preservation.

Many of the reports in low temperature biology and medicine appear in a very large number of scientific publications and this book attempts to collate some of the more recent information available. There are, however, a few journals which specialise in this field.

Journal List

(1) *Biodynamica* (now ceased publication)
(2) *Bulletin of the International Institute of Refrigeration*
(3) *Cryo letters*
(4) *Cryobiology*
(5) *Japanese Journal of Low Temperature Medicine*

General reference books in this field include the following.

Book List

Asahina, E. (ed.) (1967). *Cellular Injury and Resistance in Freezing Organisms.* Sapporo: The Institute of Low Temperature Science
Douzou, P. (1977). *Cryobiochemistry.* London: Academic Press
Elliott, K. and Whelan, J. (eds.) (1977). *The Freezing of Mammalian Embryos.* Ciba Foundation Symposium. Amsterdam: Elsevier
Fennema, O.R., Powrie, W.D. and Marth, E.H. (1973). *Low Temperature Preservation of Foods and Living Matter.* New York: M. Dekker
Holden, H.B. (ed.) (1975). *Cryosurgery.* Tunbridge Wells: Pitman Medical
Li, P.H. and Sakai, A. (eds.) (1978). *Plant Cold Hardiness and Freezing Stress. Mechanisms and Crop Implications.* London: Academic Press
Lozina-Lozinskii, L.K. (1974). *Studies in Cryobiology.* New York: John Wiley
Luyet, B.J. and Gehenio, P.M. (1940). *Life and Death at Low Temperatures.* Normandy, Missouri: Biodynamica
Meryman, H.T. (ed.) (1966). *Cryobiology.* New York: Academic Press
Rey, L. (ed.) (1964). *Researches and Development in Freeze-drying.* Paris: Hermann
Rose, A.H. (ed.) (1967). *Thermobiology.* New York: Academic Press
Smith, A.U. (1961). *Biological Effects of Freezing and Supercooling.* London: Edward Arnold
Smith, A.U. (ed.) (1970). *Current Trends in Cryobiology.* New York: Plenum
Wolstenholme, G.E.W. and O'Connor, M. (eds.) (1970). *The Frozen Cell.* Ciba Foundation Symposium. London: Churchill

There are several Learned Societies in this field.

Societies

Society for Cryobiology
Society for Low Temperature Biology
Japanese Society for Research of Freezing and Drying
Japanese Society for Low Temperature Medicine
Japanese Society for Frozen Semen Research
Japanese Society for Tissue and Organ Banking
International Institute for Refrigeration (Commission C)

Most of the contributors to this book are members of the Society for Low Temperature Biology. This society meets three times a year, usually twice in the UK and once in continental Europe. The meetings have fulfilled the hopes of the founding members in that informality prevails and that science is accompanied by good food and wine. Much of the research of the authors contributing to this book has been presented to the Society for Low Temperature Biology where, we hope, argument and fellowship have made good science in an area which is noticeable for both its intrinsic interest and its beneficial application to medicine, biology and agriculture.

Acknowledgements

The editors would like to thank Mrs Alexa Kennedy, Mrs Carol Warby and Miss Heather Lee for help in the preparation of the manuscript. We also thank Miss Maeve O'Connor and Mrs Julie Whelan of the Ciba Foundation for helpful advice.

List of Contributors

M.J. Ashwood-Smith, BSc, MSc, PhD
Senior NATO Research Fellow (1977–78), Department of Biology, University of Victoria, Victoria, British Columbia, Canada

J. Farrant, B Pharm, PhD
Clinical Research Centre, Watford Road, Harrow, Middlesex, HA1 3UJ, United Kingdom

E.R. James, BSc, MSc, PhD
Winches Farm Field Station, London School of Hygiene and Tropical Medicine, 395 Hatfield Road, St. Albans, Hertfordshire, United Kingdom

Stella C. Knight, BSc, PhD
Clinical Research Centre, Watford Road, Harrow, Middlesex, HA1 3UJ, United Kingdom

Leslie L. Lenny, MSc
Lindsley F. Kimball Research Institute, New York Blood Center, 310 East 67th Street, New York, NY 10021, USA

P. Mannoni, MD
Centre départmental de Transfusion Sanguine, Du Val-de-Marne, Hôpital Henri Mondor F94010, Creteil, France

G.J. Morris, BSc, PhD
Culture Centre of Algae and Protozoa, 36 Storey's Way, Cambridge, United Kingdom

C. Polge, BSc, PhD
ARC Institute of Animal Physiology, Animal Research Station, 307 Huntingdon Road, Cambridge, United Kingdom

R.A. Ring, BSc, PhD
Department of Biology, University of Victoria, Victoria, British Columbia, Canada

A.W. Rowe, PhD
Lindsley F. Kimball Research Institute, New York Blood Center, 310 East 67th Street, New York, NY 10021, USA

U.W. Schaefer, MD
Universitätsklinikum Essen, Innere Klinik und Poliklinik (Tumorforschung), Hufelandstrasse 55, D - 43 Essen 1, Germany

D.G. Whittingham, MA, PhD, FRCVS
MRC Mammalian Development Unit, Wolfson House, (University College London), 4 Stephenson Way, London NW1 2HE, United Kingdom

CHAPTER ONE

General Observations on Cell Preservation

J Farrant

Introduction

The reverberations of the epoch-making discovery made by Polge, Smith and Parkes in 1949 [1] that glycerol can be used to protect living cells from injury due to freezing are still with us. Workers in many medical and biological fields are increasingly using stored cells as a matter of routine. The unique ability of the preserver of cells is that of slowing or even stopping that otherwise inexorable phenomenon, time.

To those who wish to practise low temperature techniques and use stored cells in their own field of biology or medicine without themselves becoming full time low temperature biologists there are two main hazards. The first is the trusting application of a published recipe for some preservation procedure which may turn out to be inappropriate for the slightly different conditions of the particular problem in hand. The second danger becomes apparent when an attempt is made to delve into the literature of low temperature biology so that a recipe can be adapted for a different system. The frustrated seeker after a practical low temperature technique may then become enmeshed in the conflicting theories of mechanisms of injury and protection, few of which stand on solid ground.

In making a general survey of the principles underlying cell preservation, this chapter is intended to describe the solid ground of observed phenomena that have a general application. Possible hypotheses to explain the data will be discussed but the description that will be attempted is that based on general observed phenomena. These are illustrated by examples taken (where appropriate) from the personal experiences of the author. Data specific to different biological systems will be given in succeeding chapters. This chapter will be restricted to observations that are general and that apply to many or several biological systems.

Damage by cooling in the absence of freezing

Most cell types can be cooled rapidly or slowly in the absence of freezing (for example from +37°C to 0°C) without injury. However, there is a diverse collection of cell types that are injured on cooling, particularly when this is done rapidly. These systems include most cells from the pig, in particular spermatozoa and embryos [2], to a certain extent most mammalian spermatozoa [3] and some Gram-negative bacteria [4]. This injury is often termed cold shock or thermal shock and appears to be related to the rapidity of cooling (see intrinsic cold shock, Figure 1.1). It is distinct from the injuries produced by keeping

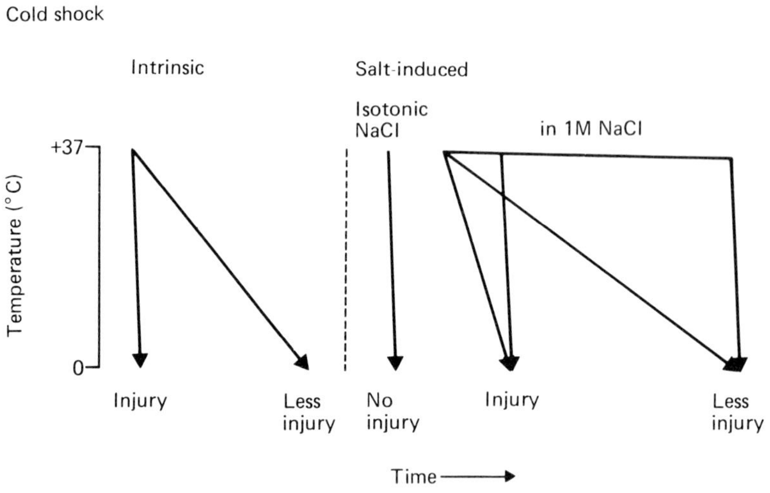

Figure 1.1 Diagram of conditions leading to injury on cooling above 0°C without freezing. With some cell types, e.g. some spermatozoa and bacteria, injury occurs particularly on rapid cooling. With red cells, NaCl (1 M) induces sensitivity to cold injury particularly following short exposure

cells at a low temperature (e.g. 0°C) when ionic imbalances may occur because of the disproportionate reduction in active ion pumps and other processes in comparison with passive ionic diffusion. Ways of avoiding cold shock include slow rather than rapid cooling and exposure before cooling either to a cryoprotective additive such as glycerol and dimethylsulphoxide (DMSO) [5] or to a specific phospholipid (e.g. phosphotidyl serine) [6]. This last finding indicates that the lesion in cold shock is located in the membrane. This is supported by the fact that ions and cytoplasmic enzymes leak through the membranes following cold shock injury. Some systems, particularly those of cells from the pig are damaged by cold shock whatever precautions are taken [2].

A second type of cold or thermal shock is that induced in red cells by high concentrations of electrolytes [7, 8] (see Figure 1.1). Sensitivity to injury on cooling occurs with concentrations of 0.8 M to 2.0 M NaCl. Above this concentration, injury will occur directly without cooling. Injury is less when cryoprotective agents are present [5]. Injury is also reduced by slow cooling or even by a period of time in the hypertonic electrolyte before rapid cooling (Figure 1.1). Damage is little affected by different cations but is very sensitive to a change in anions. In the sequence acetate, chloride, nitrate, iodide, sulphate, acetate is the least and sulphate the most injurious [9].

Mechanisms of cold or thermal shock injury may involve differential contractions of membrane components leading to mechanical fractures and conformational changes to the membrane topography [10].

Because of the association in the red cell between high salt concentration and sensitivity to injury on cooling, it has been suggested that phenomena similar to cold shock are involved in the injury to cells *during* freezing [11]. The freezing process itself provides both the increased concentration of electrolytes and the reduction in temperatures. However, this hypothesis has never been accepted generally, perhaps because many mammalian cells (e.g. lymphocytes and tissue culture cells) do not show sensitivity to cold shock in the presence of hypertonic salts above 0°C. However, there is evidence that salt-induced thermal shock becomes more severe as the temperature is lowered [7].

Thermal shock before freezing should always be remembered when examining a new system for preservation but with many systems it may not be relevant.

Before considering events during freezing in detail there is a general phenomenon illustrated in Figure 1.2 that is applicable to many biological systems. In essence, both rapid cooling and slow cooling kill cells during freezing. However, often an intermediate rate of cooling allows some survival. Each of these phenomena are now considered in turn.

Rapid cooling during freezing kills cells

This is perhaps the clearest and most unequivocal general phenomenon in the field of low temperature biology. There are certain exceptions

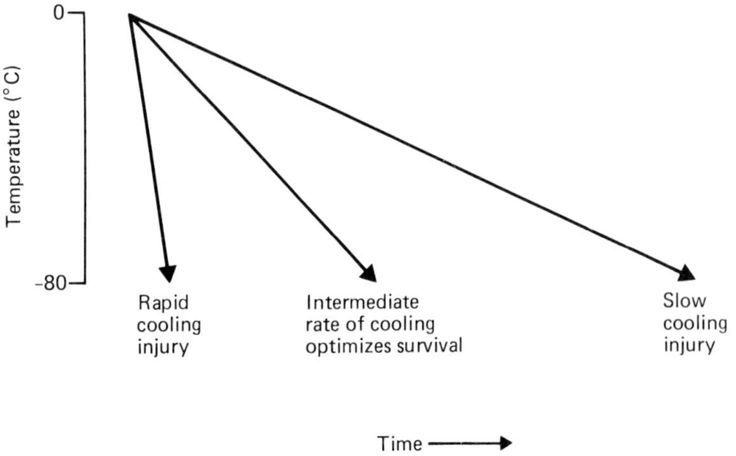

Figure 1.2 Representation of three rates of cooling. More injury occurs on thawing after cooling at both rapid and slow rates of cooling. Intermediate rates improve survival. This phenomenon occurs with many biological systems

and boundary conditions, but compared with statements to be discussed later these are few in number.

Before considering the definition of 'rapid' cooling, the onset of freezing must be described.

The initiation of ice nuclei is a random process; the first nucleus may form inside the cells or outside. If it occurs outside the cells, the freezing process can spread throughout the entire extracellular solution, a build-up in external osmolality will begin and the cells will begin to lose water by osmotically induced shrinkage. The maintenance of the osmotic gradient that may lead to cell shrinkage is only possible because the spread of ice into the cells is halted by the barriers of cellular membranes [12, 13]. If ice is present on both sides of a cellular membrane there is no driving force for water transport. If instead the first ice nucleus occurs within a cell only that cell is affected, again because of the barrier properties of cellular membranes.

What is meant by 'rapid' cooling varies from system to system. It can be defined as cooling that is sufficiently rapid that it does not allow significant osmotic shrinkage of the cells. In this definition the slowest rate of 'rapid' cooling will be determined by the water permeability of the cells [13] and to a lesser extent by the interaction with the water transport of other solutes present. What is the fate of the water within a rapidly cooled cell? Because it cannot be removed by shrinkage of the cell, the water becomes progressively more and more supercooled until intracellular freezing occurs. A rapidly cooled cell before thawing is, therefore, generally unshrunken and contains intracellular ice. As already stated, after this combination of events most cells are dead on thawing. Table 1.1 shows for several cell types the lowest rate of 'rapid' cooling that kills most of the cells.

Table 1.1 Examples of lowest rate of rapid cooling that kills most cells.

Cell type	*Cryoprotective additive*		*Cooling rate (°C/min)*	*Reference*
Human lymphocyte	DMSO	10%	1	14
	DMSO	5%	5	14
Human erythrocyte	None		5000	15
	Dextrose	10%	9000	15
	Glycerol	10%	1500	15
Chinese hamster fibroblast	DMSO	5%	200	16
Mouse marrow stem cells	Glycerol	11.5%	8	17

The actual causes of injury on thawing to cells that have been cooled rapidly will be discussed later under conditions of thawing. It is clear, however, that injury following rapid cooling becomes manifest during rewarming rather than during cooling. This is shown by the different survivals that can sometimes be obtained when thawing conditions are altered [17].

The overt cause of rapid cooling injury has been ascribed not so much to the formation of intracellular ice in cells during cooling, but to some of the consequences of having intracellular ice within cells during the rewarming process [18].

Under conditions of ultra-rapid cooling several orders of magnitude greater than those that just prevent cell shrinkage, it is possible to cool so fast that ice nuclei will not form [19]. Theoretically, this will allow preservation (provided that rewarming is equally rapid) but has no practical utility for the recovery of function. Morphologists often use ultra-rapid cooling for what they term the 'preservation' of good structure [20, 21]. This means that ice crystals are minimised by ultra-rapid cooling. The incorporation of high concentrations of cryoprotective additives combined with rapid cooling minimises the formation of obvious ice crystal artefacts in pictures obtained by techniques of freeze-etching or freeze-substitution. Unfortunately, from the functional point of view these conditions seldom allow survival, since cryoprotective agents are usually only effective during freezing at rates of cooling under which osmotic shrinkage occurs [22]. This will be discussed under slow cooling. Also very high concentrations of glycerol or other cryoprotective agents can damage cells directly. An alternative approach in morphological studies would be to tolerate some ice crystal artefacts in order to see cells that are potentially viable [22].

Slow cooling during freezing kills cells

Avoidance of the hazard of rapid cooling that has already been described as due to insufficient time for the cells to shrink thus allowing too much intracellular ice to form, may lead to another problem in that excessive exposure to conditions allowing shrinkage can also kill the cells. This is what may happen during slow cooling. The barrier properties of most cellular membranes to ice nuclei, at least at high subzero temperatures, are well attested by the general finding that slowly cooled cells will shrink in response to extracellular freezing rather than freeze internally. The lack of barrier properties of the granulocyte membrane to ice nuclei may be the reason why this cell type is so difficult to preserve in a potentially functional state. (Chapter 6, Leucocytes).

Thus, in slow cooling once extracellular freezing has begun the cells begin to shrink. To define slow cooling is as difficult as to define rapid cooling. One possibility is to relate slow cooling to that which is sufficiently slow that shrinkage can be termed 'equilibrium shrinkage', i.e. the water permeability is sufficiently high that the cell can respond rapidly to the changed osmolality imposed by each temperature reached. Under these conditions the cells are exposed to three classes of disturbance to their normal microenvironment that might cause injury. First, there is the effect of the reduction in temperature alone (as already discussed). Second, there is the presence of large amounts of extracellular ice close to or contiguous with the cells. Third, there is exposure to increased concentrations of extracellular solutes due to the removal of extracellular solvent as ice. There is little evidence that extracellular ice damages cells directly [23]; almost all attention has been given to the consequences of increased concentrations of solutes as potentially damaging agents. This might occur in several ways. There is the possibility of direct attack upon the cell by the high concentrations of solutes (including changes in pH) [24]. There is also the stress imposed on the cells by the increased extracellular osmolality, that is the stress of shrinkage [25]. Last, there is the possibility of transmembrane transport of solute species driven by their now very abnormal extracellular concentrations [24]. Which of these agencies of solutes or shrinkage is primarily responsible for the damage has occasioned some of the most controversial aspects of low temperature biology. There are data supporting the idea that injury correlates with osmolality *per se* (that is, a critical level of shrinkage beyond which the cell is injured), with the mole fraction of the major electrolyte present (sodium chloride), with pH etc. Perhaps more light will be thrown on this as yet unsolved problem when we consider the avoidance of slow cooling injury by using cryoprotective additives as discussed later.

Intermediate cooling is protective

If cooling is slow enough that intracellular ice may not form and yet rapid enough that the cells are not damaged by excessive exposure to high concentrations of solutes or by shrinkage, they may survive the processes of freezing and thawing. With many cell types these hazards overlap (at least in the absence of cryoprotective additives) and survival does not occur. However, with some cell types including red cells, bacteria and yeast, there is, as we have already seen, a 'window' on the rate of cooling scale that allows some survival [14, 15, 17] (Figure 1.2). A diagrammatic representation of cell shrinkage during cooling at a rate giving good survival is shown in Figure 1.3.

The real breakthrough into the practical preservation of cells came from the use of cryoprotective agents that widen and extend this

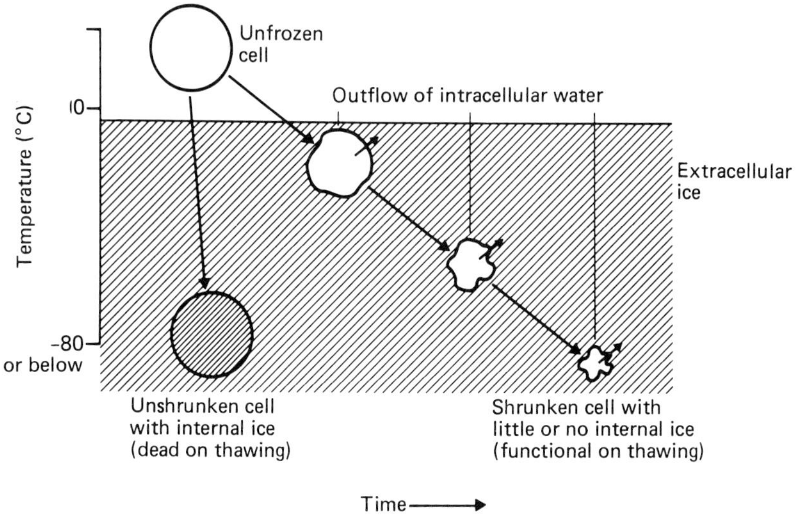

Figure 1.3 Shrinkage correlates with preservation. Extracellular freezing induces conditions that allow osmotically induced loss of water from cells during slow freezing. This correlates with survival on thawing. Rapidly cooled cells do not have time to shrink, form intracellular ice and are dead on thawing

'preservation window' particularly in the direction of overcoming injury caused by slow cooling.

Cryoprotective agents

The first cryoprotective agent, glycerol, was discovered almost by accident in the now classical work of Polge, Smith and Parkes [1]. Cryoprotective agents are of many types and are chosen for many different reasons. They do, however, have some common features and these can be discussed first. The primary property they all have is that of high solubility in water. High water solubility at room temperature also confers the related property that once freezing begins in a system containing the compound it will remain in solution and consequently reach a lower temperature in solution rather than crystallise out soon after the first ice has formed. There are several studies relating cryoprotective ability of compounds to their capacities of forming hydrogen bonds and binding water. This reflects the primary requirement of high water solubility.

A second essential property is that these compounds should have low toxicity to the cells to be preserved. These two properties of high water solubility and low toxicity to the cells are the only primary requirements for a compound to be cryoprotective. The conditions under which a particular additive will protect cells in any system will vary and will depend on other secondary factors such as the molecular weight and the permeability of the cell to the compound.

Cryoprotective agents are effective against slow cooling injury but not rapid cooling injury

This statement is the first indication in this chapter of the interactive nature of conditions for cell preservation. Rapidly cooled cells are protected far less efficiently by additives than are slowly cooled cells (Figure 1.4). There are many instances in which as the concentration of protective agent is increased, optimal preservation occurs at lower rates of cooling [14, 17].

There have been many attempts to explain the mechanism of protection given by these compounds. There is evidence that sometimes they can act directly on the cell, as for example when thermal shock of red cells in the absence of freezing is reduced by DMSO [5]. However, most of the evidence is that during freezing these compounds work by altering the physical conditions of the ice and solutions around the cells and that these altered conditions favour cellular survival. The presence

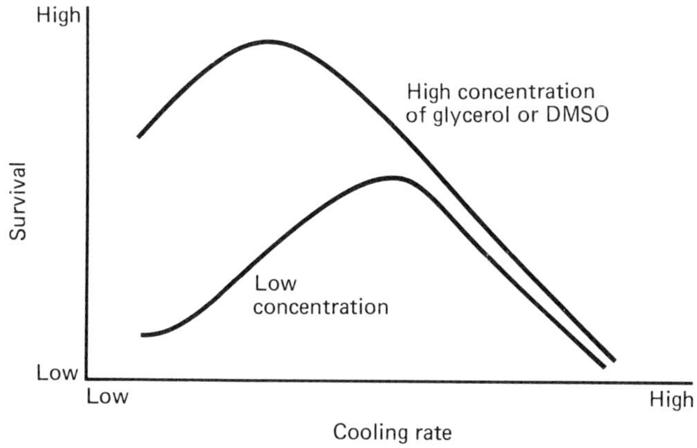

Figure 1.4 Glycerol or DMSO protect at low cooling rates but not at rapid rates

of cryoprotective additives reduces the proportion of the system that converts into ice [26]. This reduction in amount of extracellular ice at each subzero temperature may not, in itself, be the direct cause of protection since, as we have seen, extracellular ice may not be damaging. However, less extracellular ice implies alterations to the conditions in solution around the cells. Since there is less ice the solutes present, e.g. electrolytes, will be less concentrated during freezing [26]. Another way of looking at this is that part of the build-up of high concentrations of salts produced by freezing in the absence of cryoprotective agents is replaced by the build-up in the less toxic solute of the cryoprotective agent itself, e.g. DMSO or glycerol. Much has been made of the reduction in driving force for shrinkage of the cells due to the cryoprotective molecule being able to penetrate the cell membrane. However, as we have seen, some cell shrinkage correlates with protection even though excessive shrinkage is harmful. It is also possible that cryoprotective agents exert some other influences by reducing the probability of the formation of the ice nuclei [27]. There is also strong evidence that excellent protection is obtained using high molecular weight cryoprotective compounds that do not penetrate the cell membrane and therefore do not reduce the effective driving force for cell shrinkage [17, 28]. This indicates the importance of the reduction in build-up of the other intrinsic solutes (e.g. electrolytes) by the use of the cryoprotective substance. This does not mean that high concentrations of electrolytes are proved to be the sole injurious agent during slow freezing. Instead, it suggests that the use of the protective agent moderates the rise in salt concentration so that sufficient cell shrinkage can be induced to avoid problems with intracellular ice during rewarming.

This is supported by data obtained using the so-called two-step cooling procedure. The essence of this method is that cells can be protected from freezing injury by what has been termed 'prefreezing', that is cells frozen to a subzero temperature of for example –20°C for a short period can survive subsequent rapid freezing to storage temperatures [29–32]. The extracellular freezing during the 'prefreezing' stage appears to provide sufficiently high solute concentrations to shrink the cells to avoid or reduce intracellular ice formation on subsequent cooling [32, 33] (Figures 1.5 and 1.6). As with the more conventional preservation procedure involving a continuous rate of cooling, the two-step method requires both that the cell membrane acts as a barrier for ice crystals, thus allowing shrinkage to take place, and that the high concentrations around the cells are not themselves inducing injury at the prefreezing temperature. Work with protective agents and the two-step method shows that with an increased con-

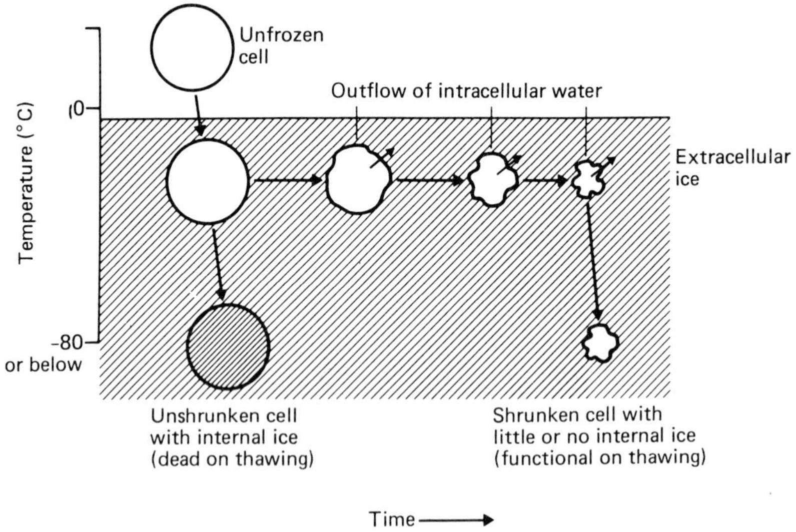

Figure 1.5 Principle of two-step freezing method. Cells held for a time at a temperature above that where intracellular ice forms can shrink in response to the conditions produced by the extracellular freezing. The cells can then be cooled rapidly to a very low temperature and survive on thawing

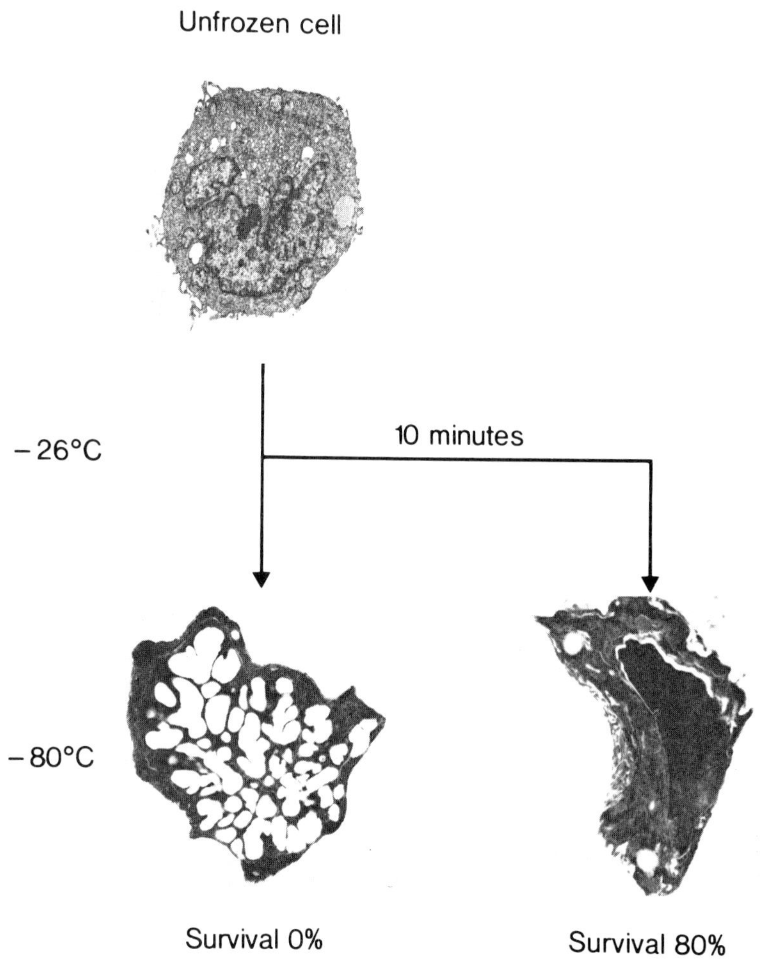

Figure 1.6 Ultrastructural representation of Chinese hamster fibroblasts in dimethylsulphoxide (5% v/v) before and after two-step freezing. Rapid cooling induces ice cavities and survival does not occur. In contrast 10 min at -26°C before cooling to -80°C allows cellular shrinkage, the absence of obvious intracellular ice and high survival. The cooled cells were examined by freeze-substitution [33] (approximate magnification X 5400)

centration of the permeating compound, e.g. DMSO, a lower prefreezing temperature is needed for optimal protection. Increasing the DMSO concentration means that a lower temperature is needed during freezing for the other solutes to reach the same high concentrations to bring about shrinkage. Not only is the process of osmotic shrinkage temperature dependent, but also toxic injury to the cells by solutes at high concentrations is reduced at low temperatures. To summarise, therefore, cryoprotective agents may work by allowing the shrinkage required to minimise intracellular ice formation to take place at lower, less toxic temperatures, thus avoiding exposure to high concentrations of other solutes, e.g. salts at higher temperatures. It is clear that protective agents are acting only by changing conditions quantitatively since several systems show good survival in their absence providing the cooling conditions are optimised. The concentration of cryoprotective agents used should be as high as possible to maximise cryoprotection without being so high that direct toxic injury occurs. This leads on to the problem of the design of procedures for the addition of the cryoprotective agent before freezing and its removal after thawing (see Chapter 12). Usually protective compounds are added to the cells at either room temperature (20–25°C) or at 0°C. Addition at 0°C minimises direct chemical toxicity and is probably to be recommended. The main reason for adding permeating protective compounds at room temperature was the previously held but erroneous belief that permeation was essential for protection. Permeation may, however, alter the optimal conditions for preservation. The problem of removal of the cryoprotective agent after thawing is dealt with later under the heading 'Post-thaw handling'.

Rapid thawing generally improves survival

The optimal conditions for cooling are frequently found using a single (usually rapid) rate of thawing. As with cooling the terms rapid and slow are both arbitrary and relative when applied to rewarming. Often the routine procedure has been to rewarm as rapidly as possible, e.g. by agitating the sample vial in a +37°C water bath. The determining factor for the rewarming rate then becomes the volume of the sample and the nature of its container. When changes in the rate of rewarming have been studied the most consistent general finding has been that slower rates worsen survival. This is especially so for samples cooled at more rapid rates (those likely to encourage the formation of intracellular ice) [17] (Figure 1.7). With slower cooling the sensitivity of survival to changes in the rewarming conditions is less marked. The finding that rapid thawing can protect cells to some extent against the injury associated with rapid cooling is a powerful argument that the injury associated with intracellular ice occurs at least partially during rewarming and not during the initial formation of ice on cooling. Many

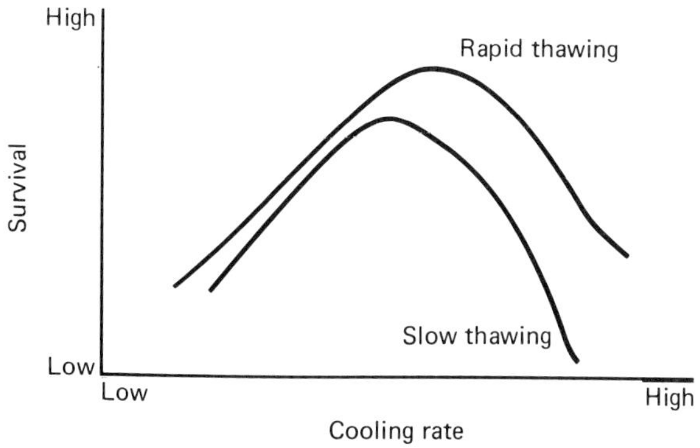

Figure 1.7 Rapid thawing often improves survival after rapid cooling

correlations have been made between increased injury on slow thawing and the observed redistribution or growth of ice during rewarming to larger intracellular crystals [32, 34]. This process is generally known as recrystallisation. It has been assumed that recrystallisation leads to an increase in the total amount of intracellular ice within the cell, although it is not clear whether instead it represents merely a redistribution of the same amount of ice from many small crystals to few large ones. Mechanistically the large size of the ice has been linked to increased injury [35] although as already discussed it has also been proposed that damage is controlled by osmotic events following the melting of the intracellular ice [18]. Recently, membrane lesions close to intracellular ice crystals have been demonstrated ultrastructurally [36].

Slow thawing is sometimes better than rapid thawing

Just as increased damage following slow thawing has been linked to the recrystallisation of intracellular ice formed during cooling, so similar survival after rapid or slow thawing has been ascribed to the absence of intracellular ice [32]. However, there are instances in which slow thawing gives much higher survival than rapid thawing. A notable example of this is described in the chapter on the preservation of mammalian embryos (Chapter 4), but it has also been reported for red cells cooled in high concentrations of permeating cryoprotective compounds, particularly glycerol [37 - 39]. Recent work with embryos described elsewhere in this book suggests that this phenomenon is that of injury caused by rapid thawing (see Chapter 4). This injury on rapid thawing can be avoided, at least with embryos, by prematurely curtailing a slow rate of cooling by plunging the sample from about –40°C into liquid nitrogen. Under these cooling conditions it is possible that a small amount of intracellular ice is formed.

It is thus possible to make some preliminary general conclusions about the effects of rates of thawing on cellular survival. When a small amount of intracellular ice is present rapid thawing is necessary to avoid the injury associated with recrystallisation. Conversely, under the less well understood conditions of rapid thawing injury, a small amount of intracellular ice may be protective. The mechanism of this last effect is unclear. It is possible that the presence of a small amount of intracellular ice seals the cell to osmotically induced damaging water movements during the initial part of thawing. Another possibility is that rapid thawing injury is brought about by transmembrane movements of solutes, including the cryoprotective agent itself, although this is harder to relate to the protection afforded by a small amount of intracellular ice. One thing is clear that if a cell contains too much intracellular ice following cooling no increase in thawing rate can avert its destruction.

The interrelationship between all the variables in a preservation procedure and survival

Several instances of this have already been mentioned. For example, with more rapid cooling, the rate of thawing must be increased to maintain survival. An increase in the concentration of cryoprotective additive often demands a slower rate of cooling. Another instance even involves such an 'inert' variable as the temperature of storage. With red cells when high concentrations of glycerol are used and low rates of cooling, storage can be successful for many months at –70 to –80°C. However, with low concentrations of glycerol and more rapid rates of cooling, giving equally high initial survival, rapid loss of function occurs during storage at these temperatures. This can be avoided by using liquid nitrogen as the storage medium. Another example is that where tissue culture cells frozen by a two-step method under certain conditions may initially survive slow thawing from a storage temperature of –80°C; this ability is lost after five weeks' storage [32]. These are only a few of the examples of the interactions between cooling conditions and survival. Different cells require different cooling conditions. A small change in a single variable, for example the stage in the cycle of a synchronised cell or the concentration of a protective agent may demand changes in *ALL* the other variables including rates of cooling and rewarming, temperatures and times of storage and conditions applied to the system after thawing for example those necessary for the removal of the added protective agent.

Post–thaw handling of cells that have been preserved

The normal procedure for verifying the usefulness of preserved cells, particularly when the cooling and thawing recipe is first designed is to assay the cells for function soon after the thawing and removal of the protective additive. Perhaps the most neglected aspect of cryobiology has been that of the conditions for either progressive damage or repair of cells after storage. There have been a few studies in this field which suggest that repair is an important factor that may be augmented by changing post–thaw conditions [40]. Rather more information is available concerning the immediate differences between thawed cells and their unfrozen counterparts from the same source. This includes not only whether the cells are functional but important other factors such as whether they have an increased sensitivity to osmotic or mechanical shock after the stresses of cooling and rewarming. Common changes to cells following thawing include an increased sensitivity to injury with high *g* centrifugation and to osmotic shock [41]. There is almost complete ignorance on how long these effects persist and on factors affecting them.

Traditionally the most important post–thaw stress has been that associated with the removal of a permeating cryoprotective additive from the system. This has usually been carried out by dilution, either continuous (e.g. by dialysis), stepwise or abruptly. Some cell types will tolerate abrupt dilution better than others, and the presence of an extracellular colloid, e.g. serum, in the diluting medium may reduce the trauma. A vital but neglected variable has been that of the temperature of the dilution. With many cell types and when diluting from several diverse solutes including sodium chloride, dimethylsulphoxide and glycerol it has now been shown that an abrupt dilution is tolerated better at room temperature or +37°C than at 0°C [41, 42]. Thus, when slow dilution out of DMSO may be necessary at 0°C the simple expedient of warming to room temperature may permit the use of a single dilution step.

It is helpful to remember that dilution of the cells after thawing is merely a continuation of the dilution that takes place during thawing as the ice melts. It is thus easy to see that post–thaw handling conditions also come into the category of variables that interact with all the other conditions of the preservation procedure. In fact, there is a direct relationship between cells tolerating an osmotic stress of dilution at a higher (room) temperature than a lower one (0°C), and the often beneficial effects of more rapid thawing which bring the cells to higher temperatures before they receive the osmotic shocks imposed by the dilution as the ice melts. In other words, thawing is a dilution procedure. A particularly clear example of the interaction between cooling conditions and post–thaw handling conditions has been observed with mouse lymphocytes cooled in DMSO over a range of rates. At the 'optimal' rate of cooling changes to post–thaw handling had little effect on functional survival, whereas cells cooled at rates both slower and faster than the optimal rate functioned well on thawing but only when 'gentle' conditions of handling were used [41, 42].

It is thus helpful to consider a preservation procedure in its entirety without neglecting the very important phase of post–thaw handling.

Problems caused by different nature of thawed material

Some of the factors under this category are blindingly obvious while others have, until recently, been totally ignored. Clearly, after attempted preservation procedures, cells may be dead or their function may be impaired. Also, as already discussed, they may have an increased sensitivity to damage when stressed. The chief problem caused by the different nature of the thawed material is that of assessing functional survival after thawing. To do this a comparison has to be made between the function recovered with that of a similar sample acting as the unfrozen control. It has been customary with many cellular systems to relate the assayed function recovered to that of the control by means of a simple percentage figure. In some systems this is justified as for example when considering haemolysis where there is good evidence that cells either contain their normal quota of haemoglobin or none. However, when injury is not 'all or none' and cells are present covering a whole spectrum of damage, attempts to obtain a single averaged figure for survival are fraught with danger. The chapter on the preservation of lymphoid cells deals with one of the most important of these circumstances. The chief problem comes when the total function of a sample of cells (see Chapter 6) is not linearly related to the number of cells present. Any trauma or injury such as that imposed by freezing may then not only damage the cells but change the nature of the sample (that is, numbers of functional cells surviving) so that it has to be compared with a *different unfrozen sample.* When cell concentration is the critical variable this difficult problem can be partially overcome as discussed in Chapter 6 by comparing function of thawed and unfrozen samples over a whole range of cell concentrations. This use of a one–dimensional rather than a single point comparison may only be the first step towards the proper conditions for comparing function of cellular samples. For example, it is likely that a two–

dimensional comparison is necessary where both cell concentration and time of assay before the response is measured are included.

The safest course is to avoid the assumption that a thawed sample of cells must be qualitatively similar to the unfrozen control sample but diluted due to the loss of functional cells during preservation. Instead, it would be better to consider the thawed sample as *potentially* different in nature from the unfrozen cells.

Conclusions

Despite many ingenious hypotheses to explain freezing injury the general results used as section headings in this chapter form the main framework of cryobiology. To most of them there are exceptions some of which are described in detail in succeeding chapters. The most important conclusion is that there is no general recipe for cellular preservation. The dogmatic application of any published protocol to a system even slightly different from that for which it was designed may lead either to poor results or to disaster. It is nevertheless hoped that familiarity with the general points considered in this chapter may help in the design of specific freezing procedures for each new circumstance in each laboratory. In Chapter 12 an attempt is made to show the approach to designing one's own freezing protocol for a new system. In this field as in several others the do-it-yourself approach is often to be preferred.

Acknowledgements
I would like to thank Dr. C. A. Walter for permission to use the electron micrographs in Figure 1.6.

References

1 Polge, C., Smith, A.U. and Parkes, A.S. (1949). *Nature* **164**, 666
2 Polge, C., Wilmut, I. and Rowson, L.E.A. (1974). *Cryobiology* **11**, 560
3 Blackshaw, A.W. (1954). *Australian Journal of Biological Sciences* 7, 573
4 Farrell, J. and Rose, A.H. (1968). *Journal of General Microbiology* **50**, 429
5 Farrant, J. and McGann, L.E. (1975). *Advances in Cryogenic Engineering* **20**, 417
6 Butler, W.J. and Roberts, T.K. (1975). *Journal of Reproduction and Fertility* **43**, 183
7 Lovelock, J.E. (1955). *Biochemical Journal* **60**, 692
8 Morris, G.J. and Farrant, J. (1973). *Cryobiology* **10**, 119
9 Morris, G.J. (1975). *Cryobiology* **12**, 192
10 Lovelock, J.E. (1954). *Nature* **173**, 659
11 Farrant, J. and Morris, G.J. (1973). *Cryobiology* **10**, 134
12 Chambers, R. and Hale, H.P. (1932). *Proceedings of the Royal Society, London, Series B* **110**, 336
13 Mazur, P. (1963). *Journal of General Physiology* **47**, 347
14 Farrant, J., Knight, S.C. and Morris, G.J. (1972). *Cryobiology* **9**, 516

15 Rapatz, G., Sullivan, J.J. and Luyet, B. (1968). *Cryobiology* **5**, 18
16 McGann, L.E. Unpublished data
17 Leibo, S.P., Farrant, J. Mazur, P., Hanna, M.G. and Smith, L.H. (1970) *Cryobiology* **6**, 315
18 Farrant, J. (1977). *Philosophical Transactions of the Royal Society London Series B* **278**, 191
19 Luyet, B. (1960). In *Recent Researches in Freezing and Drying,* p.3. A.S. Parkes and A.U. Smith (Ed.). Oxford: Blackwell
20 Stolinski, C. and Breathnach, A.S. (1975). In *Freeze-Fracture Replication of Biological Tissues. Techniques, Interpretation and Applications.* London: Academic Press
21 Bullivant, S. (1973). In *Advanced Techniques in Biological Electron Microscopy,* p.67. J.K. Koehler (ed.). Berlin: Springer
22 Farrant, J., Walter C.A., Lee, H., Morris, G.J. and Clarke, K.J. (1977). *Journal of Microscopy* **111**, 17
23 Nei, T. (1968). *Cryobiology* **4**, 303
24 Lovelock, J.E. (1953) *Biochimica et Biophysica Acta* **10**, 414
25 Meryman, H.T., Williams, R.J. and Douglas, M.J. (1977). *Cryobiology* **14**, 287
26 Lovelock, J.E. (1953). *Biochimica et Biophysica Acta* **11**, 28
27 Luyet, B. and Rasmussen, D. (1968). *Biodynamica* **10**, 167
28 Knight, S.C., Farrant, J. and McGann, L.E. (1977). *Cryobiology* **14**, 112
29 Luyet, B. and Keane, J.A. (1955). *Biodynamica* **7**, 281
30 Asahina, E. (1959). *Nature* **184**, 1003
31 Polge, C. (1957). *Proceedings of the Royal Society, London Series B,* **147**, 498
32 Farrant, J., Walter, C.A., Lee, H. and McGann, L.E. (1977). *Cryobiology* **14**, 273
33 Walter, C.A., Knight, S.C. and Farrant, J. (1975). *Cryobiology* **12**, 103
34 Asahina, E. (1965). *Federation Proceedings* **24**, S183
35 Mazur, P. (1966). In *Cryobiology,* p.213. H.T. Meryman (Ed.). New York: Academic Press
36 Fujikawa, S. (1978). *Cryobiology* **15**, 707
37 Rapatz, G., Luyet, B. and MacKenzie, A. (1975). *Cryobiology* **12**, 293
38 Miller, R.H. and Mazur, P. (1976). *Cryobiology* **13**, 404
39 Farrant, J., Lee, H. and Walter, C.A. (1977). In *The Freezing of Mammalian Embryos,* p.49. K. Elliott and J. Whelan (Ed.). A Ciba Foundation Symposium. Amsterdam: Elsevier, Excerpta Medica, North Holland
40 McGann, L.E., Kruuv, J. and Frey, H.E. (1972). *Cryobiology* **9**, 496
41 Thorpe, P.E., Knight, S.C. and Farrant, J. (1976). *Cryobiology* **13**, 126
42 Woolgar, A.E. and Morris, G.J. (1973). *Cryobiology* **10**, 82

CHAPTER TWO

Low Temperature Preservation of Cells, Tissues and Organs

M J Ashwood-Smith

Introduction

Attempts to preserve organs at low temperatures in the frozen state have met, in general, with no success with the exception of skin, cornea and certain fetal tissues. When the different requirements for freezing and thawing rates of the constituent cells that comprise organs are considered the largely negative results are not surprising. Compounding an already difficult situation are the thermal gradients present in tissues during freezing and thawing, the diffusion of the protective molecules and the possible requirements for differing concentrations and perhaps varieties of cryoprotective agents for optimum cell survival.

In this chapter consideration will be given to some of the more recent and interesting aspects of the cryoprotection of isolated mammalian cells before a more detailed discussion of the freeze-preservation of selected systems such as the cornea. A recent account of the problems of the long term preservation of cells and tissues has been given by Pegg [1]. An earlier review is that of Sell [2]. The attempted, and unsuccessful, preservation of mammalian kidneys and adult hearts is discussed only very briefly.

Tissue culture cells

There are few problems associated with the successful low temperature preservation of mammalian cells from tissue culture. The majority of researchers use either glycerol or dimethylsulphoxide as the cryoprotective agent of choice, usually at a concentration of between 5 and 10% (v/v) added to standard tissue culture media in the presence of 10 or 15% serum. The efficacy of the added serum has been discussed [3] and the type of buffer system used, namely bicarbonate or organic buffers, has been investigated [4]. Cooling rates normally have been within the range of 1–5°C/min and thawing rates, usually rapid, have been of the order of 150–200°C/min. Microwave thawing has been shown to be no more effective for cell systems than conventional thawing methods [5].

Theoretical models for freezing damage in mammalian cells in tissue culture have their origin in elegant experiments by Mazur and his colleagues [6–11] in the USA and by Farrant and his research group in the UK [12–16]. Research by Kruuv and others [17–19] in Canada has concentrated on the sensitivity of mammalian cells as a function of the cell cycle, on the possibility of the cellular repair of freezing damage and on the influence on survival of changes in membrane composition. Synchronised Chinese hamster cells, for example, were frozen at a rate of 5°C/min for 30 minutes and then plunged into liquid nitrogen and kept at –196°C for one week before thawing rapidly. The highest survival was achieved with cells frozen in M and late S. Cells frozen in G2 were 2.2 times more sensitive [20]. Terasima and Yasukawa [21] using *HeLa* cells have also reported that changes in cell survival after freezing and thawing in DMSO were related to the cell cycle.

Repair of freezing damage in diploid Chinese hamster lung cells has been shown to occur when cells were incubated for 3 hours at 37°C in spinner flasks prior to plating [19]. Ultrastructural changes in hamster ovary cells after freezing and thawing in DMSO at 1.7°C/min and followed by rapid thawing have been observed and studied by Hunt *et al* [22]. The main areas of damage were the cell surface and the cytoskeletal framework of the cell (microtubules and microfilaments). A delay in the return of the frozen cell to a normal morphology may, in the opinion of these authors, be related to a need for repair.

The kinetics of cell growth after freezing cells to –196°C in 15% fetal calf serum added to Eagle's Basal Medium with Hanks' salt solution (cooling rate 8°C/min; thawing rate 485°C/min) have recently been studied by Frim *et al* [17]. Division delay was noted (Figure 2.1) and was dependent on cellular age. The time of the first generation was significantly prolonged but in subsequent generations cell cycle times were comparable to controls. In plateau phase cells, mitosis was delayed 7 hours and it is clear that repair mechanisms are involved in the reported phenomenon. Frim *et al* [17] noted that some cells were metabolically active, as observed microscopically, for at least 35 hours after thawing although they failed to divide. The implications of these findings to viability assays are noteworthy.

The effect of three membrane lipid perturbers (BHT, 2,4-ditertiary

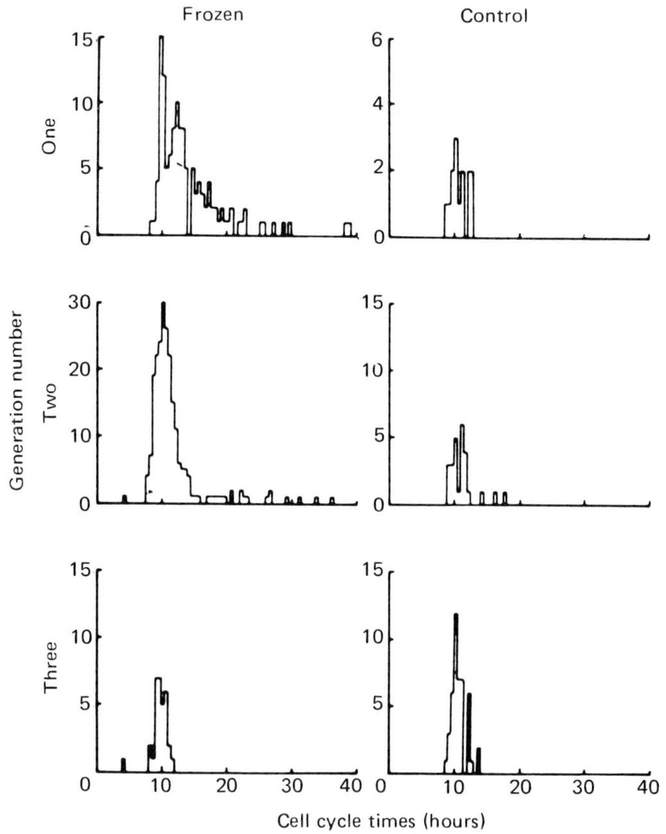

Figure 2.1 Histograms of cell cycle times. Frequencies of occurrence (histograms) of various cell cycle times are shown for both unfrozen (control) and frozen-thawed (experimental) populations as functions of cell generation number. (From Frim *et al.*, 1978 [17] by permission of the authors and publishers).

butyl 5-hydroxytoluene; 2-adamantanone and adamantane) on the survival of Chinese hamster cells exposed to cold has recently been reported by Rule and his colleagues [23]. BHT produced significant protection against damage associated with exposure of cells to +5°C for up to 12 days. All phases of the cell cycle were equally well protected with BHT.

An interesting experiment, which should be repeated using a better assay for viability than the eosin exclusion technique, was published by Holečková and Cinnerová [24]. These workers demonstrated a difference in the survival of DMSO protected mouse L cells as a function of storage time at –70°C depending on whether or not the cells had been 'cold adapted' to +4°C prior to freezing and thawing. The 'cold adapted' subline survived better than the control cells that had not been 'cold adapted'.

Most observers have confirmed that the long term storage of tissue culture cells in either DMSO or glycerol does not result in decreased viability, provided that the temperatures are below about –130°C. Biochemical and morphological markers are not altered and chromosomes are not damaged. Forty-two mouse neoplasms stored for periods up to 6 years at –195°C after freezing in 5% glycerol (cooling rate 1°C/min until –25°C; 5°C/min from –25°C to –70°C; thawing rate was rapid) retained all of their biological and biochemical characteristics; no loss of viability as a function of storage time was noted [25]. Resistance to 14 chemotherapeutic agents was not affected. Ashwood-Smith and Grant [26] concluded, however, that indefinite preservation at –196°C is a theoretical impossibility as there can be no escape from the slow accumulation of radiation damage, albeit reduced by both the presence of DMSO or glycerol and the low temperature. A figure of approximately 32 000 years is estimated for frozen cells to accumulate the equivalent of 600 rads of acute ionizing radiation damage!

Before discussing attempts to preserve specialised mammalian cells and tissues several important caveats should be noted with respect to the use of the two standard cryoprotective agents, DMSO and glycerol. The following facts should not, however, detract from the overwhelming usefulness of either cryoprotectant.

Lyman *et al* [27] showed that DMSO, possibly through its actions on cellular membranes, induced the differentiation of Friend leukaemic cells along the erythroid pathway. It has been suggested [28] that this effect of DMSO is at the transcription level. Travers [29] has shown that both DMSO and glycerol at concentrations up to 20% (v/v) stimulate RNA synthesis *in vitro (E. coli* system). The stimulation is due to an increase in initiation with the targets for glycerol and DMSO being the DNA template rather than RNA polymerase. Two other recent papers concerning molecular and biological effects of DMSO which are pertinent are those of Krystosek and Sachs [30] on differentiation of myeloid leukaemic cells and of Young and Wright [31] on an RNA polymerase B' mutation in *Salmonella typhimurium.* A separate and probably unrelated property of DMSO to these molecular attributes is its action in mitigating NaCl hypertonicity effects on the cell surfaces of tissue culture cells recently disclosed in a series of beautiful electron micrographs published by Mironescu and Seed [32]. These and other studies [33] have been aimed at an understanding of hypertonic effects associated with the stresses of freezing and thawing and their possible modifications by cryoprotective agents.

Surprisingly little attention has been focused on methanol as a cryoprotective additive even though early studies indicated its usefulness in the preservation of blood [34, 35], of heart [36] and trypanosomes [37]. Ashwood-Smith and Lough [38] demonstrated, with mammalian tissue culture cells, good protective action of methanol and these results are summarised in Figure 2.2. Methanol is surprisingly non-toxic to a

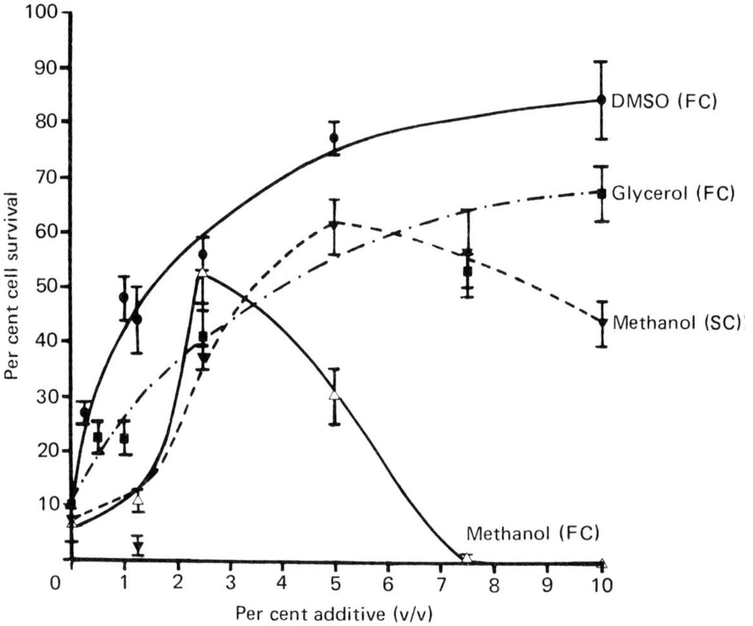

Figure 2.2 Cryoprotection of mammalian cells in tissue culture by methanol: effect of concentration and comparison with glycerol and dimethylsulphoxide. Vertical bar represents standard deviation. FC, fast cooling rate (20°C/min); SC, slow cooling rate (1°C/min). Cells thawed at approximately 115°C/min. ●, DMSO; ■, glycerol; ▼, methanol(SC); △, methanol(FC). (From Ashwood-Smith and Lough, 1975 [38], by permission of the authors and publisher).

number of isolated cells. However, Rajotte *et al* [39] found methanol to be incapable of affording cryoprotection to fetal mouse hearts.

Other cryoprotective agents which have been used with varying degrees of success include polymers such as dextrans, hydroxyethyl starch and polyvinylpyrrolidone (PVP). This later compound should be dialysed before use. The molecular weight descriptions associated with

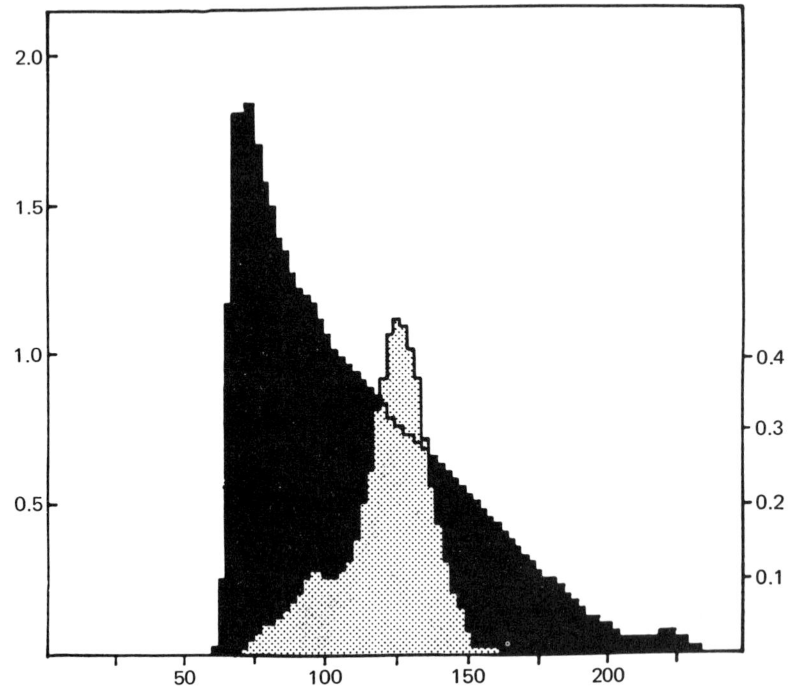

Figure 2.3 Elution pattern of Plasdone C (K30) on Sephadex G 200. Column 45 X 2.5 cm. Sample (20 mg) dissolved and eluted with phosphate buffer (pH 7.4; 0.01 M). Fractions collected every 2.5 ml. Flow rate 7.5 ml per hour. Ordinate: (left) absorbance X 10^{-1} (PVP) at 203 nm; (right) absorbance (protein) at 280 nm. Abscissa, volume of eluate (ml). Void volume 60 ml. Black, PVP; hatched grey, bovine serum albumin. (From Ashwood-Smith, M.J. and Warby, C. (1971). *Cryobiology* **8**, 453, by permission of the authors and publisher).

its commercial packaging are often far from accurate (Figures 2.3 and 2.4). Combinations of small quantities of either DMSO or glycerol (at levels which are essentially without osmotic effects) with dextrans have been shown to be remarkably effective in the cryopreservation of mammalian cells [40]. Results are illustrated in Figures 2.5 and 2.6.

Different mechanisms for the cryoprotective actions of intracellular (DMSO) and extracellular agents (hydroxyethyl starch) have been postulated [41] and thus the results obtained by using combinations of either glycerol or DMSO with dextrans (probably acting in a similar manner to hydroxyethyl starch) could be explained.

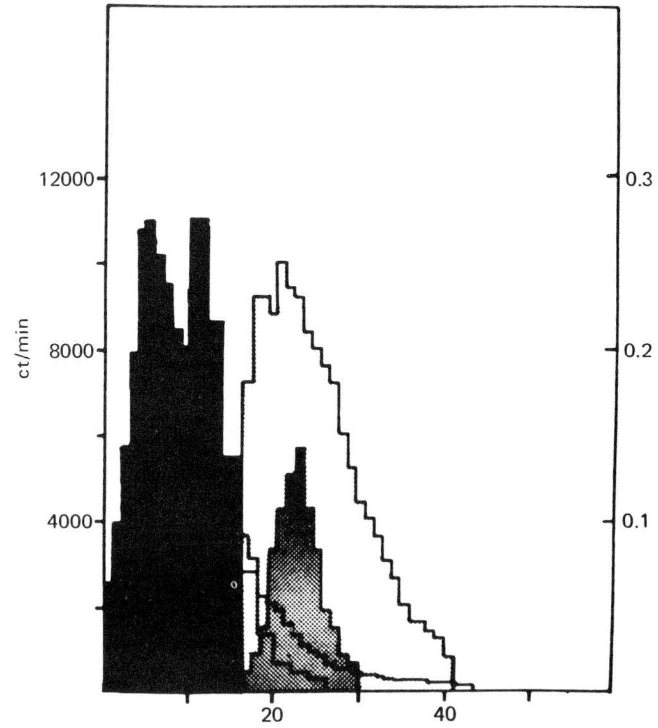

Figure 2.4 Density gradient centrifugation of PVP. Behaviour of PVP ^{14}C (K33), black; lysozyme, (hidden) bovine serum albumin, cross-hatched grey; and haemoglobin, white; on sucrose gradient (5–20%) subjected to a centrifugal field of 2.5×10^5 g for 18 h at 4°C. Beckman ultracentrifuge model 65B with a SW 56 swing-out rotor. Ordinate: (left) ^{14}C content in counts per minute; (right) absorbance (protein) at 280 nm. Abscissa: fraction no. (each fraction 0.1 ml). 5% sucrose at fraction 1 and 20% sucrose at fraction 43. (From Ashwood-Smith, M.J. and Warby, C. (1971). *Cryobiology* **8**, 453, by permission of the authors and publisher).

Results of Mazur's theoretical approach to the freezing and thawing of mammalian cells

The successful preservation of mammalian embryos (see Chapter 4) owes much to a theory of cryobiological damage developed by Mazur. Critical factors associated with survival of cells after freezing and thawing are (1) cooling rate, (2) warming rate, (3) concentration of protective additive, (4) permeation of additives into cells before freezing commences, (5) osmotic problems associated with the addition and removal of the additive.

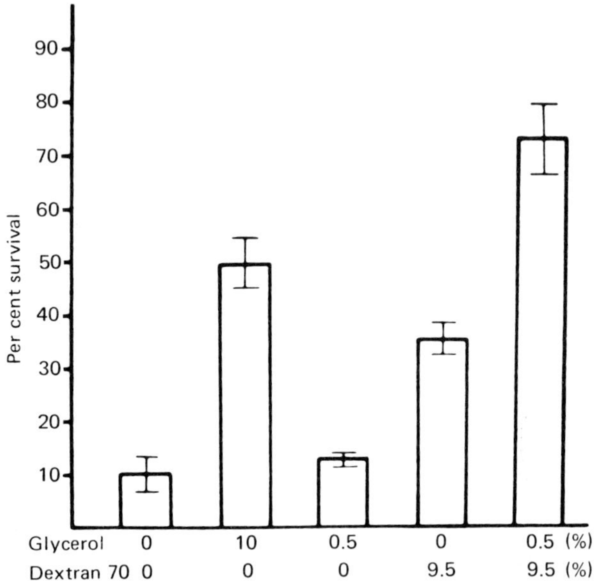

Figure 2.5 Effect of glycerol, dextran 70 (mol. wt. 70000 daltons), and combinations of glycerol and dextran on the survival of Chinese hamster cells frozen and thawed in tissue culture medium that contained 10% fetal calf serum. (From Ashwood-Smith, 1975 [40], by permission of the author and publisher).

The critical cooling rate for a cell is reached when a cell is cooled too rapidly and intracellular ice (usually but not always fatal, see Chapter 1) is formed. The basic premises of a critical cooling rate have been confirmed by experiments with red blood cells, yeast, embryos and tissue culture cells [9, 42–44]. To prevent the formation of intracellular ice, cells must be cooled slowly so that equilibrium (in terms of the chemical potential of the intracellular and extracellular water) is achieved.

The rate of osmosis during freezing is defined by Mazur by four simultaneous equations:

$$dV/dt = (Lp\,ART \ln p_e/p_i)\nu_1{}^\circ \qquad (2.1)$$

V is the volume of cellular water, t is time, Lp is hydraulic conductivity, A is cell surface area, $\nu_1{}^\circ$ is the molar water volume, p_e and p_i are the vapour pressure of extra- and intracellular water, when

$$d\ln(p_e/p_i)/dT = L_f/RT^2 - [n_2\nu_1{}^\circ\ (V+n_2\nu_1{}^\circ)\ V]\ dV/dT \qquad (2.2)$$

T is temperature, n_2 is osmoles of solute and Lf is the latent heat of

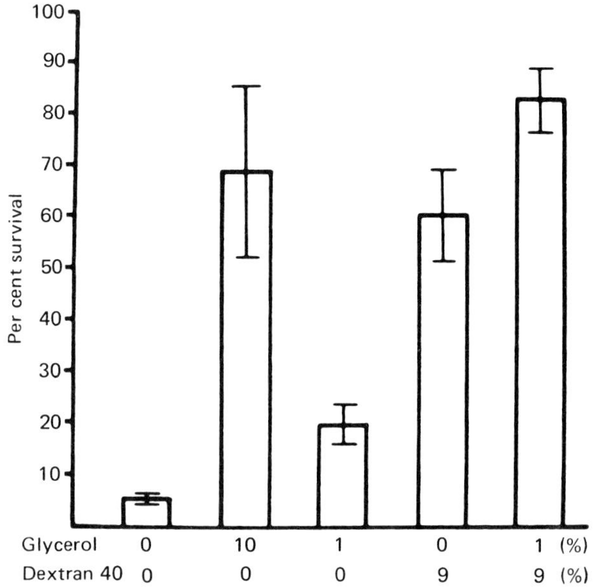

Figure 2.6 Effect of glycerol, Rheomacrodex dextran 40 (mol. wt. 40000 daltons), and combinations of glycerol and dextran on the survival of Chinese hamster cells frozen and thawed in tissue culture medium that contained 10% fetal calf serum. (From Ashwood-Smith, 1975 [40], by permission of the author and publisher).

fusion, R is the gas constant. Time/temperature are related by the cooling rate, which when linear is

$$dT/dt = B \tag{2.3}$$

and finally Lp is related to temperature by

$$Lp = (Lp)_g \exp\,[b(T-T_g)] \tag{2.4}$$

where $(Lp)_g$ is the permeability coefficient to water at a known temperature, T_g, and b is the temperature coefficient.

In order to avoid intracellular freezing, the cellular water content given by equations 2.1 to 2.4 must be reduced to the equilibrium water content before the intracellular nucleating temperature is reached. This equilibrium water content is

$$\ln[V/(V+n_2 v_1{}^\circ)] = (Lf/R)\,(1/273 - 1/T) \tag{2.5}$$

These equations are taken from the paper of Mazur *et al* [45] and are based on equations originally proposed by Mazur in 1963 [6].

Two further equations used for constructing the successful experimentation for pancreas (see later) and embryo preservation are related to additive permeability (still a controversial issue!) when P the permeability constant is

$$ds/dt = PA[m_s{}^e/1000 - S/V] \qquad (2.6)$$

where S is moles of additive and $m_s{}^e$ is the molarity of the external additive.

A final equation gives the water content (V) that cells *must* maintain to remain in osmotic equilibrium with the extracellular fluid

$$V = [M^i{}_{iso} + 1000\, \varphi\, {}^i{}_s S]/(M_n{}^e + \varphi\, {}_s{}^e m_s{}^e) \qquad (2.7)$$

where $M^i{}_{iso}$ is the osmolality of the isotonic cell, $\varphi^i{}_s$ and $\varphi\ {}^e{}_s$ the osmotic coefficients of the internal and external electrolytes respectively, and $M^e{}_n$ is the external electrolyte osmolality.

Pancreas

Considerable activity has been focused on various aspects of pancreatic islet functions in the last year or so and the papers documenting these facts are noteworthy for their overall excellence and will, therefore, be dealt with in some detail. Mazur's work in this area relating to the preceding equations will also be considered.

'Human pancreatic islet transplantation could be a successful technique in the treatment of some cases of diabetes.' Thus starts the first paragraph of a paper by Yudilevich *et al* [46]. In this study, experiments were performed with rat pancreas to investigate the optimum conditions for obtaining and preserving pancreatic material for the later isolation of viable islets. The authors found that as little as 30 minutes of warm ischaemia resulted in a poor yield of viable, isolated islets. The use of cold, bicarbonate buffered medium plus HEPES (an organic buffer) and a proteolytic enzyme inhibitor, Trasylol, resulted in islet activity demonstrable after more than 8 hours of storage.

Mazur *et al* [45] demonstrated the survival of fetal rat pancreases after freezing to –78°C and –196°C. Optimal conditions included suspension of tissue (16.5- to 17.5-day intact fetal pancreases) in 2 M DMSO (15.6% DMSO w/v). Cooling rates less than 1°C/min were necessary and were combined with thawing rates of 4°C/min and 15°C/min (these are slow thawing rates for most mammalian cells) for successful cryopreservation. The large amounts of DMSO used required its gradual removal by dilution after thawing. Success was measured by the ability of the preserved pancreases to synthesise protein and to yield viable allografts. Results are shown in Figures 2.7 and 2.8.

Initial experiments by this group indicated that the survival of pancreases was not appreciably dependent on the time of DMSO exposure or the temperature and thus intracellular permeation of DMSO was not critical. Experiments with glycerol yielded results which were far less good than with DMSO. The excellence of Mazur's results which were based on the theoretical predictions obtained from equations 2.1 to 2.7 and also successfully applied to the preservation of embryos suggested, in Mazur's words that '. . . . the approaches used here in freezing pancreases will prove to be useful guides to the successful freezing of other mammalian organs'.

Ferguson *et al* [47] studied the cryopreservation (as well as simple cold storage and organ culture) of pancreatic islets obtained from mouse, rat, guinea pig and human pancreas. Isolated animal islets in 10% DMSO (Earle's solution + 10% fetal calf serum) were frozen at 1°C/min to -50°C, followed by rapid freezing to -187°C. Thawing was rapid. Some cells were cooled at 5°C/min and in all instances cells were kept at -187°C for between 1 and 2 weeks. Viability was

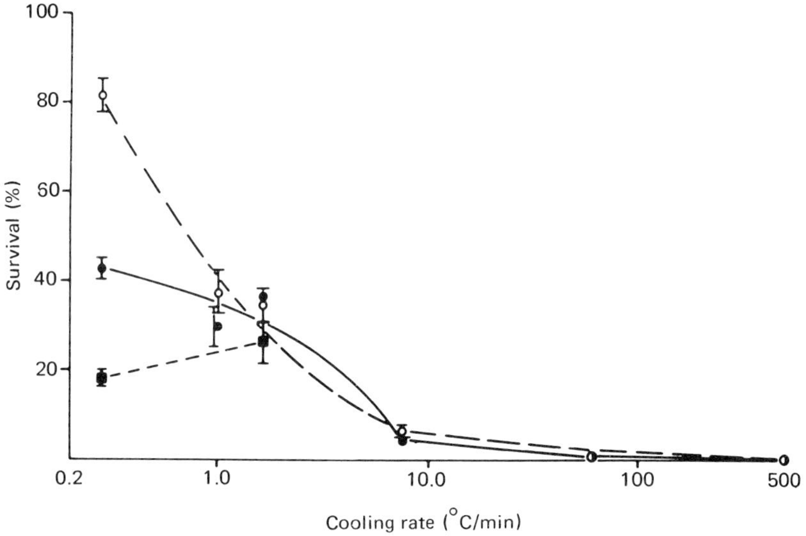

Figure 2.7 Specific (○) and absolute (●, ■) survivals of frozen-thawed 16.5 to 17.5 day fetal rat pancreases as a function of the rate of cooling. Most were frozen to −78°C; some were frozen to −196°C. Warming was at 4 or 15°C/min. Prior to freezing, pancreases were either subjected to standard procedures—1 M DMSO, 60 min, 22°C + 2 M DMSO, 15 min, 0°C (○, ●)—or to 1 M DMSO, 60 min, 22°C (■). (From Mazur *et al.*, 1976 [45], by courtesy of the authors).

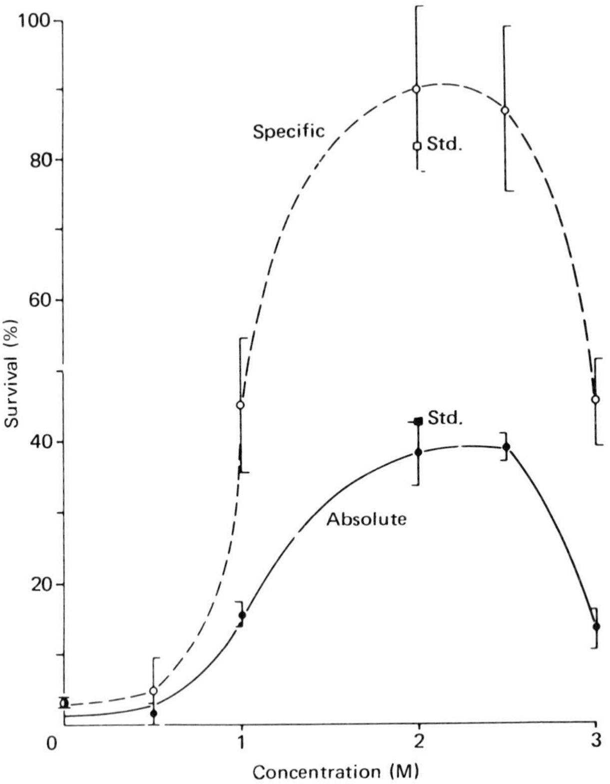

Figure 2.8 Survival of frozen-thawed 16.5 to 17.5 day fetal rat pancreases as a function of the DMSO concentration in the suspending medium. Pancreases were suspended in the indicated concentrations for 60 min at 22°C prior to initiation of freezing. The cooling rate was 0.28°C/min to −78 or −196°C. Warming rates were 4 - 15°C/min. □ and ■ refer to the survival obtained with the standard additive treatment. (From Mazur *et al.,* 1976 [45], by courtesy of the authors).

assessed by light microscopy and insulin assays (radioimmune assay). Function was absent after the 1°C/min cooling procedures for mouse pancreas and was poor for rat pancreas. With the higher cooling rate of 3°C/min mouse pancreas was not functional but rat pancreas survived and functioned reasonably well. In the light of Mazur's work it is apparent that better results would, perhaps, have been obtained by Ferguson *et al* [47], had different rates of cooling and thawing been used.

Studies by Rajotte *et al* [48] have concerned themselves with the cryopreservation of rat Islets of Langerhans. Viability of cryopreserved cells was assessed by their ability to secrete insulin when thawed and cultured, by electron microscopy and by transplantation into diabetic rats. Separated islet cells were suspended in DMSO (7.5–10% v/v) in Eagle's minimal essential medium containing 10% fetal calf serum. DMSO was added to give the required concentration, slowly over 30 minutes at +4°C. Cells were cooled to –100°C at between 0.5 and 0.7°C/min before storage at –150°C for 1–7 days. The thawing rate was about 150°C/min and then DMSO was removed slowly by stepwise dilution. Unfortunately, in this study, only one diabetic rat was injected with cryopreserved islet cells and only 400 cells were injected. However, the response was remarkable in that the mean daily urine volume of the diabetic animal fell from 70 ml/day to a normal level of 12–15 ml/day, 16 weeks after treatment. The hypoglycaemia was reduced to normal after 13 weeks. Histological examination of the sacrificed rat (after 16 weeks) showed dead beta cells in the original islets. Rat islets lodged in the blood vessels and liver showed the stained secretory granules (aldehyde-fuchsin). This paper is tantalising and hopefully more work on the *in vivo* responses to cryopreserved islets will be pursued.

Thyroid tissue

Mateeva [49, 50] investigated the cryoprotective action of glycerol, DMSO and polyvinylpyrrolidone (PVP) on rat thyroid gland fragments stored for several days at –196°C after cooling slowly at 1°C/min to –40°C. Thereafter samples were plunged into liquid nitrogen for storage; thawing was achieved by warming in a water bath at +40°C. Parker '199' tissue culture medium, supplemented with 10% bovine serum was used as a suspending medium to which the three cryoprotective agents had been added separately. Viability was estimated by cellular proliferation in diffusion chambers implanted intraperitoneally in white rats. DMSO (10%) was the best agent; PVP was very poor.

Human thyroid tissue has been preserved at –196°C for up to 8 weeks in 15% glycerol (rate of cooling and thawing not controlled). Thawed tissue was viable as judged by the stimulation of cyclic-AMP production by the long acting thyroid stimulator (LATS) and the human thyroid stimulator (HTS). Thus thyroid slices could be satisfactorily employed for bioassay [51].

Parathyroid

Rat parathyroid tissue has been successfully preserved [52] at –196°C with 10% DMSO in Waymouths medium containing 10% syngeneic rat serum. Parathyroid tissue functioned normally for 9 months and in some instances 12 months after transplantation of the thawed material. Samples were cooled at 1°C/min to -80°C and then cooled rapidly to -200°C, prior to storage at -196°C. Samples were thawed rapidly by immersion in water at 37°C. Before transplantation into hypocalcaemic rats the freshly thawed cryopreserved glands were washed three times with Waymouths medium at +4°C to remove DMSO.

Liver cells

There is an increasing need for isolated animal hepatocytes in a variety of *in vitro* systems for the study of liver metabolism and the action of

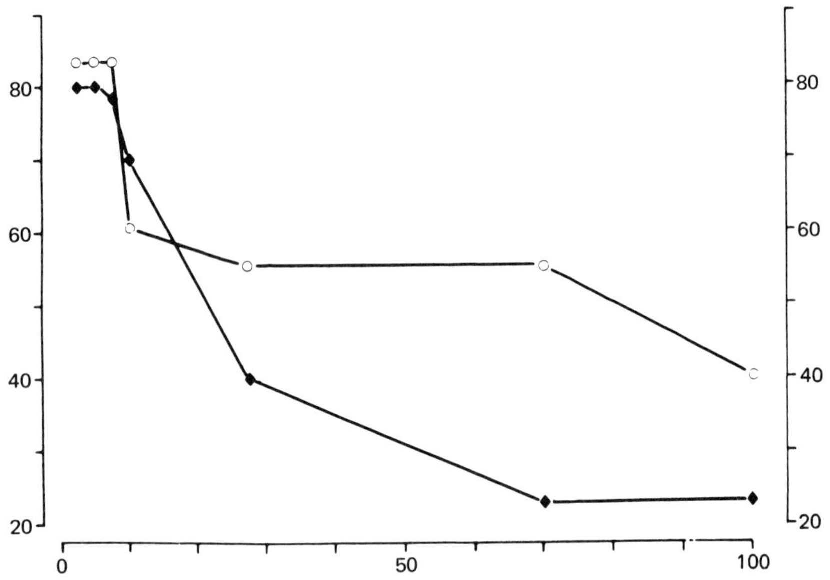

Figure 2.9 The effect of the cooling rate on liver cell survival, and ability to bind insulin. Cells isolated by method 2, resuspended in KRb-4% BSA supplemented with tryptose phosphate and DMSO, were cooled at the indicated rates. After thawing and washing, the viability (♦) of the frozen cell suspensions corresponding to each cooling rate was assessed using a trypan blue test and the ability to bind ^{125}I insulin at 0.1 nM (○) was measured as described above. Similar experiments (control) with respect to cell concentration, buffer conditions, but with cells that had not been frozen were simultaneously carried out. The data are expressed as a percentage of the controls. Abscissa: cooling rate (°C/min); ordinate: (left) survival of cells (% of control); (right) binding of ^{125}I insulin (% of control). (From Cam *et al.*, 1976 [53], by permission of the authors and publisher).

hormones. The cryopreservation of adult rat liver cells has been reported by Le Cam [53]. Cells were suspended in Eagle's basal medium containing 10% DMSO (v/v) and 10% (v/v) calf serum or else in Krebs buffer containing 4% BSA, 4% tryptose phosphate and 10% (v/v) DMSO. Viability of cells after freezing and thawing was assessed using trypan blue exclusion and the ability of cells to bind ^{125}I insulin. Cooling rates were varied from 2°C/min to 100°C/min; the thawing rate was not stated but was rapid and probably of the order 150–200°C/min. Two washes to remove DMSO were employed before viability assays. Results are shown in Figure 2.9 and demonstrated an optimum survival of between 80 and 85% of control values at cooling rates between 2 and 7°C/min. Ability to bind insulin (Figure 2.10) was

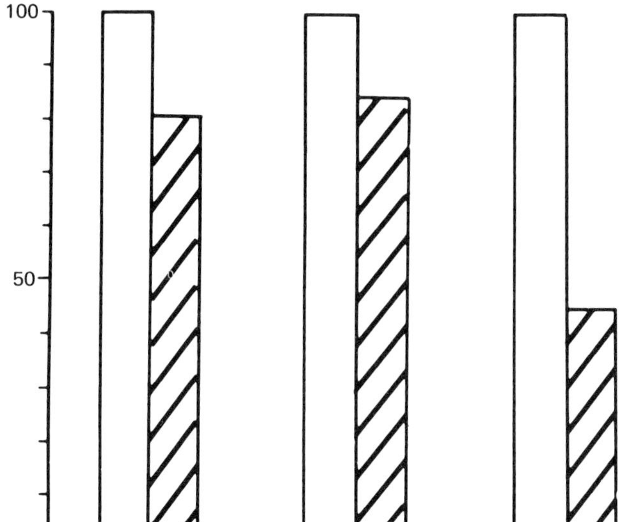

Figure 2.10 Cell survival, insulin-binding ability and gluconeogenic capacity in cells that had been frozen and in control cells. Similar sets of cells were treated as in Figure 2.9; one set was frozen (hatched bars) at a cooling rate of 7°C/min; the other set (open bars) was maintained at 4°C during the same period. After thawing and washing cells of both sets were assessed for viability (trypan blue test), ability to bind ^{125}I insulin (see legend to Figure 2.9) and gluconeogenic capacity (the glucose production was measured for a 2 h incubation time in the presence of 10 mM pyruvate). The data are expressed as a percentage of the controls. Abscissa: (a) cell survival; (b) insulin-binding ability; (c) gluconeogenic capacity; ordinate: % of the controls. (From Le Cam, *et al.*, 1976 [53], by permission of the authors and publisher).

decreased by about 15% after freezing and thawing compared with unfrozen controls. An examination of Figure 2.10 reveals, however, a disturbing fall, commented on by the authors, of the gluconeogenic activity of hepatic cells after freezing; the activity was decreased by 55%. Fine structure, as revealed by electron microscopy, and the insulin receptors were not extensively damaged by freezing and thawing procedures. However, mitochondrial and microsomal synthetic mechanisms were damaged and it is clear that further research is necessary before completely acceptable hepatocyte preservation is a routine procedure.

Recent studies by Idoine and her colleagues [54] have been concerned with the long term culture and cryopreservation of rat liver cells in culture. These authors were interested in test and model systems relating cell culture to the biochemistry of carcinogenesis and as the ability to metabolise carcinogens is largely centred on the liver such studies are apposite. The cell lines were 'epithelial like' and derived from the liver of young rats. Levels of aryl hydrocarbon hydrolase (AHH), an enzyme associated with the activation of a number of carcinogens, were monitored. Cells were frozen in Williams D medium supplemented with 15% fetal calf serum and containing 9% DMSO. Freezing and thawing rates, unfortunately, were not recorded in these studies but storage temperatures were -70°C for 4 months and then -100°C for extended periods. Morphology of cells before and after freezing and storage for 17 months at low temperature was not changed. Activity of AHH enzyme remained unaltered and a second enzyme, tyrosine aminotransferase (TAT), was not decreased in activity as a function of freezing and thawing or associated cryopreservation. A very important adjunct to this study was the finding that no evidence for malignant transformation (as indicated by the ability of cells to grow in soft agar) was seen. Thus it may be inferred that there was genetic stability and no selection either as a result of freezing or thawing or as a result of storage. Research with Chinese hamster ovary cells by Ashwood-Smith and Grant [26] and with mouse embryos by Lyon *et al* [55] supports the view that genetic effects are not produced by normal cryopreservation methods. Idoine *et al* [54] state that 'the response of a cell (or organism) depends to a great extent on the levels of metabolizing enzymes; if these enzyme levels change with the cell generation, the apparent response to a test material may not be due to its carcinogenicity but to a gradual aging of the cells'. Therein is the advantage of cryopreservation.

Microsomal enzymes P450, P448 are stable, *in vitro*, to freezing and thawing and preservation is not enhanced by the addition of either

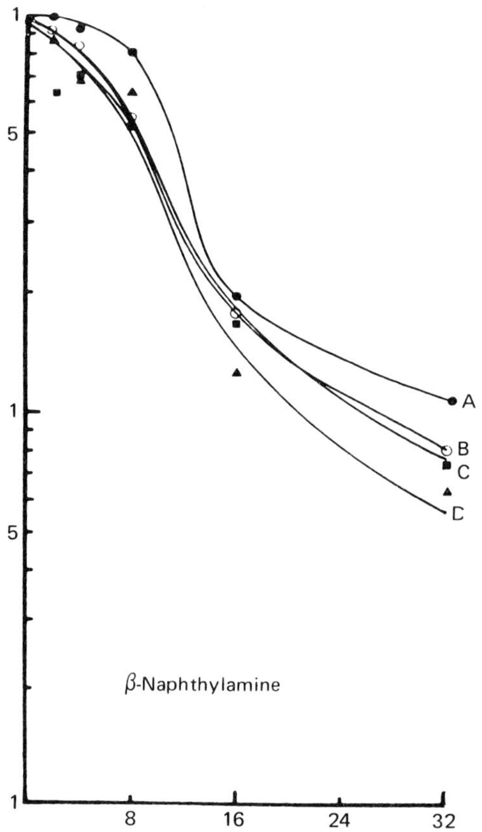

Figure 2.11 Effect of freezing and thawing liver microsomes on the detection of histidine mutants of *Salmonella typhimurium* (TA 1535) in response to the carcinogen β-naphthylamine (100 μg). A, unfrozen control; B, frozen and thawed, no additives; C, frozen and thawed, 10% DMSO; D, frozen and thawed, 10% glycerol. Ordinate: percentage of decrease in detection of mutants (log base 10); abscissa: microsomes diluted from 1 (no dilution) to 32 (1/32). (From Ashwood-Smith, 1977 [56], by permission of the author and publisher).

glycerol or DMSO [56]. Results are illustrated in Figure 2.11. These enzymes are necessary for the formation of active mutagens from carcinogens such as β-naphthylamine and aminoanthracene. The Ames test for carcinogens as mutagens depends on the activity of these microsomally associated enzymes.

Heart and kidney

There have been a number of reports on the ability of fetal animal hearts to resume some form of activity after freezing and thawing and these early studies have been reviewed by Luyet [57]. Some activity after freezing and thawing in 10% DMSO was observed by Rajotte *et al* [39] when fetal mouse hearts were reimplanted into mouse ears. Larger hearts had to be cut in half for protection to be achieved! The heart beat strengths were, however, approximately half that of control, non-frozen hearts. Adult mammalian hearts have not been successfully frozen and thawed and the immediate prospects are not hopeful unless variations and applications of Mazur's theoretical equations to cell freezing could be the key. The cryopreservation of kidneys has been equally unsuccessful although considerable information on perfusion physiology has been gained (see, Pegg [1]). The successful freezing of dog kidneys to -22°C in 12.5% DMSO has been reported by Rebelo *et al* [58]. The holding temperature of -22°C was far too high for 'useful' long term storage, however.

Offerijns *et al* [59] and Offerijns and Ter Welle [60] have devoted much of their energies to cryobiological studies of isolated and cultured heart cells with special reference to the effects of cryoprotectants and hypotonic stress on contractility. It was hoped that such cells would offer a good model system for cryobiological studies of highly differentiated mammalian cells. The hopes, were in fact, borne out and a detailed series of papers by Alink and his colleagues was included in a thesis presented by Alink to the University of Amsterdam *(Heart Cell Preservation by Freezing, a Cryobiological Study,* G.M. Alink, 1977). Ultrastructural changes in rat hearts after freezing and thawing with and without DMSO have been carefully studied by Karow and Schlafer [61] and correlations made between size of ice crystals and cooling rates. A cellular approach to the possibilities of kidney preservation has not been attempted and such an attempt might, in fact, be impossible given the varied histology of the kidney.

Preservation of vein grafts

Weber *et al* [62, 63] have made a series of studies on the long term preservation of canine jugular veins frozen to −196°C in 15% DMSO solution. Homologous vein grafts in man are used for peripheral vascular reconstruction and immunological reactions are generally regarded as being weak. 15% DMSO was shown to be superior to other cryoprotective agents for the preservation of vein smooth muscle contractile ability (freezing rates were surprisingly high; 5°C/s; thawing rates were also 5°C/s). Samples of canine vein were kept at −120°C for up to 28 days before thawing and autografting in the carotid arteries.

Frozen, preserved grafts were regarded as identical in function to fresh veins with patency rates of between 62.5 and 87.5% and which, after grafting, continued for a year. Scanning electron microscopy showed intact endothelium. Veins frozen in the absence of DMSO had a high rate of thrombosis and only one of these grafts lasted a year and scanning electron microscopy indicated disrupted endothelium.

Viable heart valves

Heart valves preserved with DMSO at −196°C have been shown to be viable [64]. These experimental studies on the dog have been applied to human valves. The use of frozen, x-irradiated aortic valve homografts was investigated by Beach *et al* [65] who claimed considerable success.

Teeth

The cryopreservation of developing teeth has been studied by Bartlett and Reade [66]. Mouse 2 days old neonatals were used as donors into adult males. The tooth germs were suspended in phosphate buffered saline containing 20% inactivated fetal calf serum and 10% DMSO (v/v). One hour after incubation of the samples at +4°C they were cooled at approximately 2–5°C/min to −90°C before immersion directly into liquid nitrogen at −196°C and left for periods of up to three months. Samples were thawed rapidly (rates not given) and after removal of DMSO isologously transplanted. Histology and mitotic activity were assayed in the transplants at periods of 1–20 days after transplantation. It was estimated that 86% of the cryopreserved developing teeth commenced to grow and produce normal dental hard and soft tissue structures.

Low temperature preservation of cornea

It is now considered that the cornea is not immunologically privileged and that some late developing failures in corneal preservation, whether with fresh or preserved material can be attributed to immune reactions (see discussions in CIBA Foundation Symposium, 1973 [67]). Bourne [68] has shown that the antigenicity of dog corneal tissue used for interlamellar xenografts in rabbits was not changed by the cryopreservation method originally described by Capella *et al* [69]. The resistance of cells to antigenic change as a result of freezing and thawing is a widely observed phenomenon.

The early literature describing the successful preservation of full thickness rabbit and human corneas [82–85] has, surprisingly, not been sufficiently exploited clinically. Recent transplantations of human corneas are largely based on the techniques of Kaufman and Capella [70].

A useful and critical review of this area has been prepared by Van Horn and Schultz [71]. Methods for intermediate term storage as well as long term storage in organ culture are surveyed by these authors and the review contains a useful description of various methods used for the critical evaluation, *in vitro*, of corneal functions prior to keratoplasty. The assessment, *in vitro*, of corneal functions is a difficult and controversial area. There appears little doubt, however, that successful cryopreservation requires an intact and functioning endothelium and nearly all the workers on corneal preservation are agreed on this point (see Stocker [72]).

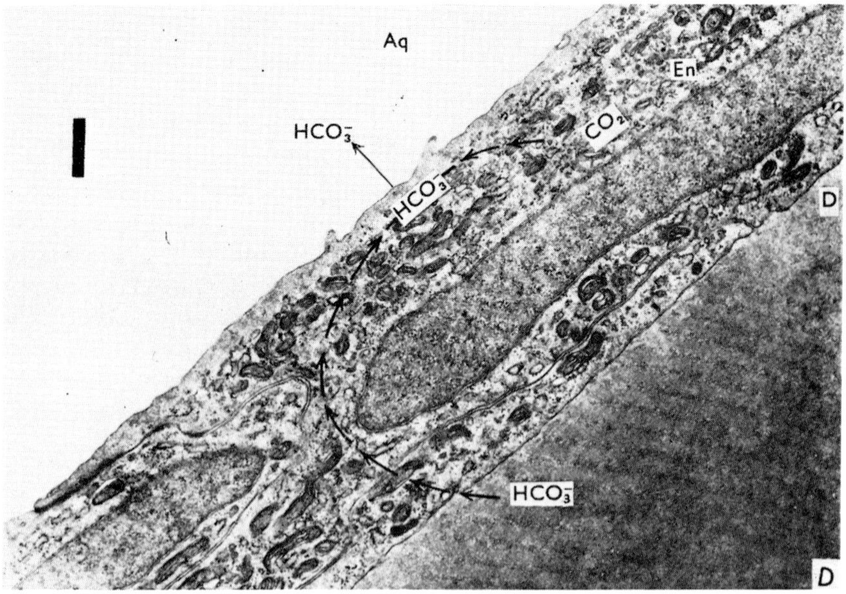

Figure 2.12 Electron micrograph of rabbit corneal endothelium. Vertical bar equals 1 μm. Aq is aqueous humour, En is the endothelium, D is Descemet's membrane. The proposed scheme of the bicarbonate ion transport system which regulates corneal hydration. The pump is supposed to be located at the posterior membrane of the endothelium, two-thirds of its substrate is supplied by HCO_3^- entering from the stromal side through Descemet's membrane (D), one-third is supplied by the intracellular conversion of CO_2 to HCO_3^- by carbonic anhydrase located just underneath the membrane. (From Hodson and Miller, 1976 [73], by permission of the authors and publisher).

The controversies on assessment centre on the measurement of endothelial function. The movement of water into the corneal stroma from the aqueous humour is controlled by the endothelium which also acts as a water and ion pump. Most methods for evaluating endothelial function destroy the cells so that grafts cannot be performed on the actual cornea which is assessed. Techniques which have been used include histochemical staining (nitro-blue tetrazolium) to measure dehydrogenases and diaphorase activity and trypan blue staining. Perhaps specular microscopy 'should prove invaluable in evaluating the state of the endothelium after penetrating keratoplasty' (Van Horn and Schultz [71]).

Interesting and fundamental studies by Hodson and his colleagues in Cardiff have considerable bearing on the problems of assessment of corneal viability *in vitro*. Hodson and Miller [73] have proposed a model of corneal water relations in which the endothelium pumps an anion into the aqueous humour; bicarbonate or hydroxyl ions are suggested as substrates for the 'anion pump' as chloride, phosphate and sulphate were not detected as part of the 'pump'. Results indicated that about 70% of the endothelial pump capacity is associated with the translocation of HCO_3^- from corneal stroma to aqueous humour. The localization of the scheme is illustrated in Figure 2.12.

The importance of these fundamental studies is illustrated in their application to the estimation of the viability of corneal storage in 'MK medium' (TC 199 medium with 5% dextran and antibiotics, McCarey and Kaufman [74]). Periods of storage of up to 2 weeks were obtainable, without freezing, with this method (see Mayes *et al* [75]). The relatively low concentration of bicarbonate in MK medium (4.2 mEq $.l^{-1}$ bicarbonate) is a factor in good corneal survival (see Figure 2.13). Interestingly, transendothelial potentials across rabbit corneal endothelium still occurred at 4°C (Figure 2.14).

No statistical differences between three different corneal storage methods have been observed by McCarey *et al* [74] when comparisons of human postoperative thickness measurements were made. Storage in these studies was accomplished in traditional moist chambers, 'MK' medium and by cryopreservation [70].

The success of the various methods for the long term cryopreservation of corneas has had important implications in transplantation medicine. Unfortunatley attempts with other tissues of medical importance, with the exception of skin, have not been so successful.

Skin

Clinical experience with viable frozen human skin has been documented

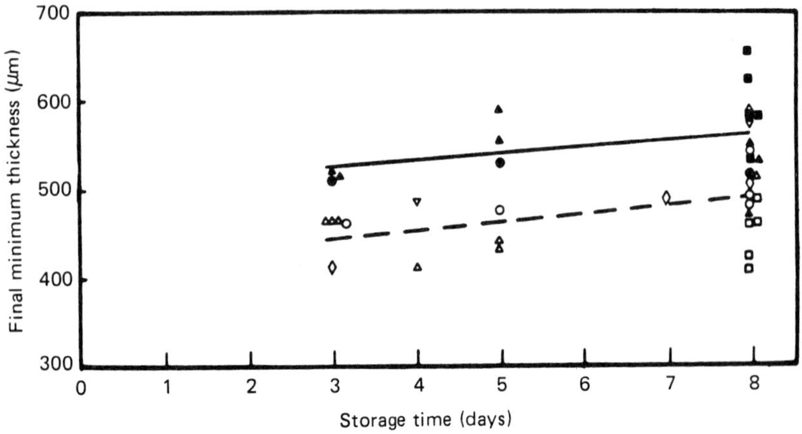

Figure 2.13 Final minimum thickness achieved after deturgescence at 35°C by rabbit corneas stored by refrigeration in various media. High bicarbonate media: (▲) aqueous humour (whole eyes in moist chamber); (●) immersed in bicarbonate–glutathione Ringer (Dikstein and Maurice, 1972 [81]) with 5% dextran added; (■) immersed in MK medium with 41 mM sodium bicarbonate added. Low bicarbonate media: (○) immersed in unmodified MK medium; (□) immersed in MK medium with 41 mM sodium chloride added; (△) suspended in air after washing with bicarbonate and carbon dioxide-free Ringer [73]; (◇) suspended in air after washing with bicarbonate and carbon dioxide-free Ringer plus 5% dextran; (▽) immersed in bicarbonate and carbon dioxide-free Ringer plus 5% dextran. Regressions of final minimum thickness on storage time: continuous line, high bicarbonate storage (slope not significant); broken line, low bicarbonate storage (slope significant at the 5% level only). (From Mayes *et al.*, 1978 [75], by permission of the authors and publisher).

by Bondoc and Burke [76]. In these studies human skin was cryopreserved with 15% glycerol in Ringer's solution and cooled to −70°C at an approximate rate of 1°C/min, a rate which was originally used, with success, by Billingham and Medawar as early as 1952. Samples were then transferred to a storage temperature of −160°C, cooling was rapid and not stated but probably of the order of 80–100°C/min. Bondoc and Burke concluded that '. . . frozen skin grafts subject to slow freezing and storage at liquid nitrogen temperatures possess the same biologic and clinical properties of freshly harvested skin grafts'. Rabbit skin treated with DMSO (10%) and cooled to −80°C at a rate of 5–10°C/min has been successfully preserved by Nathan *et al* [77]. Thawing was fast (not stated) and *in vitro* tissue culture on pig skin was used prior to autografting.

Changes in the metabolism of rat skin after cryopreservation at −196°C in the presence of different amounts of glycerol in combination with different cooling rates have been observed by de Loecker *et al* [78]. Both protein and DNA synthesis were lower than with fresh skin and indications were that some of the damage resulted from the initial contact of the skin with the buffer solution containing the cryoprotectant.

The freeze-drying of diploid mammalian cells

Several unsuccessful attempts have been made in efforts to lyophilise

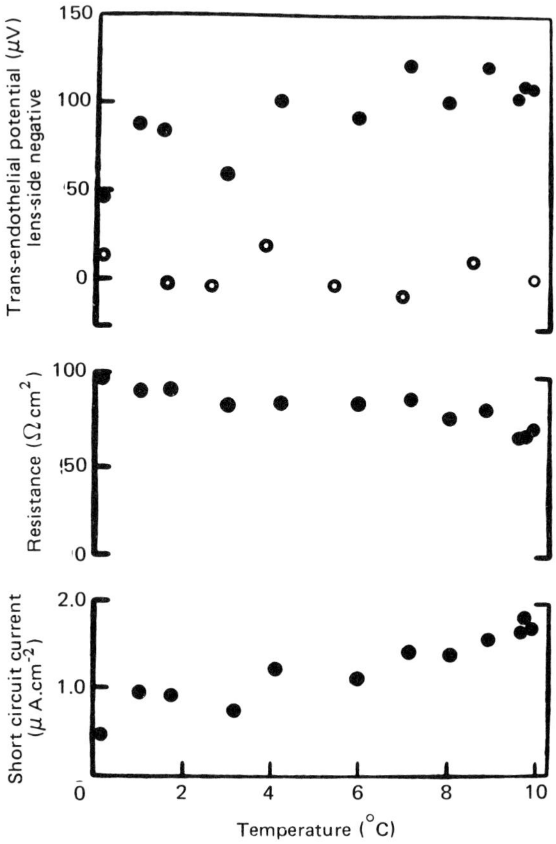

Figure 2.14 Transendothelial potentials, resistances and short circuit currents across rabbit corneal endothelium at refrigerated temperatures incubated in: (●) bicarbonate-glutathione Ringer; (○) bicarbonate and carbon dioxide-free Ringer. (From Mayes *et al.*, 1978 [75], by permission of the authors and publisher).

diploid mammalian cells including attempts by the author of this chapter! The usefulness of methanol [38] as a cryoprotectant was revealed during these studies as it was considered essential to use a volatile agent which would be removed during the drying process, to protect cells against damage produced during the initial freezing. However, Thomas *et al* [79] and Damjanovic *et al* [80] have reported the successful freeze-drying of mouse spleen cells from PVP solutions. Success was measured in terms of their usefulness as target cells for certain immunological reactions which required surface receptor molecules to be intact. It would appear, however, extremely doubtful if any of the cells were 'alive' in terms of their ability to divide. Freeze-dried cells were 'totally unable to bind either PHA or concanavalin A and did not form rosettes when mixed with sheep red blood cells' [79]. The trypan blue viability tests, however, yielded values indicating 86% viable cells. Non-staining of cells in this test equals viability and the test is notoriously capricious. Staining of cells correlates well with cellular death, however.

At the present time evidence for the lyophilisation of diploid mammalian cells is very unconvincing although Greiff (personal communication) has preliminary data suggesting some success with lymphocytes.

Conclusions

The successful application of the cryobiological principles developed by Mazur resulting in the low temperature preservation of the embryos of a number of mammalian species suggests that organ preservation may not be such a dream as it seemed ten years ago. The extensive work on perfusion physiology by Pegg and his colleagues and the MRC Cryobiology Group has proved to be a very necessary and basic prerequisite in the quest for organ preservation.

The pioneering studies on the effects of low temperatures on cells and tissues by Dr Audrey Smith are the foundation on which much of modern cryobiology rests. Her book *Biological Effects of Freezing and Supercooling* published by Edward Arnold in 1961, should be consulted for many key references.

References

1 Pegg, D.E., (1973). In *Organ Preservation.* Edinburgh and London: Churchill Livingstone
2 Sell, K.W., (1971). *Transplantation Proceedings* **3**, 274
3 CIBA Foundation Symposium 52 (new series) (1977). In *The Freezing of Mammalian Embryos,* pp. 45, 122, 225. Amsterdam, Oxford, New York: Elsevier, Excerpta Medica
4 Ashwood-Smith, M.J., (1973). *Cryobiology* **10**, 502

5 Voss, W.A.G., Warby, C., Rajotte, R. and Ashwood-Smith, M.J., (1972). *Cryobiology* **9**, 562
6 Mazur, P. (1963). *Journal of General Physiology* **47**, 347
7 Mazur, P. (1970). *Science* **168**, 939
8 Leibo, S.P. and Mazur, P. (1971). *Crybiology* **8**, 447
9 Bank, H. and Mazur, P. (1973). *Journal of Cell Biology* **57**, 729
10 Bank, H. and Mazur, P. (1972). *Experimental Cell Research* **71**, 441
11 Mazur, P., Leibo, S.P. and Chu, E.H.Y. (1972). *Experimental Cell Research* **71**, 345
12 Farrant, J. (1965). *Nature* **205**, 1284
13 Farrant, J., Walter, C.A. and Armstrong, J.A. (1967). *Proceedings of the Royal Society, London Series B* **168**, 293
14 Farrant, J. and Woolgar, A.E. (1972). *Cryobiology* **9**, 9
15 Farrant, J. and Woolgar, A.E. (1972). *Cryobiology* **9**, 16
16 Farrant, J. and Morris, G.J. (1973). *Cryobiology* **10**, 134
17 Frim, J., Snyder, R.A., McGann, L.E. and Kruuv, J. (1978). *Cryobiology* **15**, 502
18 McGann, L.E., Kruuv, J., Frim, J. and Frey, H.E. (1975). *Cryobiology* **12**, 530
19 McGann, L.E., Kruuv, J. and Frey, H.E. (1972). *Cryobiology* **9**, 107
20 Koch, G.J., Kruuv, J. and Bruckschwaiger, C.W. (1970). *Experimental Cell Research* **63**, 476
21 Terasima, T. and Yasukawa, M. (1977). *Cryobiology* **14**, 379
22 Hunt, C.J., Beadle, D.J. and Harris, L. (1977). *Cryobiology* **14**, 135
23 Rule, G.S., Frim, J., Thompson, J.E., Kepock, J.R. and Kruuv, J. (1978). *Cryobiology* **15**, 408
24 Holečková, E. and Cinnerová, O. (1972). *Experimental Cell Research* **72**, 600
25 Wodinsky, I., Meaney, K.F. and Kensler, C.J. (1971). *Cryobiology* **8**, 84
26 Ashwood-Smith, M.J. and Grant, E.L. (1977). In *The Freezing of Mammalian Embryos*, p. 251. CIBA Foundation Symposium 52 (new series). Amsterdam, Oxford, New York: Elsevier, Excerpta Medica
27 Lyman, G.H., Preisler, H.D. and Papahadjopoulos, D. (1976). *Nature* **262**, 360
28 Preisler, H.D., Houseman, D., Scher, W. and Friend, C. (1973). *Proceedings of the National Academy of Sciences* **70**, 2956
29 Travers, J. (1974). *Biochemistry* **47**, 435
30 Krystosek, A. and Sachs L. (1976). *Cell* **9**, 675
31 Young, B.S. and Wright, A. (1977). *Molecular and General Genetics* **155**, 191
32 Mironescu, S. and Seed, T.M. (1977). *Cryobiology* **14**, 575
33 Mironescu, S. (1978). *Cryobiology* **15**, 178
34 Hudita, H. (1959). *Low Temperature Science, Series B* **17**, 85
35 Meryman, H.T. (1968). *Nature* **218**, 333
36 Rapatz, G. (1973). *Cryobiology* **10**, 181
37 Polge, C. and Soltys, M.A. (1960). In *Recent Advances in Freezing and Drying*, p. 87. A.S. Parkes and A.U. Smith. (ed) Oxford: Blackwell
38 Ashwood-Smith, M.J. and Lough, P. (1975). *Cryobiology* **12**, 517
39 Rajotte, R.V., Dossetor, J.B. and Voss, W.A.G. (1976). *Cryobiology* **13**, 609
40 Ashwood-Smith, M.J. (1975). *Annals of the New York Academy of Sciences* **243**, 246
41 McGann, L. (1978). *Cryobiology* **15**, 382
42 Diller, K.R. (1975). *Cryobiology* **12**, 480
43 McGrath, J.J., Cravalho, E.G. and Huggins, C.E. (1975). *Cryobiology* **12**, 540
44 Leibo, S.P., Magrath, J.J. and Cravalho, E.G. (1975). *Cryobiology* **12**, 579
45 Mazur, P., Kemp, J.A. and Miller, R.H. (1976). *Proceedings of the National Academy of Sciences* **73**, 4105
46 Tellez-Yudilevich, M., Tyhurst, M., Howell, S.L. and Sheldon, J. (1977). *Transplantation* **33**, 217
47 Ferguson, J., Allsopp, R.H., Taylor, R.M.R. and Johnston, I.D.A. (1976). *British Journal of Surgery* **63**, 767
48 Rajotte, R.V., Stewart, H.L., Voss, W.A.G., Shnika, T.K. and Dossetor, J.B. (1977). *Cryobiology* **14**, 116

49 Mateeva, G. (1970). *Folia Medica* **12**, 309
50 Mateeva, G. (1974). *Folia Medica* **16**, 201
51 Knox, A.J.S., Westarp, C., Row, V.V. and Volpe, R. (1977). *Cryobiology* **14**, 543
52 Wells, S.A. and Christiansen, C. (1974). *Surgery* 75, 49
53 Le Cam, A., Guillouzo, A. and Freychet, P. (1976). *Experimental Cell Research* 98, 382
54 Idoine, J.B., Elliott, J.M., Wilson, M.J. and Weisburger, E.K. (1976). *In Vitro* 12, 541
55 Lyon, M.F., Whittingham, D.G. and Glenister, P. (1977). In *The Freezing of Mammalian Embryos.* p. 273. CIBA Foundation Symposium 52 (new series). Amsterdam, Oxford, New York: Elsevier, Excerpta Medica
56 Ashwood-Smith, M.J. (1977). *Cryobiology* **14**, 240
57 Luyet, B. (1971). *Cryobiology* 8, 190
58 Rebelo, A.E., Graham, E.F., Crabo, B.G., Lillehei, R.C. and Dietzman, R.H. (1974). *Surgery* 75, 319
59 Offerijns, F.G.J., Freud, G.E. and Krijnen, H.W. (1969). *Nature* **222**, 1174
60 Offerijns, F.G.J. and Ter Welle, H.F. (1974). *Cryobiology* **11**, 152
61 Karow, A.M. and Schlafer, M. (1975). *Cryobiology* **12**, 130
62 Weber, T.R., Dent, T.L., Salles, C.A., Ramsburgh, S.R., Fonseca, F.P. and Lindenauer, S.M. (1975). *Surgical Forum* **26**, 291
63 Weber, T.R., Lindenauer, S.M., Dent, T.L., Allen, E., Salles, C.A. and Weatherbee, L. (1976). *Annals of Surgery* **184**, 709
64 Mermet, B., Angell, W.W. and Dor, V. (1971). *Annales de Chirurgie Thoracique et Cardio-vasculaire* **10**, 463
65 Beach, P.M., Bowman, F.O., Kaiser, G.A. and Malm, J.R. (1973). *New York State Journal of Medicine* 73, 651
66 Bartlett, P.F. and Reade, P.C. (1972). *Cryobiology* 9, 205
67 Lyon, M.F., Whittingham, D.G. and Glenister, P. (1973). In *Corneal Graft Failure.* CIBA Foundation Symposium 15 (new series). Amsterdam, Oxford, New York: Elsevier, Excerpta Medica
68 Bourne, W.M. (1975). *Archives of Ophthalmology* **93**, 215
69 Capella, J.A., Kaufman, H.E. and Robbins, J.E. (1965). *Cryobiology* 2, 116
70 Kaufman, H.E. and Capella, J.A. (1968). *Journal of Cryosurgery* **1**, 125
71 Van Horn, D.L. and Schultz, R.O. (1977). *Survey of Ophthalmology* **21**, 301
72 Stocker, F.W. (1953). *Transactions of the American Ophthalmological Society* **51**, 669
73 Hodson, S. and Miller, F. (1976). *Journal of Physiology* **263**, 563
74 McCarey, B.E., Meyer, R.F. and Kaufman, H.E. (1976). *Annals of Ophthalmology* 8, 1488
75 Mayes, K.R., Graham, M.V. and Hodson, S. (1978). *Experimental Eye Research* **26**, 555
76 Bondoc, C.C. and Burke, J.F. (1971). *Annals of Surgery* **174**, 371
77 Nathan, P., Robb, E.C., Harper, A.D. and Ballantyne, D.L. (1978). *Cryobiology* **15**, 133
78 De Loecker, W., De Weever, F., Jullet, R. and Stas, M.L. (1976). *Cryobiology* **13**, 24
79 Thomas, D., Edwards, D.C. and Damjanovic, V. (1976). *Cryobiology* **13**, 191
80 Damjanovic, V., Edwards, D.C. and Thomas, D. (1975). *Nature* **253**, 116
81 Dickstein, S. and Maurice, D.M. (1972). *Journal of Physiology* **221**, 29
82 Smith, A.U., Ashwood-Smith, M.J. and Young, M.R. (1963). *Experimental Eye Research* **2**, 71
83 Mueller, F.O. and Smith, A.U. (1963). *Experimental Eye Research* **2**, 237
84 Mueller, F.O., Casey, T.A. and Trevor-Roper, P.D. (1964). *British Medical Journal* **2**, 473
85 Mueller, F.O., (1964). *British Journal of Ophthalmology* **48**, 377

CHAPTER THREE

Freezing of Spermatozoa

C Polge

Introduction

The young science of cryobiology is growing to adolescence and there is still much to be learned, but studies on the freezing of spermatozoa have played a significant part in the development of our present concepts of the biological effects of low temperature. Even in what might be considered the preconceptual era, the observation by Mantegazza in 1866 that human spermatozoa survived freezing in semen frozen to −17°C must rank as one of the earliest reports of the recovery of mammalian cells after exposure to a temperature below their freezing point [1]. It was a long time, however, before Jahnel noted in 1938 that a proportion of human spermatozoa survived after storage for comparatively long periods in semen frozen to the temperature of solid CO_2 or in the liquid gases [2]. Further experiments on human spermatozoa by Parkes in 1945 [3] revived interest in this subject and a few years later in 1949 one of the most important advances was made by Polge, Smith and Parkes, namely the discovery of the protective effect of glycerol on the spermatozoa of several species during freezing and thawing [4]. This opened up the field tremendously, not only to further work on spermatozoa, but also to experiments on a wide variety of other cells and tissues. The book by Luyet and Gehenio (1940), *Life and Death at Low Temperatures,* compiled much of the information available before that time and tentative hypotheses were developed as to the mechanisms involved in the preservation of life or the causes of death at low temperatures [5]. Outstanding among the new generation of cryobiologists, however, was Audrey Smith whose monograph *Biological Effects of Freezing and Supercooling* published in 1961 records the great progress that was made during the next decade [6].

The impetus to work on spermatozoa received its biggest boost when bull spermatozoa were successfully frozen and thawed by Smith and Polge [7] and when the first calves were produced following the insemination of semen stored at −79°C [8, 9]. The fundamental importance of these experiments was to introduce the concept that 'slow' freezing in the presence of glycerol was less damaging to the cells than 'very fast' freezing. However, the practical implications of being able to store semen for prolonged periods of time generally took precedence over the more basic aspects of the biological effects of freezing and thawing because artificial insemination was becoming extensively used for the breeding of farm animals, particularly cattle. Thus, there followed many experiments which in essence recorded developments and modifications of the technique devised by Smith and Polge [7] and extended by Polge and Rowson [9]. A basis for semen preservation at very low temperatures was established, at least as far as bull spermatozoa were concerned, and the technique was soon adopted as routine practice in cattle breeding. Nevertheless, it is hardly surprising that many of the modifications which were studied led to the improvement of a method which in the first instance was developed empirically. Of particular importance have been the use of liquid nitrogen for freezing and long-term preservation, the development of small plastic 'straws' for containing the semen [10] coupled with more rapid freezing, or alternatively the method of freezing semen in small 'pellets' [11]. Work on spermatozoa of species other than cattle was also extended and revealed important differences which have necessitated modifications in the methods used for their preservation at low temperatures. It is probably true to say that more has been written about the freezing of spermatozoa than any other cells and a comprehensive review of many of the techniques used for semen preservation has recently been published by Watson [12]. It is certainly not intended in this chapter to cover this ground again or to provide recipes for freezing the spermatozoa of any particular species. Rather, an attempt will be made to review some of the work which points towards or underlines basic principles.

With the growth of experimentation, numerous theories have emerged concerning factors affecting the survival of cells during freezing and thawing and these are discussed in more detail in Chapter 1. Briefly, however, most evidence now supports the concept that intracellular crystallisation, at least of a large part of the water, is incompatible with cell survival. Very rapid cooling from temperatures above the freezing point increases the likelihood of intracellular ice formation. This may be avoided by slower cooling whereby ice formation outside the cells permits cellular dehydration and shrinkage. Under these conditions, however, the cells may become exposed to potentially damaging environmental hazards due to increased concentrations of solutes, and agents such as glycerol may protect against these effects. Nevertheless, it should be stressed that crystallisation of some intracellular water may not necessarily be damaging and cryoprotective agents do not necessarily have to permeate the cells. The conditions required to obtain maximum survival of different cells vary enormously and depend on individual characteristics. These obviously include size and permeability to water or to various solutes, but also other less well defined characteristics are probably involved which may be associated with the nature of the cell membrane. How do some of the observations which have been made on spermatozoa fit in with these concepts and how have they helped to mould ideas? In considering these questions it should be borne in mind that the experiments on spermatozoa have spanned many years and what might seem well accepted today was probably far from obvious at the time when much of the work was done.

Cooling rate and cryoprotective agents

In the absence of cryoprotective agents, very few spermatozoa of any species studied have been found to survive freezing to very low temperatures. The optimal cooling rate compatible with the survival of some of the cells during the freezing of neat semen has therefore not been determined. Human spermatozoa are interesting in that none survive when thin films are plunged directly into liquid nitrogen, but some survival can be obtained if undiluted semen is cooled more slowly. This was achieved as early as 1945 by plunging semen contained in ampoules directly into solid CO_2 and alcohol [3]. During ultra-rapid cooling the cells are probably damaged by the formation of intracellular ice, but within the centre of the ampoule the initial cooling rate of the semen should be retarded which perhaps allows the formation of sufficient extracellular ice to reduce the amount of intracellular nucleation. If this is the case, it is also evident that human spermatozoa possess considerable tolerance to the transient solution effects caused by the formation and dissolution of ice during freezing and thawing, a view substantiated by the very early observation relating to their survival after freezing at $-17°C$. Although the spermatozoa of different species withstand freezing for short periods in unprotected semen at various subzero temperatures and Lovelock and Polge showed that there are differences in their resistance to increased concentrations of electrolytes [13], none so far examined appear to be so hardy as those of the human. Methods of freezing and thawing which might enable some cells to survive in the absence of cryoprotective agents therefore become extremely critical and they are very hard to achieve in practice.

The addition of cryoprotective agents to the semen greatly extends the tolerance of spermatozoa to freezing at slower rates and it is now clear that the optimal cooling rate depends on the nature and concentration of the cryoprotectant that is used. Initially, most of the experiments were carried out with glycerol. Fowl semen diluted to contain 15–20% glycerol resumed full motility of the spermatozoa after thawing when ampoules containing the semen had been plunged directly from room temperature into alcohol at −79°C [14]. Bull spermatozoa required slower rates of cooling and the rate widely adopted for semen diluted in a yolk-citrate medium containing 7–10% glycerol was that shown in Figure 3.1 [15]. It soon became evident, however, that the semen could be frozen much more rapidly without causing a reduction in cell survival as when ampoules were placed

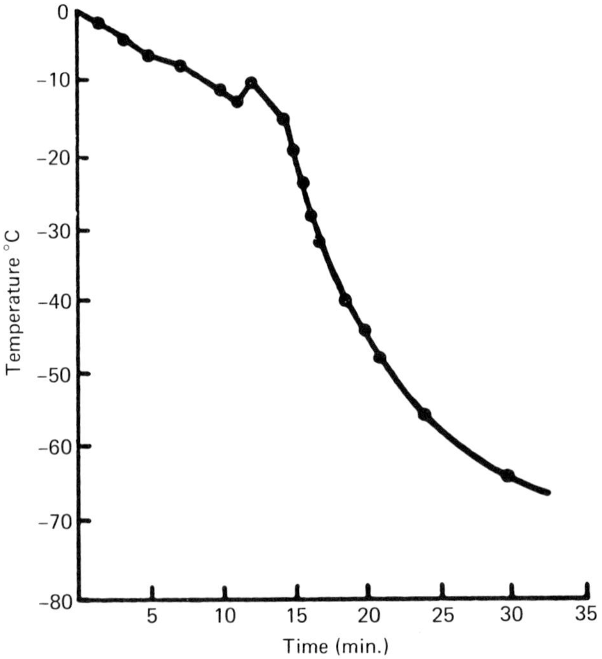

Figure 3.1 The cooling rate of 1.0 ml of bull semen in 15% glycerol contained in an ampoule of 1 cm diameter by immersion in an insulated vessel cooled in a bath at −79°C (Polge and Lovelock, 1952 [15])

directly from 0°C into crushed solid CO_2 [16, 17]. The current theory at that time was that the spermatozoa of many mammals, including those of the bull, had been shown to be irreversibly damaged by sudden chilling from body temperatures to 0°C and it was thought that this sensitivity to thermal shock might extend to temperature ranges well below 0°C. This necessitated some degree of slow cooling and the protective action of glycerol was explained by its ability to reduce the concentration of electrolytes to which the cells were exposed during the freezing process [13].

Some important experiments were then carried out in 1955 by Luyet and Keane [18] during an investigation of the effects of cooling velocity, within various temperature intervals, on the survival of spermatozoa in frozen bull semen. It was found that bull spermatozoa diluted in a yolk-citrate buffer containing 7% glycerol could be successfully cooled from 0 to −195°C in two abrupt steps. If tubes containing the semen were placed for 5 min in a bath at −27°C and then transferred directly to liquid nitrogen, about three-fourths of the spermatozoa survived. There was practically no survival when the terminus of the first step was −20°C. In this procedure the average cooling velocity was of the order of 20°C/min in the upper range from 0 to −25°C, and of some 15°C/s in the lower range from −27° to −195°C. Luyet and Keane concluded that the critical temperature, in the region of −27°C, seemed to be characterised by a sensitivity of bull semen to a too rapid cooling [18]. Stopping or slowing the cooling process in the critical range apparently afforded protection against freezing injury at lower temperatures. These results were confirmed by Polge [19], but it was also noted that when bull spermatozoa in 7.5% glycerol were cooled abruptly from 0 to −20°C in one step and then in another step directly to −79°C instead of −195°C, a very large proportion of the spermatozoa survived. By contrast, few spermatozoa survived abrupt cooling from −15 to −79°C. A full description of the results of experiments undertaken at that time is shown in Table 3.1. Clearly, abrupt cooling from 0°C to temperatures below −30°C caused a progressive reduction in sperm survival, but little damage was caused by freezing abruptly to temperatures above −30°C. In two-step freezing, as the terminus of the first step was shifted from 0 towards −30°C, so the velocity of freezing that could be tolerated in the second step was increased.

The fundamental importance of these observations in explaining the mechanisms of cell injury or survival was not fully appreciated at that time, but the first clue was provided somewhat later. Asahina found in 1959 that certain nematodes, chrysalises and overwintering pupae could

Table 3.1 Motility of bull spermatozoa after 'two-step' freezing

				First step			
	°*C*	*0*	*–10*	*–15*	*–20*	*–25*	*–30*
	–10	8					
	–15	7	7				
	–20	7	7	7			
	–25	7	7	7	7		
Second	–30	7	7	7	7	7	
step	–40	5	7	7	7	7	7
	–50	3	6	6	7	7	7
	–60	0	6	6	7	7	7
	–79	0	0	2	6	7	7
	–196	0	0	0	0	2	6

The semen was diluted 1/50 with medium containing 2% w/v sodium citrate, 25% v/v egg yolk and 7.5% v/v glycerol. After 6 h at 0°C, 0.2 ml in glass ampoules was transferred abruptly to the temperatures indicated and thawed 5 min later by immersion in a water bath at 37°C. In the two-step procedure, the semen was first frozen for 5 min at the temperature indicated on the top of the table, and then transferred abruptly to one of the lower temperatures before thawing. The motility of the spermatozoa was scored on a scale of 0 – 10 where 10 = 100% of the spermatozoa showing progressive motility. The initial motility of the spermatozoa = 8.

survive freezing in liquid oxygen provided they had been prefrozen at −30°C [20]. The hypothesis was developed that if animals are prefrozen extracellularly at −30°C, hardly any water crystallises in their tissue cells even in liquid oxygen, and in hardy organisms death by extracellular freezing takes a long time. The two-step freezing procedure has now been extended to studies on other cells and it has been possible to correlate cell survival with the amount of intracellular ice that is formed [21]. As far as bull spermatozoa in 7% glycerol are concerned, it would appear that the cooling velocity during abrupt transfer from 0 to −30°C (about 20°C/min) is probably sufficiently slow to permit extracellular crystallisation and consequent dehydration of the cells. By −30°C the cells have lost enough water so that very fast cooling thereafter, 15°C/s, is not injurious. Similarly, when the terminus of the first step is −20°C, the cells would not be dehydrated to the same extent and subsequent cooling at 15°C/s results in poor survival. By contrast, cooling at 1-2°C/s, as is obtained by plunging the ampoules directly from −20 to −79°C, allows the cells to lose more water before they reach very low temperatures. Whether any or how much intracellular ice is formed when bull spermatozoa are frozen under these conditions has not been determined. Larson and Graham found that when bull spermatozoa in 7% glycerol were plunged from −25 to −79°C, the proportion of motile spermatozoa decreased significantly during 2 weeks' storage at −79°C, which would suggest that some nucleation had taken place and crystallisation was developing [22]. In their experiments, however, the ampoules were contained in a bath which was cooling at a rate of 5°C/min and they were plunged to −79°C from progressively lower temperatures. It was noted that, although there was a marked increase in survival between samples plunged at −20 or at −30°C, there was a continued increase in survival if freezing at 5°C/min was continued to −45°C before very rapid cooling. The samples cooled at 5°C/min may have had insufficient time to reach equilibrium at the higher temperatures. Martin subjected bull spermatozoa in 7.5% glycerol to freezing rates of 2, 6 or 18°C/min to −25°C followed by transfer to −79°C [23]. It was found that motility after thawing was significantly improved by holding the semen at −25°C for 10 min before rapidly cooling to −79°C. However, the proportion of spermatozoa unstained by congo-red after thawing was reduced by holding for this time at −25°C after initial freezing at 2 or 6°C/min, but it was increased by holding at −25°C in samples frozen at 18°C/min.

Early in the studies on freezing bull spermatozoa in the presence of glycerol it had been noted that there was a critical temperature zone between −15 and −30°C in which injury to the spermatozoa was increased when they were held at these temperatures for progressively longer periods [19, 24]. It is within this temperature zone that the cells would be exposed to increased concentrations of solutes and therefore it might be considered advantageous to pass through this zone as quickly as possible. By contrast, it is within the same temperature range that the cells require sufficient time to lose water so that they can withstand cooling to lower temperatures. It is therefore a reconciliation of these two factors which determines optimal cooling rate in the presence of different cryoprotective agents or in the presence of different concentrations of a particular cryoprotective agent. This point is well demonstrated in experiments in which bull spermatozoa diluted in a yolk–citrate medium containing different concentrations of glycerol have been frozen by the two-step method [25]. The temperature from which the spermatozoa could be plunged directly into liquid nitrogen without causing a further reduction in motility was −15°C when the semen contained 1.75% glycerol, but was −25°C when the semen contained 7% glycerol (Figure 3.2). However, 1.75% glycerol provided much less protection to the cells than 7% during the initial step of freezing.

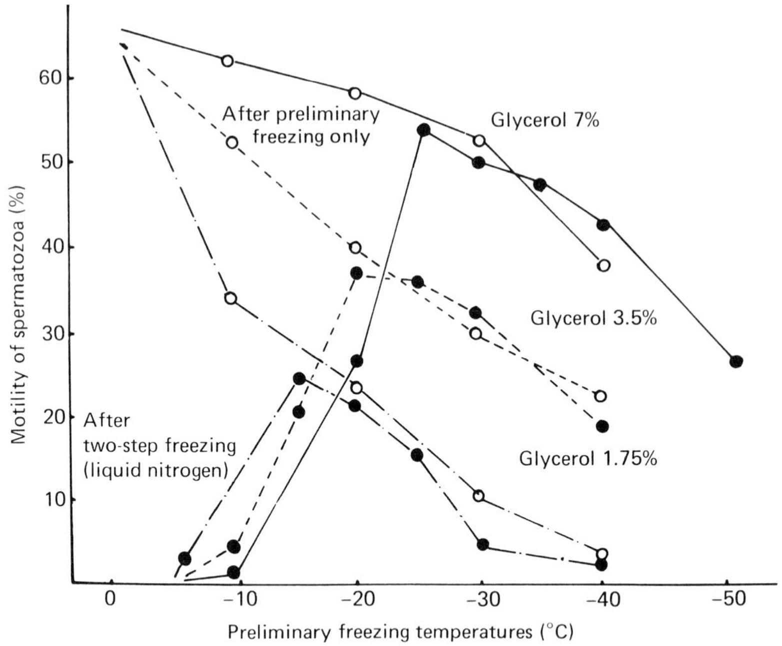

Figure 3.2 Post-thaw motility of bull spermatozoa in egg yolk-citrate diluent containing different glycerol levels. ○, after preliminary freezing only; ●, after two-step freezing; first at the preliminary temperature and then in liquid nitrogen (Nagase, 1972 [25])

One of the consequences of the experiments on two-step freezing was to allay the concept that slow cooling to very low temperatures was essential in order to avoid thermal shock. Very fast cooling rates could obviously be tolerated within certain temperature ranges. This opened up the possibility of re-examining protective effects of a number of neutral solutes other than glycerol during freezing and thawing. In early experiments by Lovelock on red blood cells, protection against haemolysis by freezing and thawing was obtained only with those solutes which were able to permeate the cells and which were non-toxic [26]. Smith and Polge examined the protective action of a number of polyhydric alcohols, including ethylene glycol and propylene glycol, during the freezing and thawing of avian and mammalian spermatozoa [14]. These substances permeated the cells and offered some protection, but Emmens and Blackshaw found that sugars gave very poor protection to bull spermatozoa during slow freezing [27]. Although it seemed clear at that time that a permeating solute was required to protect the cells during slow cooling, the question arose as to whether this also applied during much more rapid freezing. In 1960, Polge and Soltys compared the protective effect of six different solutes during the freezing of bull spermatozoa by the two-step method [28]. Three were permeating solutes, methanol, acetamide and glycerol, and three which did not readily permeate the cells were also chosen, xylose, glucose and sucrose. Permeating solutes were added to a solution of 2.2% sodium citrate containing 25% egg yolk, but because the addition of sugars to an isotonic medium rapidly reduced the motility of the spermatozoa, the non-permeating solutes were dissolved in water and egg yolk at a concentration which supported good sperm motility. After freezing to −79°C by the two-step method, it was found that the sugars provided excellent protection and the motility of the spermatozoa after freezing in the medium containing xylose was only slightly less than that in the medium containing glycerol (Table 3.2).

Table 3.2 Bull spermatozoa. Proportion surviving freezing when cooled abruptly to −79°C after preliminary freezing for 2 min at the temperatures indicated

Diluent	*Temperature drop (°C)*							
	0→	*−79*	*−10→*	*−79*	*−20→*	*−79*	*−30→*	*−79*
Yolk-citrate solution	(100)	0	(0)	0	(0)	0	(0)	0
5% methanol solution	(100)	0	(90)	0	(70)	0	(50)	0
7.5% acetamide solution	(100)	5	(70)	5	(50)	40	(60)	50
10% glycerol solution	(100)	5	(90)	5	(70)	50	(80)	70
7.5% xylose solution	(100)	25	(80)	50	(60)	55	(70)	60
7.5% glucose solution	(100)	15	(80)	40	(60)	50	(60)	50
10% sucrose solution	(100)	10	(70)	30	(60)	40	(60)	40

From Polge and Soltys (1960) [28]

Moreover, the temperature from which the spermatozoa could be frozen rapidly to −79°C was higher in the samples diluted in the sugars than in those containing glycerol. These experiments provided for the first time a method for freezing bull spermatozoa in the absence of permeating cryoprotective agents. They cast doubt on the idea that protection both within and without the cell was required during freezing; solutes such as sugars could provide adequate protection when they were used as diluents for the semen in the absence of extra salts and when the cooling rate was relatively rapid.

During the next few years, extensive studies, principally by Nagase and his colleagues [25], were made on bull spermatozoa frozen by the two-step method in yolk–glucose diluents containing various con-

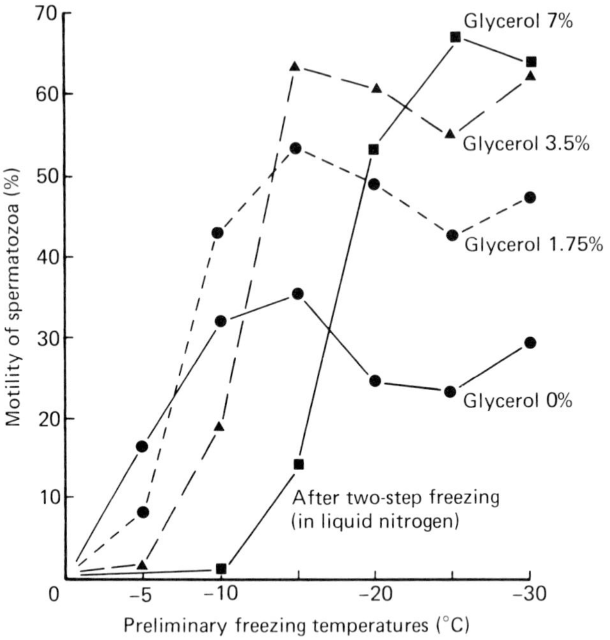

Figure 3.3 Post-thaw motility of bull spermatozoa after two-step freezing in egg yolk-glucose diluents containing different concentrations of glycerol (Nagase, 1972 [25])

centrations of glycerol (Figure 3.3). It was confirmed that a relatively high survival of spermatozoa could be obtained after freezing in yolk–glucose medium alone, although survival was increased by the addition of glycerol. As the level of glycerol in the medium was increased, however, lower temperatures of preliminary freezing were required in order to give the best motility after plunging into liquid nitrogen.

These observations led Nagase to develop the 'pellet' method of freezing [11] in which small volumes of semen (0.2 – 0.013 ml) were frozen by dropping them onto the surface of solid CO_2. The freezing rate of a pellet was considerably faster than the two-step method and it took about 4–5 min to cool from +5 to −75°C, but the time spent between 0 and −40°C was about 1 min. Quite good survival of spermatozoa in a yolk–citrate diluent containing 7% glycerol was obtained by this method, but the survival was better in yolk–glucose diluents. Moreover, the survival was equally good in yolk–glucose media containing only 1.75% glycerol. In the next few years, the protection afforded to bull spermatozoa by a wide variety of sugars and polyols during freezing by the pellet method was examined [29–32]. Disaccharides (maltose, sucrose, lactose) and trisaccharide (raffinose) appeared to provide better protection than pentose (xylose) and hexoses (glucose, galactose, fructose). When different polyols were tested, it was noted that the effect of these substances on the initial motility of the spermatozoa varied according to their molecular weight and their ability to permeate the cells. Inositol was relatively impermeable and supported good motility at a concentration of 0.3 M, but xylitol, erythritol and adonitol did not support motility in the absence of other solutes to increase the osmolarity. The postfreezing motility of spermatozoa was very low when egg yolk–polyol solutions were used as diluents without added solutes. By contrast, all the polyols tested provided some protection during pellet freezing when they were added to isotonic yolk-citrate or TRIS diluents.

The various cryoprotective agents studied were compared with glycerol and ethylene glycol during freezing at three different rates, slow freezing (5–7°C/min), pellet freezing (about 25°C/min) and very fast freezing of pellets dropped directly into liquid nitrogen. The results of this investigation (Figures 3.4 and 3.5) summarise very well the interactions between the type of cryoprotective agent and freezing rate [25]. What stands out most clearly is that glycerol and ethylene glycol are by far the most effective cryoprotective agents during slow freezing, but they are relatively ineffective during very rapid freezing. There is no doubt that non-permeating solutes provide good protection during rapid freezing. It is interesting that it has more recently been shown that a significant improvement in the motility of bull spermatozoa after pellet freezing can be achieved if the time allowed for glycerol to permeate the cells is severely curtailed. Post-thaw motility was better after glycerol exposure of 10 s at 5°C before freezing as compared with 2 min exposure or longer. Post-thaw motility was not influenced by the length of prefreezing exposure to lactose at 5°C [33, 34].

Today the most widely adopted method of packaging and freezing bull semen is in small plastic straws. The straws used initially contained 1.2 ml of diluted semen [35], but their size has now been reduced to a 'mini-straw' of very small diameter containing only 0.2 ml [36]. Jondet's method of freezing is very simple and consists of lowering a basket containing the straws in a vertical position onto the surface of liquid nitrogen. As the bottom of the basket comes into contact with the liquid nitrogen the vapour shoots up and envelops the straws ensuring a consistent and even freezing rate throughout each individual dose of semen. The straws are left to freeze in the vapour phase and the

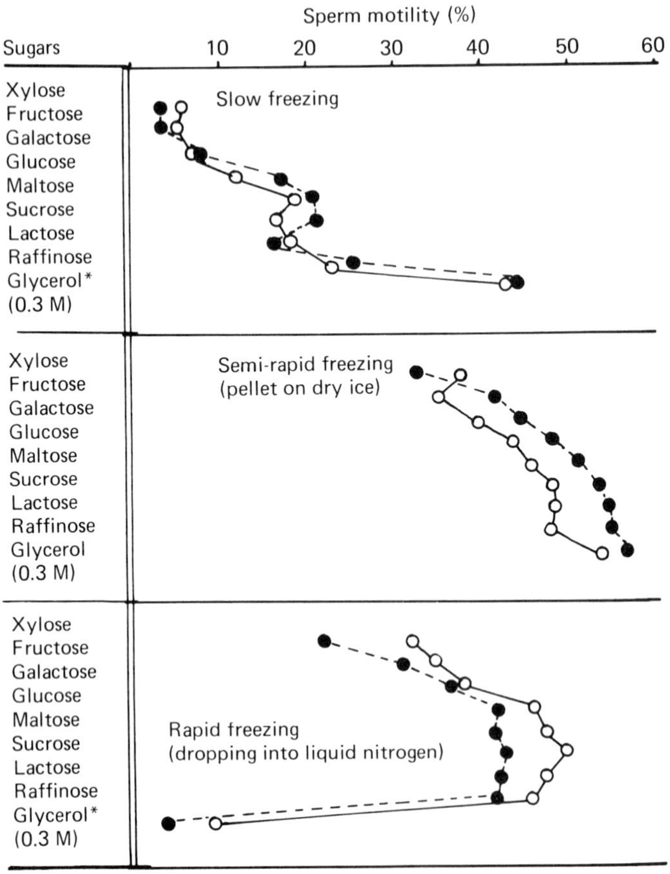

Figure 3.4 Protective effect of sugars on the motility of bull spermatozoa after freezing at different rates (Nagase *et al.,* 1968 [32]). ○, no addition of thawing media; ●, addition of thawing media to thawed samples; *, glycerol in egg yolk-citrate medium

time taken to cool from 0 to −196°C is about 3 min. By this method it is claimed that only 1.45% of the spermatozoa are destroyed by the freezing and thawing operations. Semen has been frozen successfully in a number of different media including egg yolk–citrate, egg yolk–lactose, milk or media buffered with TRIS. The most commonly used concentration of glycerol in the diluents is 7%, but equally good survival rates have been reported in glycerol concentrations ranging from 4 to 9%.

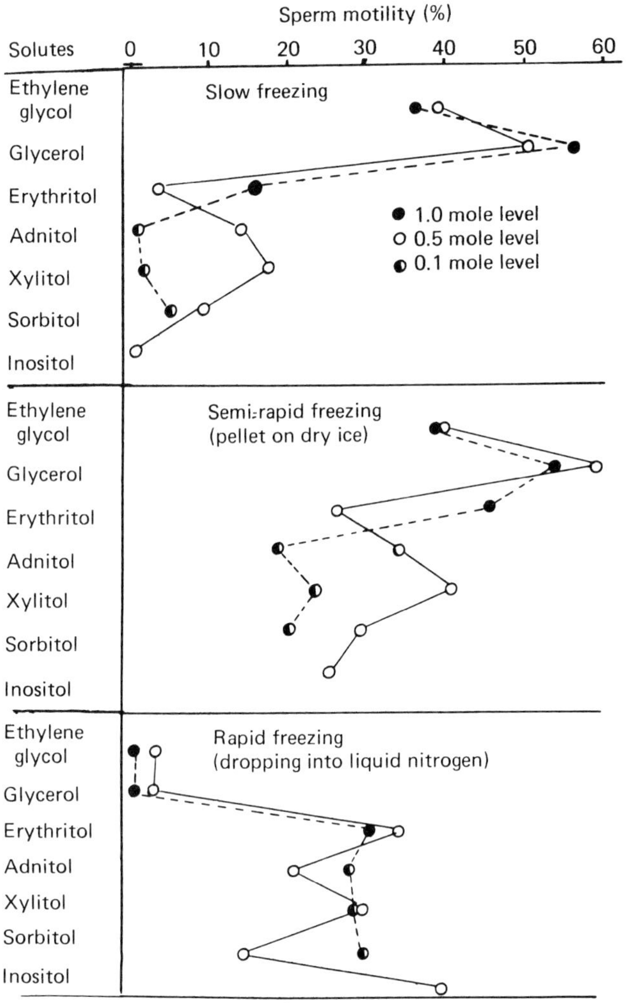

Figure 3.5 Protective effect of polyols in egg yolk-citrate diluent on the motility of bull spermatozoa after freezing at different rates (Nagase and Tomizuka, 1968 [31])

Other diluent components

Egg yolk has been used very extensively as a component of diluents for spermatozoa ever since its beneficial effects on the preservation of bull spermatozoa at temperatures above 0°C were first demonstrated in

1940 [37]. It has also been included in many of the diluents used for freezing spermatozoa. However, the important role of egg yolk in protecting the spermatozoa during freezing and thawing has perhaps not been fully appreciated. When bull spermatozoa were pellet frozen in TES buffer [38], a medium containing glucose, neither buffer alone nor buffer plus glycerol protected the spermatozoa [39]. By contrast, semen diluted in buffer containing egg yolk in the absence of glycerol had 24% motility after thawing. The components of egg yolk which afforded protection to the spermatozoa appeared to be associated mainly with a lipoprotein fraction of low density. There was a synergistic effect between the concentrations of glycerol and egg yolk in the diluent which was also noted in some of the very early experiments on freezing bull spermatozoa [40]. Further observations on the role of the low density lipoprotein fraction of egg yolk have been made by Watson [41] and it is assumed that it plays an important part in protecting the cell membrane.

Milk is another commonly used diluent either for storage of spermatozoa above 0°C or in combination with glycerol as a freezing medium. Some spermatozoa may survive freezing in milk alone, perhaps due to the presence of lactose or other beneficial components, but combination with glycerol and added fructose increases survival rate [42, 43].

There have been numerous experiments on the effects of diluent composition on the survival of spermatozoa of different species during freezing and thawing and a full discussion of these may be found in the recent review by Watson [12]. There are obviously complex interactions between the type of diluent, cryoprotective agent, freezing rate and thawing rate. Nevertheless, the principles underlying the successful preservation of spermatozoa are now reasonably well defined.

'Equilibration' period

In the very early experiments on bull semen by Polge and Rowson [9] it was noted that the motility of spermatozoa was increased in samples that had been diluted and stored at 2°C in the presence of glycerol for 18 h before freezing as compared with samples frozen soon after collection and dilution. This period of storage before freezing was referred to as the 'equilibration' period. Since then there have been very many experiments on varying the equilibration time; although there have been some conflicting results, the main body of evidence supports the beneficial effect of storage before freezing, but the time can be reduced to just a few hours. However, it soon became obvious that the spermatozoa did not require a long period to reach equilibrium with the glycerol, since motility was not reduced when very high concentrations of glycerol were added directly to the semen [13]. An improvement in motility after freezing was also obtained in samples which had been held at 2°C for 8 h in the absence of glycerol and when the glycerol was added just before freezing [17]. Spermatozoa thus acquired an increased resistance to freezing and thawing simply by storing the semen for a period of time at temperatures above 0°C. This phenomenon has not yet been explained, but it is probably associated with changes in the cell membranes which make them more hardy. Certainly the rate at which the capacity for motility of the spermatozoa is destroyed during freezing and storage within the critical temperature range of −15 to −25°C is considerably reduced by 'equilibration' before freezing [19]. Increased resistance to solution effects might be expected to be most beneficial during conditions of very slow freezing, but an advantage of equilibration has also been noted in samples frozen rapidly in straws and in this case there was an interaction between equilibration time and rate of thawing [44].

It is interesting that a storage period at temperatures above 0°C has also been shown to have a marked effect on the resistance of spermatozoa to thermal shock. This has been shown most clearly in boar spermatozoa which are particularly sensitive to cooling to temperatures below 15°C in undiluted semen. Both the motility of the spermatozoa and the integrity of the acrosomes are destroyed by cooling. By contrast, if boar semen is incubated for several hours at 25°C, the detrimental effects of subsequent cooling can be almost completely abolished [45]. Obviously the stresses due to freezing and thawing are different from those due to cooling under conditions where no ice is formed, but the increased hardiness acquired by spermatozoa during storage appears to be beneficial in both circumstances.

Thawing rate

In all the early experiments in which semen in media containing glycerol was frozen relatively slowly in ampoules, there was little evidence that the speed of thawing was a critical factor affecting survival of spermatozoa, although the ampoules were generally thawed by transferring them directly to water at about 30°C. Today, however, the majority of semen is frozen much more rapidly either in straws or by the pellet method. These conditions may be more conducive to the formation of some intracellular ice and the speed of thawing may then become more critical. With semen frozen in straws, rapid thawing rates obtained by immersing them for a few seconds in very hot water, up to 90°C, have generally been found to result in a slight but significant improvement in sperm motility and less damage to the acrosomes than when the straws are thawed in water at 5°C [46, 47]. In bull semen frozen by the pellet method there has been shown to be a very marked effect of thawing rate on the recovery of sperm motility and there is an interaction between thawing rate and the concentration of glycerol in which the semen is frozen (Figure 3.6) [25]. In semen containing 0 – 3.5%

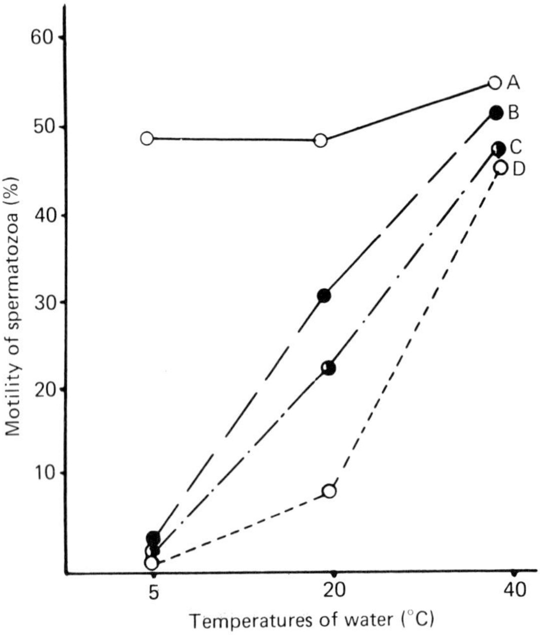

Figure 3.6 The effect of thawing temperature on the motility of bull spermatozoa frozen by the pellet method in egg yolk-glucose diluent containing different levels of glycerol (Nagase, 1972 [25]). A, Glycerol, 7.0%; B, glycerol, 3.5%; C, glycerol, 1.75%; D, glycerol, 0%

glycerol, slow thawing in dry test tubes at 5°C was very detrimental, but motility was increased by thawing at higher temperatures. When the semen was frozen in 7% glycerol there was little effect of thawing rate under these conditions. In practice, pellets are usually thawed by transferring them from liquid nitrogen to a warm thawing medium.

Fertility and variations between species

The spermatozoa of some 50 different species have now been frozen

[12]. There are variations in the success that has been achieved in preserving them at very low temperatures, but as the biological effects of freezing and thawing have become better understood, the methods employed for freezing spermatozoa have been extended. In many species the fertilising capacity of frozen semen has not been examined, but extensive studies have been made particularly in farm animals and to a lesser extent in laboratory species. It is in experiments on fertilising capacity that the greatest differences between species have become apparent. Fortunately for the cattle industry, in which artificial insemination is very extensively used, the fertility obtained with frozen-thawed bull semen is generally very good and the conception rate in routine A.I. compares favourably with fresh semen. Thus, laboratory observations on techniques affecting the motility or integrity of the spermatozoa have a good correlation with fertilising capacity. Nevertheless, individual variations in fertilising capacity of semen from different bulls have been observed and these are not necessarily reflected in differences in sperm survival as measured *in vitro*. In competitive fertilisation studies in which thawed semen from several bulls has been mixed before insemination, the spermatozoa of certain individuals have been shown to be far more effective in achieving fertilisation than others [48].

The results with bull semen contrast dramatically with those obtained in experiments with boar semen. The fertilising capacity of boar spermatozoa is severely reduced simply by exposure to more than 5% glycerol at ambient temperatures, although motility is not affected [49, 50]. Acceptable levels of fertility following normal techniques of artificial insemination were not achieved with frozen-thawed boar spermatozoa until methods were employed for freezing them in the absence of glycerol or in media containing very low concentrations of glycerol [51, 52]. The semen is generally frozen by the pellet method in media containing egg yolk and sugars. Although a slight improvement in the motility of the spermatozoa can be obtained by including a low concentration of glycerol in the medium, fertility is not necessarily increased [53]. Obviously motility of the spermatozoa is a very poor criterion of the functional integrity of boar spermatozoa at least in media containing glycerol. Severe damage to the sperm acrosome appears to be caused by freezing and thawing in the presence of glycerol or other permeating cryoprotective agents [54]. Exposure to glycerol also causes an increased loss of certain enzymes [55]. The differences in fertility of frozen-thawed semen from individual males has been found to be greater in the boar than in the bull. These differences seem to be associated with variations in the ability of spermatozoa to survive in different regions of the female reproductive tract and may reflect subtle changes at the sperm surface which have not as yet been determined [56].

The fertility of ram semen after freezing and thawing is also generally lower than of fresh semen although acceptable levels of fertility have been obtained in some experiments with semen frozen either by the pellet method [57] or in straws [58]. Sperm motility is also a poor criterion of fertilising capacity in this species as fertility is rapidly reduced during storage at temperatures above zero under conditions in which motility is well maintained.

These examples and others [12], not discussed here, show some of the diversity that has been observed in the spermatozoa of different species. They emphasise the necessity to determine more precisely some of the changes induced by freezing and thawing which are associated with the maintenance of fertilising capacity in the female reproductive tract. Cryobiology has come a long way since the early observations on freezing semen were first made. The type of cryoprotectant that might be used and the methods of freezing and thawing that might be applied in experiments today offer a wide variety of choice that should help to solve outstanding problems. The experiments on spermatozoa reflect much of the recent history of cryobiology and in some ways they have come full circle. It is amusing to consider what course events would have taken if the experiments on freezing spermatozoa rapidly in the presence of sugars had proved successful before those in which glycerol was used.

References

1 Mantegazza, P. (1866). *R.C. Ist lombardo* **3**, 183
2 Jahnel, F. (1938). *Klinische Wochenschrift* **17**, 1273
3 Parkes, A.S. (1945). *British Medical Journal* **ii**, 212
4 Polge, C., Smith, A.U. and Parkes, A.S. (1949). *Nature* **164**, 666
5 Luyet, B.J. and Gehenio, P.M. (1940). Life and death at low temperatures. *Biodynamica*, Normandy, Missouri
6 Smith A.U. (1961). *Biological Effects of Freezing and Supercooling*. London: Edward Arnold Ltd
7 Smith, A.U. and Polge, C. (1950). *Veterinary Record* **62**, 115
8 Stewart, D.L. (1951). *Veterinary Record* **63**, 65
9 Polge, C. and Rowson, L.E.A. (1952). *Proceedings of the 2nd International Congress on Animal Reproduction and Artificial Insemination (Copenhagen)* **3**, 90
10 Cassou, R. (1964). *Proceedings of the 5th International Congress on Animal Reproduction and Artificial Insemination (Trento)* **4**, 540
11 Nagase, H. and Niwa, T. (1964). *Proceedings of the 5th International Congress on Animal Reproduction and Artificial Insemination (Trento)* **4**, 410
12 Watson, P.F. (1979). In *Oxford Reviews of Reproductive Biology*. C Finn (ed.) Oxford: Oxford University Press

13 Lovelock, J.E. and Polge, C. (1954). *Biochemical Journal* **58**, 618
14 Smith, A.U. and Polge, C. (1950). *Nature* **166**, 668
15 Polge, C. and Lovelock, J.E. (1952). *Veterinary Record* **64**, 62
16 Bruce, W. (1953). *Veterinary Record* **65**, 562
17 Polge, C. and Jakobsen, K.F. (1959). *Veterinary Record* **71**, 928
18 Luyet, B.J. and Keane, J. (1955). *Biodynamica* 7, 281
19 Polge, C. (1957). *Proceedings of the Royal Society, London Series B* **147**, 498
20 Asahina, E. (1959). *Nature* **184**, 1003
21 Farrant, J., Walter, C.A., Lee, H. and McGann, L.E. (1977). *Cryobiology* **14**, 273
22 Larson, G.L. and Graham, E.F. (1959). *Artificial Insemination Digest* 7, (6), 6
23 Martin, I.C.A. (1962). *Australian Veterinary Journal* **38**, 414
24 Polge, C. (1953). *Veterinary Record* **65**, 557
25 Nagase, H. (1972). Reproduzione animale e fecondazione artificiale, p. 189. *Societa Italiana per il progresso della Zootecnica.* Bologna: Edagricole
26 Lovelock, J.E. (1953). *Biochimica et Biophysica Acta* **11**, 28
27 Emmens, C.W. and Blackshaw, A.W. (1950). *Australian Veterinary Journal* **26**, 226
28 Polge, C. and Soltys, M.A. (1960). In *Recent Research in Freezing and Drying,* p. 87. A.S. Parkes and A.U. Smith (eds.) Oxford: Blackwell
29 Nagase, H. and Niwa, T. (1964). *Proceedings of the 5th International Congress on Animal Reproduction and Artificial Insemination (Trento)* **4**, 498
30 Nagase, H. (1968). *Proceedings of the 6th International Congress on Animal Reproduction and Artificial Insemination (Paris)* **2**, 1103
31 Nagase, H. and Tomizuka, T. (1968). *Proceedings of the 6th International Congress on Animal Reproduction and Artificial Insemination (Paris)* **2**, 1107
32 Nagase, H., Yamashita, S. and Irie, S. (1968). *Proceedings of the 6th International Congress on Animal Reproduction and Artificial Insemination (Paris)* **2**, 1111
33 Berndtson, W.E. and Foote, R.H. (1969). *Cryobiology* **5**, 398
34 Berndtson, W.E. and Foote, R.H. (1972). *Cryobiology* **9**, 57
35 Jondet, R. (1964). *Proceedings of the 5th International Congress on Animal Reproduction and Artificial Insemination (Trento)* **4**, 463
36 Jondet, R. (1972). *Proceedings of the 7th International Congress on Animal Reproduction and Artificial Insemination (Munich)* **2**, 1348
37 Phillips, P.H. and Lardy, H.A. (1940). *Journal of Dairy Science* **23**, 399
38 Graham, E.F., Crabo, B.F. and Brown, K.I. (1972). *Journal of Dairy Science* **55**, 372
39 Pace, M.M. and Graham, E.F. (1974). *Journal of Animal Science* **39**, 1144
40 Saroff, J. and Mixner, J.P. (1955). *Journal of Dairy Science* **38**, 292
41 Watson, P.F. (1976). *Journal of Thermal Biology* **1**, 137
42 O'Dell, W.T. and Almquist, J.O. (1957). *Journal of Dairy Science* **40**, 1534
43 Amann, R.P. and Almquist, J.O. (1957). *Journal of Dairy Science* **40**, 1542
44 Gilbert, G.R. and Almquist, J.O. (1978). *Journal of Animal Science* **46**, 225
45 Pursel, V.G., Schulman, L.L. and Johnson, L.A. (1973). *Journal of Animal Science* **37**, 785
46 Rodriguez, O.L., Berndtson, W.E., Ennen, B.D. and Pickett, B.W. (1975). *Journal of Animal Science* **41**, 129
47 Senger, P.L., Becker, W.C. and Hillers, J.K. (1976). *Journal of Animal Science* **42**, 932
48 Beatty, R.A., Stewart, D.L., Spooner, R.L. and Hancock, J.L. (1976). *Journal of Reproduction and Fertility* **47**, 377
49 Polge, C. (1956). *Veterinary Record* **68**, 62
50 Wilmut, I. and Polge, C. (1974). *Journal of Reproduction and Fertility* **38**, 105
51 Graham, E.F., Rajamannan, A.H.S., Schmehl, M.K.L., Makilaurila, M. and Bower, R.E. (1971). *Artificial Insemination Digest* **19**, 12
52 Pursel, V.G. and Johnson, L.A. (1975). *Journal of Animal Science* **40**, 99
53 Polge, C. (1976). *Proceedings of the 8th International Congress on Animal Reproduction and Artificial Insemination (Krakow)* **4**, 1061
54 Wilmut, I. and Polge, C. (1977). *Cryobiology* **14**, 471
55 Bower, R.E., Crabo, B.G., Pace, M.M. and Graham, E.F. (1973). *Journal of Animal Science* **36**, 319

56 Polge, C. (1978). *Journal of Reproduction and Fertility* **54**, 461
57 Salamon, S. (1971). *Australian Journal of Biological Sciences* **24**, 183
58 Colas, G. (1975). *Journal of Reproduction and Fertility* **42**, 277

CHAPTER FOUR

Principles of Embryo Preservation

D G Whittingham

Introduction

A major breakthrough in low temperature biology occurred in 1949 when it was discovered that glycerol provided protection to avian spermatozoa during freezing to −79°C [1]. Subsequently, this discovery led to the successful preservation of many different types of cells and tissues and today banks of frozen tissue culture cells, blood, corneas, cartilage and semen are commonplace in biological research, clinical medicine and animal breeding. Because mammalian embryos have the potential to develop into complete new individuals when transferred from donor to recipient foster mothers, prior to implantation in the uterus, the practical advantages of embryo storage in mammalian genetics and animal breeding were immediately foreseen. It is not surprising too, that attempts were made to freeze embryos shortly after the discovery of the protective action of glycerol. In 1952 Dr Audrey Smith [2] first demonstrated that the exposure of fertilised mammalian eggs to low temperature (−79°C and −196°C) was not incompatible with further development. Although only 1% of the fertilised 1-cell rabbit eggs continued to divide after freezing and thawing in the presence of 15% glycerol, Dr Smith predicted that, with the appropriate modifications in technique, high rates of survival should be possible. However, for the next 20 years further attempts to freeze mammalian eggs and embryos met with little or no success [3]. but during this time significant advances were being made in mammalian embryology and low temperature biology which were eventually to contribute to the development of a suitable technique for embryo storage.

In 1972, two independent reports appeared [4, 5] which described the first successful freezing of mammalian embryos to sufficiently low temperatures to allow prolonged storage in the frozen state. Mouse embryos at all stages of preimplantation development survived freezing to −196°C and live mice were born after transfer of embryos previously frozen to temperatures as low as −269°C [4]. Survival was achieved with relatively low rates of cooling (0.2 - 2.0°C/min) and thawing (4 - 25°C/min) with dimethylsulphoxide (DMSO) as the cryoprotective agent. These initial observations have now been repeated in many laboratories and the technique has been applied to the preservation of embryos from several other mammals, e.g. cattle, sheep, goats, rabbits and rats [6]. The main features of the technique for all species are based upon those originally devised for mouse embryos (Figure 4.1).

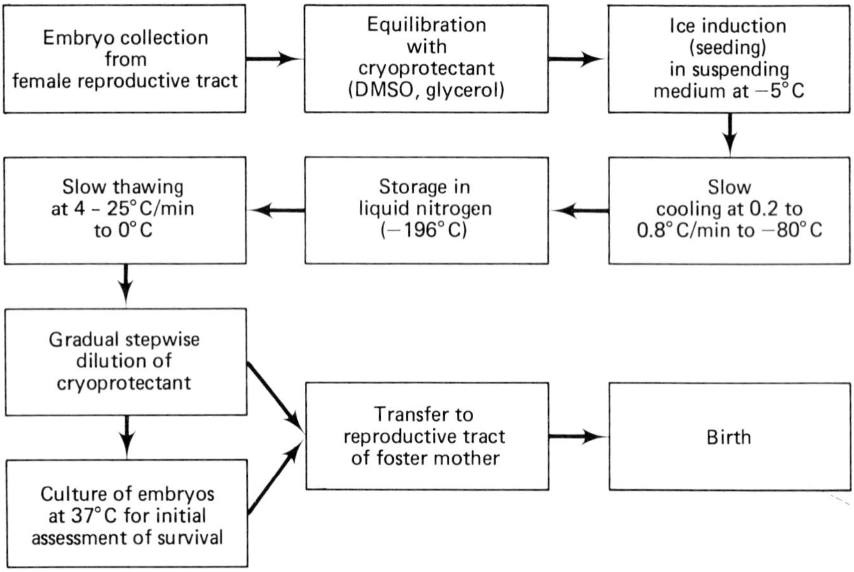

Figure 4.1 Steps in low temperature embryo preservation (original method)

Recent modifications of the cooling procedure have shown that embryos can survive relatively rapid rates of thawing and dilution [7, 8] and these are discussed in detail in this chapter. Since the experimental procedures for embryo collection, culture and transfer are covered adequately elsewhere [9, 10] only those aspects which apply specifically to embryo preservation are referred to in the text. Further reference to the source of technical detail is given in Chapter 12.

Collection of embryos

Preimplantation stage embryos (1-cell to blastocyst) are flushed from the oviduct or uterus at the appropriate time intervals after mating or ovulation (for details see refs 9, 10, 11). Unlike the vast numbers of sperm contained in each single ejaculate, the number of eggs ovulated in each oestrous cycle is small even in polyovular species such as the rat, hamster and mouse. Prior treatment of females with gonadotrophic hormones (pregnant mares' serum gonadotrophin, PMSG, and human chorionic gonadotrophin, HCG) can increase the number of eggs ovulated as much as tenfold (superovulation) in both farm and laboratory animals [9–11]. In the mouse, groups of females can be synchronised to ovulate when treated with gonadotrophins, thus providing adequate material for experimental purposes. The viability of embryos obtained from hormone treated and naturally mated females is similar after transfer to foster mothers [11, 12] and so far there is no indication that superovulated embryos are more susceptible to injury during freezing and thawing [13]. One of the disadvantages of hormone treatment is the great variability in the response of the various species, breeds and strains to the gonadotrophins. Some females fail to respond and in others where large numbers of eggs are released, few become fertilised after mating. The hyperstimulation of the ovary produces high concentrations of oestrogens which interfere with the transport of sperm through the female reproductive tract [11].

Media for collecting and preserving embryos

Embryos are best collected in a medium with a stable pH of between 7.2 and 7.4 in air, as prolonged exposure of embryos to the uncontrolled pH of bicarbonate buffered medium during collection and manipulation can adversely affect further embryonic development [14]. Originally a simple physiological phosphate buffered saline solution (PBS) was used for collecting and freezing mouse embryos [4] but in most of the subsequent studies with the mouse and other species, it has been supplemented with pyruvate, glucose and bovine serum albumin. This medium is generally referred to as PBI [15]. The presence of the nutrients helps prevent the depletion of metabolic pools within the embryos when they are held at room temperature for any length of time. In addition, the albumin acts as a membrane stabiliser and prevents the embryos sticking to plastic and glassware. PBI medium is only able to support development of the later preimplantation stages (morula and blastocyst) in culture for a limited period (up to 24 h).

The possibility of avoiding any adverse effects of this medium on embryonic viability by the use of suitably buffered embryo culture media, e.g. with Hepes, for collecting and suspending embryos during freezing and thawing has not been explored. At present such investigations are limited by the fact that, apart from the mouse and rabbit, knowledge of the *in vitro* requirements of other mammalian embryos is still fragmentary [16].

Embryonic stages

From fertilisation to implantation, the mammalian embryo remains 'free' or unattached to maternal tissue within the lumen of the female reproductive tract. It is only during this period that embryos can be removed and returned to the female tract to continue normal development to term. In the mouse, live offspring have been obtained from frozen-thawed embryos at all stages of preimplantation development [4, 17] but, for practical purposes, the most convenient stage for low temperature storage is the 8-cell embryo. The 8-cell embryos are easily collected from the oviduct, early abnormalities can be recognised and discarded, the freezing and thawing procedures are less complicated than for the blastocyst stage, the initial survival after thawing can be easily assessed *in vitro* when the culture requirements are less restrictive than for earlier stages and finally, if one or two blastomeres of an embryo are damaged there is still a reasonable chance of the embryo developing to a live offspring.

The choice of embryonic stages for freezing is limited in some of the farm animals by their sensitivity to cooling to 0°C. No pig embryos have survived cooling below +15°C. Before the late morula stage is reached in the sheep (∿32 cells) and the blastocyst stage in the cow (64 – 128 cells) the embryos are damaged by cooling to 0°C [18] but the embryos of both species can be successfully stored at these later stages. The embryos of all three species contain large amounts of lipid material and a temperature-induced phase change in the lipids during cooling may be responsible for destroying the physical integrity of the early embryonic cells.

Cryoprotective compounds

Since glycerol was first shown to protect spermatozoa from freezing damage a wide variety of compounds have been discovered with similar protective properties, e.g. DMSO, ethylene glycol, sucrose, polyvinylpyrrolidone (PVP), hydroxyethyl starch and certain proteins. Whether the cryoprotectants are permeable (DMSO, glycerol and ethylene glycol) or non-permeable (PVP, sucrose) it is generally accepted that they protect cells from the damaging effects of the high concentrations of solutes produced in the external medium as the water freezes during cooling [19]. However, the exact mechanisms by which they protect cells against the injurious effects of low temperature are not well understood. Not all of these compounds are equally as effective in protecting a given cell type. So far, only DMSO, glycerol and ethylene glycol have been shown to protect adequately mammalian embryos from freezing injury and of the three, DMSO has been the most effective and widely used.

Addition prior to freezing

Similar to other cell types the survival of embryos is influenced by the concentration of the cryoprotectant as well as its rate of addition and the temperature and duration of exposure prior to freezing. High rates of survival have been achieved with mouse embryos frozen in concentrations of DMSO ranging between 1 and 2 M [4, 20] and, for practical purposes, 1.5 M DMSO is preferable. In other species the optimal concentration appears to be somewhere between 1.5 and 2.0 M [6]. From other cell systems, e.g. mouse bone marrow stem cells, there is evidence that the cooling rate required for optimal survival varies with the concentration of cryoprotectant [21]. So far, this has not been examined in detail with mammalian embryos.

DMSO, at the required concentration, can be added rapidly in a single step to the embryos of the mouse [4], rat [22] and rabbit [23] up to the morula stage, without any adverse effect on post-thaw survival. However, the survival of blastocysts from some species (sheep and cow, [7] mouse, [17] is reported to improve after the gradual stepwise addition of DMSO. It is possible that the blastocyst is unable to withstand the major osmotic stress caused by the single rapid addition of DMSO.

The cryoprotectants used for embryo freezing are able to penetrate cells and the amount of permeation required for protection is reflected to some extent by the temperature and duration of exposure before freezing. Obviously, this will also be influenced by differences in the permeability characteristics of embryos from various species and at various stages of development. DMSO and ethylene glycol are more freely permeable than glycerol. With 8-cell mouse embryos optimal survival rates are reached after exposure for only 5 min to DMSO at 0°C [24] but longer exposure times are required to achieve comparable results with glycerol [25]. The survival of 8-cell rabbit embryos increased from 0 to 55% when exposure to DMSO at 0°C was increased from 1 to 30 min but there was only a slight improvement in the survival of rabbit morulae [23]. Either the morula stage is more permeable to DMSO or the intracellular concentration of DMSO required for protection is reached more quickly because of the reduction in cell size at the morula stage. Late morula and blastocyst stages appear to require longer exposure times to DMSO at temperatures above 0°C. For sheep and cattle, the total exposure time at room temperature to DMSO including the slow stepwise addition is 40 min or longer [7]. It is not known whether these longer exposure times are necessary for the DMSO to permeate into the blastocoele cavity to provide protection to the inner cell mass cells and the surface of the trophoblast cells lining the cavity. From the late morula stage tight junctional complexes have formed between the outer trophoblast cells [26]. These would prevent DMSO reaching the inner cells by simple diffusion through the intercellular spaces and thus the time taken for it to reach the inner cells would be increased.

The empirical basis on which many of the procedures for addition of the cryoprotectant have been worked out makes it impossible to determine the exact roles temperature and permeation play in the protection of cells from freezing injury.

Dilution of cryoprotectant after thawing

Generally, it is accepted that cryoprotectants should be removed gradually after thawing to reduce injury from osmotic shock. A solution containing 1.5 M DMSO has an osmolality of 1.5 osmol or higher and during dilution this is reduced to normal physiological levels of around 0.3 osmol. The degree of cellular swelling during dilution depends upon the amount of cryoprotectant within the cell and the rapidity with which it can move out of the cell. As with the addition of the cryoprotectant, temperature influences the rate of transport from the cell. In most instances, DMSO has been diluted from embryos after thawing in a slow stepwise procedure either at 0°C (mouse, [4] rat, [22]), 20°C (mouse, [5] cow, [20] sheep, [7]) or 37°C (mouse, [17] rabbit, [23]) to avoid osmotic stress during its removal. Various other approaches have been used to reduce the effects of osmotic shock. The DMSO has been removed from 8-cell rabbit embryos by gradually reducing the osmotic strength of the suspending medium [23] and mouse embryos have been held in isotonic sucrose at room temperature for 30 min after thawing to permit the gradual diffusion of DMSO from the embryos [27]. However, in several other cell systems abrupt osmotic stresses can be tolerated after thawing when the temperature of dilution is room temperature or above [28, 29]. Recently, this was also shown to be true for 8-cell mouse embryos and blastocysts; survival after rapid dilution of the DMSO at 20°C was similar to that after slow dilution at either 0°C or 20°C [8]. It is possible that the amount of osmotic stress that can be tolerated depends upon the extent of injury incurred during freezing and thawing; i.e. cells cooled and thawed at optimal rates can withstand abrupt osmotic stress.

Ice nucleation in the suspending medium during cooling

The freezing point of the medium in which the embryos are suspended during cooling varies with the concentration of the cryoprotectant. For PB1 medium containing 1.0, 1.5 and 2.0 M DMSO, the freezing points are approximately −2.5, −3.5 and −4.5°C respectively. When these media are cooled at the slow rates required for embryo preservation ice nucleation rarely occurs at the freezing point of the medium and samples may supercool to temperatures as low as −21°C. Excessive degrees of supercooling have a deleterious effect on embryonic survival when they are left to continue cooling in the same vessel after ice formation. The survival of 8-cell mouse embryos cooled at 0.5°C/min falls dramatically when ice forms in the suspending medium below −7°C and none survives supercooling to −12°C or below [30]. A schematic outline of these events is shown in Figure 4.2. At ice formation, the latent heat of fusion causes the sample temperature to rise close to that of the melting point of the medium. Simultaneously the bath temperature is continuing to fall at a constant rate and therefore the greater the degree of supercooling, the larger the temperature difference that will exist between the cooling vessel and the sample immediately after ice formation. This, in turn, will increase the cooling rate of the sample above the normal optimal rate for embryo survival, until thermal equilibrium is re-established with the cooling bath. The examples illustrated in Figure 4.2 show that embryos seeded at −5°C and −12°C cool at rates of <2°C/min and >6°C/min, respectively, before regaining the temperature of the cooling bath. Embryos seeded with ice at −12°C are still viable when thawed after reaching the temperature of the cooling bath but they do not survive after continuous slow cooling to −80°C and transfer to liquid nitrogen unless the rapid temperature drop is prevented by transfer to another cooling bath immediately after seeding (Figure 4.2(b)). Embryo survival depends upon the amount of shrinkage (dehydration) that takes place in

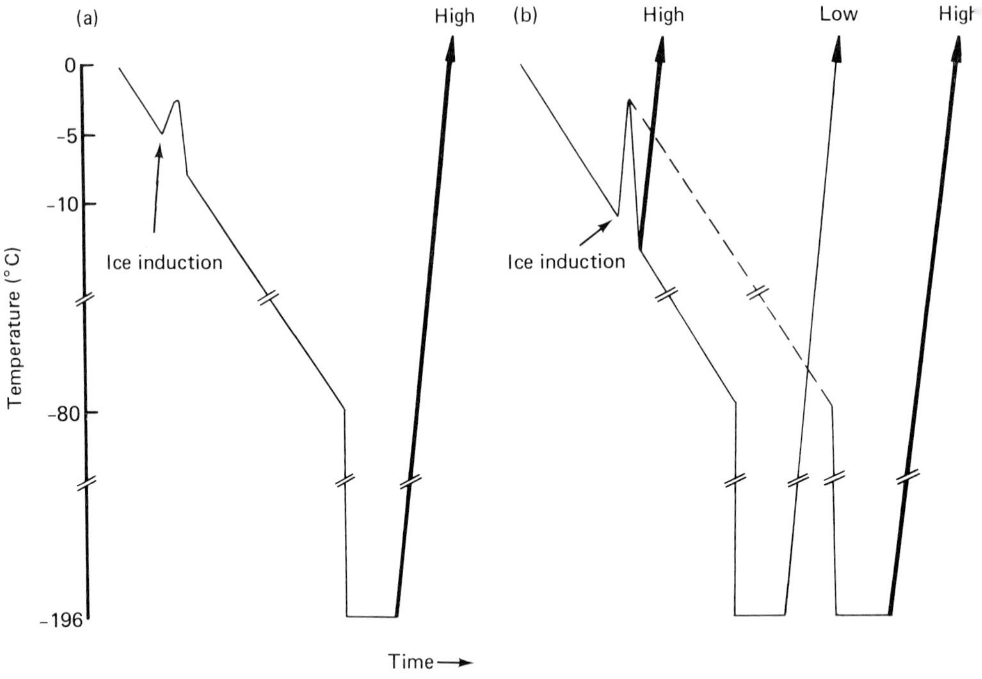

Figure 4.2 Illustration of typical cooling curves showing the elevation of temperature resulting from ice induction at −5°C (a) and −12°C (b) and the subsequent increases in cooling velocities (< 2 and > 6°C/min respectively) as the samples re-equilibrate with the temperature of the cooling bath which is cooling at 0.5°C/min. Broken line (b) represents cooling rate from elevated temperature if sample transferred to another cooling bath, cooling at 0.5°C/min from the higher temperature. Qualitative indication of survival after thawing is given

response to the increasing concentrations of solutes as water freezes out of the suspending medium during cooling (see next section). Supercooled samples cannot dehydrate until ice forms and subsequently the embryos have insufficient time to dehydrate adequately before reaching the temperature where ice forms intracellularly, i.e. if they remain in the same cooling bath after ice induction (Figure 4.2(b)).

In practice, to overcome the damage that may result from excessive degrees of supercooling, embryos are transferred, after equilibration with the cryoprotectant, to a constant temperature bath (seeding bath) a few degrees below the freezing point of the suspending medium. Several minutes later, the samples are induced to freeze, i.e. 'seeded' by one of the following procedures:

(1) touching the surface of the suspending medium with either ice contained in a pasteur pipette or a wire previously cooled in liquid nitrogen.
(2) touching the outside of the ampoule containing the embryos with dry ice or a cold metal bar.
(3) sharply tapping the samples (for this method to be effective the bath temperature should be about 4°C lower than the freezing point of the medium).

Samples seeded by methods (1) and (2) are usually left for 5 minutes to re-equilibrate with the temperature of the 'seeding' bath before transfer to a cooling vessel which is at the same temperature as the seeding bath. Samples seeded by method (3) are transferred immediately to a cooling vessel about one degree below the freezing point of the suspending medium to avoid the larger temperature drop after seeding.

Rates of freezing and thawing

In addition to the choice of a suitable cryoprotective agent, the two most important factors which contributed to the first successful attempts to preserve mammalian embryos at low temperatures were the velocities at which the embryos were cooled and rewarmed. Survival of mouse embryos was achieved with relatively slow rates of cooling (0.2 - 2.0°C/min) and rewarming (4 - 25°C/min) with DMSO as the cryoprotective agent (Figure 4.3). For 8-cell mouse embryos stored at −196°C, the critical ranges of temperature over which these low rates were required for optimal survival were from −4 to −60°C during cooling and from −70 to −20°C during rewarming [24].

Freezing

The response of mammalian embryos to different rates of cooling is not unique since this is well documented for many other cell types [33–36]. It is generally accepted that the formation of intracellular ice during cooling is the major factor contributing to cell death on thawing. Damage to the cells may occur either by the recrystallisation of the ice during thawing [37] or by the osmotic stress imposed on the cells once the ice has melted [38]. During cooling, water moves out of the cells in response to the increasing concentrations of solutes as more and more water freezes in the suspending medium. If the cooling velocity is too rapid, insufficient water is removed before the temperature is reached when the cells freeze internally. Obviously, to prevent

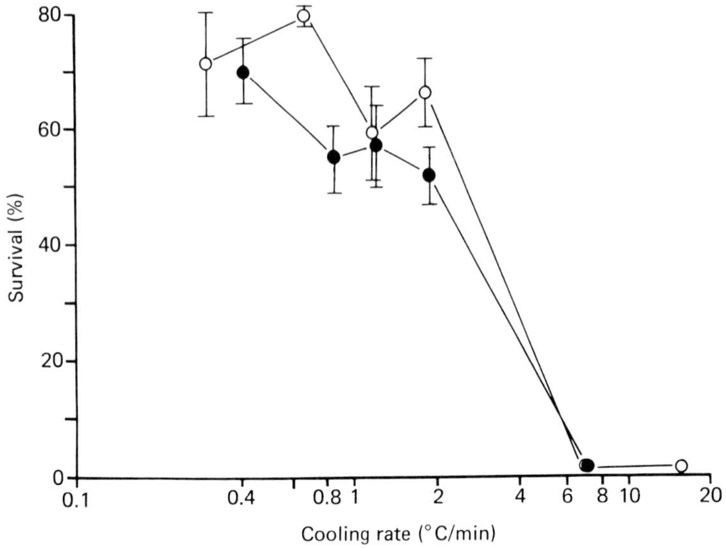

Figure 4.3 Survival of 8-cell mouse embryos after cooling at various rates in the presence of 1.0 M DMSO. Samples thawed at 4 or 25°C/min. ○——○ samples thawed from −78°C; ●——● samples thawed from −196°C. (From Whittingham *et al.*, 1972 [4]

or limit the amount of intracellular ice formation, the cooling rate must provide enough time for dehydration to take place. For a given cell type the rate of dehydration depends upon the volume of the cell, the surface area of the cell, its permeability to water and the temperature coefficient of that permeability. The total amount of dehydration that can occur will also depend on the temperature when intracellular freezing occurs. Mouse embryos cooled in I M DMSO freeze intracellularly between −40 and −50°C and the nucleation temperature is independent of the cooling rate at least for cooling rates which have not produced adequate dehydration – 2 - 30°C/min [32].

Originally, the cooling rates chosen for mouse embryos were based upon calculations previously made by Mazur [39] for sea urchin eggs (comparable in overall size to mouse eggs), since at that time the water permeability characteristics of mouse embryos were unknown. Although these calculations predicted that intracellular ice would form if the embryos were cooled at 1°C/min or above, mouse embryos survived cooling at ~2°C/min but none survived cooling at 7°C/min or above (Figure 4.3). This discrepancy was recently explained by Leibo [32] who determined the water permeability of mouse ova and the temperature coefficient of that permeability. He showed by means of the cryomicroscope that the amount of dehydration occurring in an embryo cooled at various rates to a given subzero temperature was similar to the calculated value using Mazur's mathematical model (see Chapter 2). Moreover, he found that the decrease in survival as the cooling rate increased coincided with the increased incidence of embryos containing intracellular ice (Figure 4.4). Furthermore, the probability of ice forming within an embryo cooled at a specific velocity can be determined with a reasonable degree of accuracy. Such an approach will be valuable and less time consuming when determining the optimal cooling rates for embryos of species where little is known about their sensitivity to cooling.

It is surprising that during preimplantation development the mouse embryo shows no apparent change in sensitivity to the rate of cooling since there is a marked reduction both in the size of the individual embryonic cells as the egg divides to form a blastocyst consisting of 64–128 cells and in the total protein content of the embryo (~25% decrease). The embryo appears to respond as a single unit mass throughout preimplantation similar to a small organ or piece of tissue. Where red cells are clustered together it has been shown that the efflux of water from the interior cells lags behind that of the exterior cells [41], and this may be the reason why the cooling rate optimal for embryos does not change during preimplantation. A further restriction on water movement from the inner cells of the late morula and early blastocyst may result from the tight junctional complexes which form between the outer trophoblast cells [26]. Clearly the mouse embryo is a suitable model for the study of these phenomena.

The optimal cooling rates for all the mammalian embryos studied so far are similar although there is considerable variation in the initial size of the 1-cell ovum, e.g. diameters of mouse, 70 μ; rat 70 μ; rabbit, 120 μ; sheep and cow, 140 μ. The lack of variation in cooling rate optima may reflect differences in the water permeability characteristics between the embryonic cells of various species or the temperature when intracellular ice nucleation takes place. As yet, these aspects of embryo freezing have not been determined.

In most studies, embryos have been slowly cooled to temperatures of −60°C or below before direct transfer to liquid nitrogen for storage. From the observations of Leibo [31, 32] it is now known that adequate dehydration will have occurred by the time the temperature of −60°C is reached. The possibility of dehydrating embryos by a two-step method of freezing described by Farrant *et al* [42] has now been explored for mammalian embryos. Preliminary results are promising (Maureen Wood, personal communication).

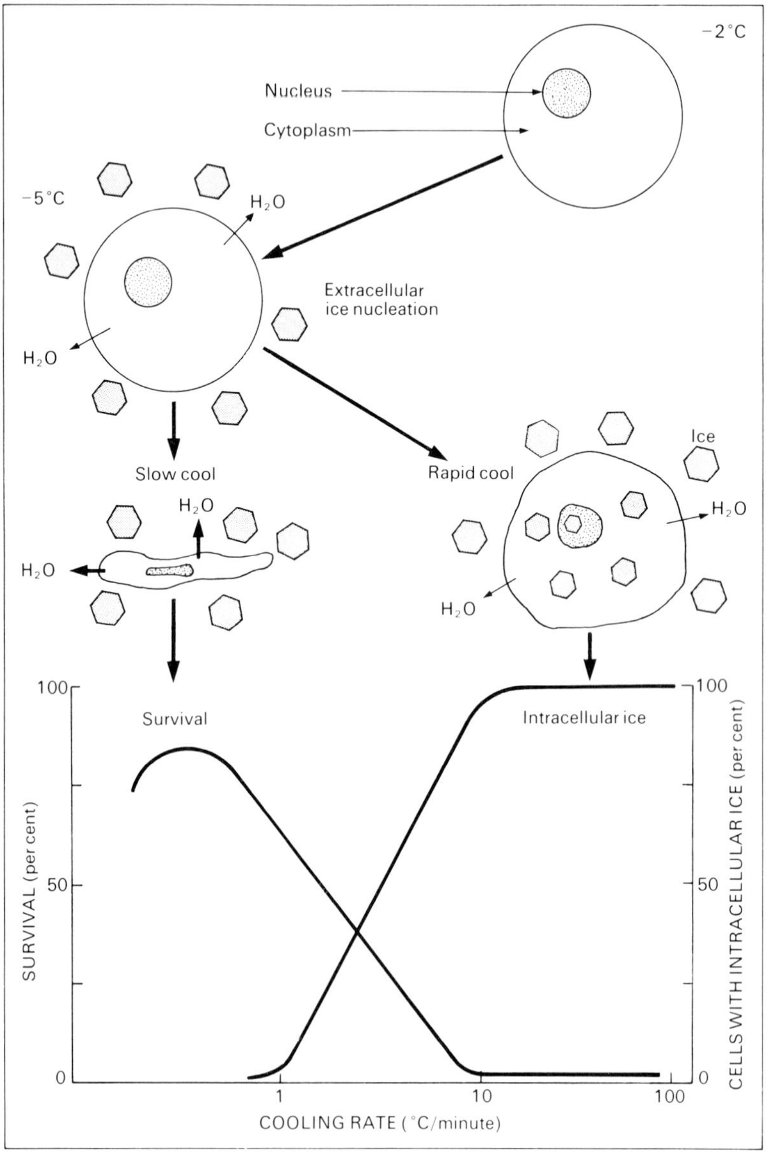

Figure 4.4 The illustration shows the behaviour of mouse ova at various subzero temperatures when cooled slowly and rapidly. The graph illustrates the decrease in survival which coincides with the increase in the number of embryos containing intracellular ice. (From Leibo, 1977 [31, 32] and Mazur, 1977 [40])

Thawing

Apart from the choice of a suitable cooling rate to minimise the possibility of intracellular ice formation, probably the most important observation which contributed to the initial success of the embryo freezing technique was the extreme sensitivity of slowly cooled embryos to the rate of thawing [4, 5]. Optimal survival of mouse embryos was achieved with warming rates of between 4 and 25°C/min. Previously, it was generally accepted that the survival of other types of slowly cooled mammalian cells was independent of the rate of thawing. However, an earlier report in the literature demonstrated that slowly cooled (0.3°C/min) human red blood cells showed less haemolysis on slow thawing than on rapid thawing [43] and since then this observation has been confirmed by other workers [44, 45]. It is difficult to explain why rapid thawing of slowly cooled embryos is lethal, for there is no intracellular ice present, unless slow rehydration is necessary for the re-assembly of the subcellular structures within the embryo which may be disrupted by rapid osmotic changes.

It is well known that some cell types such as red blood cells can survive when a small quantity of intracellular ice is present provided that thawing is rapid and the amount of intracellular ice does not exceed a critical amount [46, 47]. Until recently it was thought that rapid warming of mammalian embryos was detrimental to survival whatever the rate of cooling. However, earlier studies with mouse embryos showed that a small percentage (~10%) of 8-cell embryos [4] and blastocysts [5] can survive rapid thawing (360 - 450°C/min) when cooled at rates ranging between 1 and 10°C/min. The first successful storage of blastocysts from the cow was obtained after slow cooling (0.22°C/min) and rapid thawing (360°C/min) but again survival was less than 10% [20]. The possibility that a small amount of intracellular ice was present in the embryos which survived rapid thawing cannot be eliminated. A further indication that mammalian embryos can survive rapid thawing was demonstrated by Willadsen and Polge [7, 48, 49] with cattle and sheep embryos. When slow cooling at 0.3°C/min is terminated between −30 and −36°C by direct transfer to liquid nitrogen these embryos can survive rapid but not slow thawing.

Whittingham *et al* [8] have recently examined the effects of rapid thawing on the survival of mouse embryos. Figure 4.5 illustrates the design of the experiment. After seeding and slow cooling, embryos are transferred to liquid nitrogen by direct immersion at temperatures ranging between −10 and −80°C. The survival after slow and rapid warming is shown for 8-cell embryos in Figure 4.6 and blastocysts in Figure 4.7. Embryos survive rapid thawing only when slow cooling

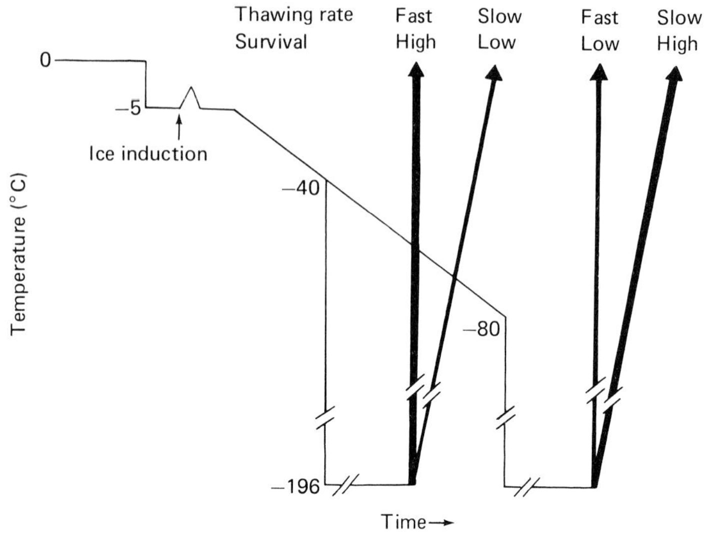

Figure 4.5 A comparison of the effect of the rate of thawing on the survival of 8-cell mouse embryos previously cooled slowly (0.5°C/min) to −40 or −80°C in DMSO (1.5M) before direct transfer to liquid nitrogen

is terminated at relatively high subzero temperatures (−10 to −50°C). The highest levels of survival *in vitro* of rapidly thawed 8-cell embryos occurs after transfer to liquid nitrogen from −35 and −40°C (72 - 88%, Figure 4.6) and of rapidly thawed blastocysts after transfer from −25 to −50°C (69 - 74%, Figure 4.7). By contrast, for embryos to survive slow thawing (8°C/min) slow cooling to lower subzero temperatures (−60°C and below) is required before transfer to −196°C. These results provide indirect evidence that embryos transferred directly to liquid nitrogen from high subzero temperatures do contain intracellular ice since survival can only be obtained by rapid thawing. However, the shape of the curves in Figures 4.6 and 4.7 indicates that the amount of intracellular ice compatible with survival is critical. Proof of the presence of intracellular ice in embryos transferred from the higher subzero temperatures to −196°C awaits investigation with the cryomicroscope and the electron microscope. It is interesting to note that the optimal range for the survival of rapidly warmed blastocysts is greater than for 8-cell embryos (Figure 4.7). Possibly, the smaller cells of the blastocyst can dehydrate faster and may be subjected to less osmotic stress during rehydration when thawed rapidly.

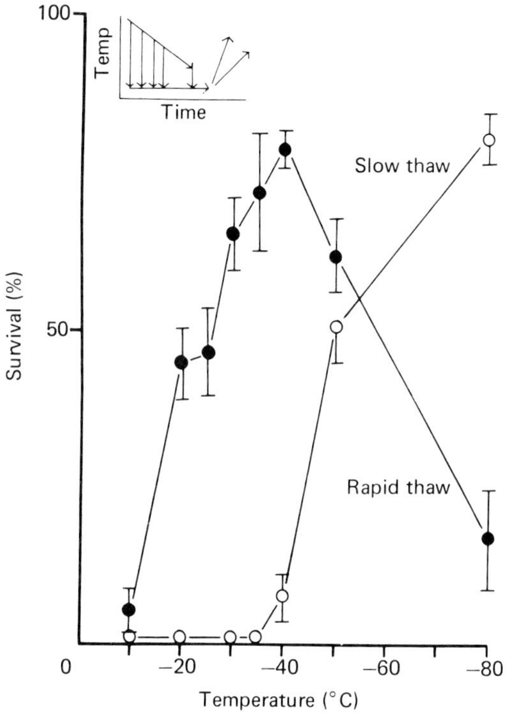

Figure 4.6 A comparison of the survival of 8-cell mouse embryos cooled in 1.5 M DMSO at 0.30 - 0.57°C/min to various temperatures between −10 and −80°C before rapid cooling to −196°C by direct immersion in liquid nitrogen and recovered after slow (20°C/min) or rapid (500°C/min) thawing from −196°C (as indicated in the inset diagram). (From Whittingham *et al.*, 1979 [8])

The ability of mammalian embryos to survive both slow and rapid thawing when frozen under the appropriate conditions indicates that their response to the stresses of freezing and thawing is similar to that of other mammalian cells. Practically, the procedure for preserving mammalian embryos is simpler and quicker when slow cooling can be terminated at relatively high subzero temperatures and the embryos can be recovered after rapid thawing.

Length of storage

Theoretically, cells can exist indefinitely in a state of suspended animation if they are able to withstand freezing to temperatures where there is no further biological activity. Liquid nitrogen at −196°C satisfies this condition because the only types of reaction that take place at that

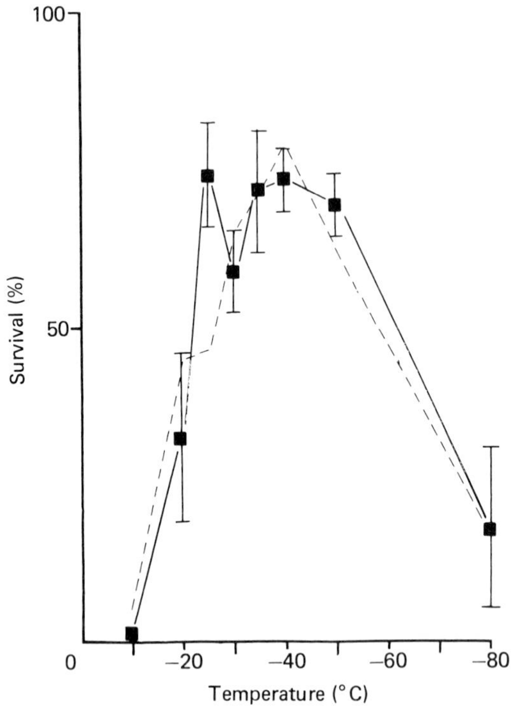

Figure 4.7 A comparison of the survival of mouse blastocysts frozen under identical conditions to those outlined for 8-cell embryos in Figure 4.6 and warmed rapidly (500°C/min) from −196°C. Dotted lines represent data from rapidly thawed 8-cell embryos in Figure 4.6. (From Whittingham *et al.*, 1979 [8])

temperature are photophysical, such as ionising radiation. This type of radiation causes genetic damage and during storage at −196°C the damage will accumulate since the normal enzymatic mechanisms for repair are no longer functioning. The natural level of background radiation is low (0.5 – 1.0 rad/yr) and since less than 1% of spontaneous mutations are attributable to background radiation at normal physiological temperatures, storage times of somewhere between 200 and 1000 years would be required to reduce survival appreciably [50]. Some protection against radiation during storage is provided by DMSO, glass ampoules and low temperature itself. So far, no additional loss of embryos or increase in mutation rate has been observed with mouse embryos exposed to the equivalent of 200 years' background radiation during storage for over 2 years at −196°C [51]. Thus, there appears to be no obvious barrier to the prolonged storage of mammalian embryos at −196°C. The subject of genetic stability at low temperatures and the effects of ionising radiation are discussed in Chapters 2 and 10.

Survival assays

A brief mention of the methods available for assaying survival is pertinent to this chapter although more detailed accounts of the technical aspects of these methods can be found elsewhere [9–11, 52]. Obviously, the ultimate test of survival of frozen-thawed embryos is the production of viable young after transfer to foster mothers. But for testing the initial viability of embryos after storage a suitable *in vitro* assay is advantageous because of the protracted period of gestation in some mammalian species (for example, 9 months for the cow). The *in vitro* assay for the survival of mouse embryos is the percentage of frozen-thawed embryos that develop to the blastocyst stage relative to the percentage of unfrozen controls developing to a similar stage. In the mouse, there is a high correlation between the numbers of frozen-thawed embryos which (1) appear morphologically normal immediately on recovery, (2) continue to develop to the blastocyst stage in culture, and (3) continue to develop to live offspring *in vivo* [4, 13, 51]. The development of embryos of other species e.g. cow, sheep, over such a prolonged period in culture is not as satisfactory as for the mouse because of the limited amount of information that is known on the nutritional requirements for the growth of these embryos *in vitro* [16]. Nevertheless development through one or two cleavage divisions *in vitro* can provide sufficient information on the success of the freezing and thawing procedures. An alternative to the *in vitro* assay in farm animals is the transfer of embryos to the rabbit oviduct where development can continue to the blastocyst stage.

Recently another technique previously used for assaying the viability of other mammalian cells [53] has been adapted for assaying the viability of frozen-thawed unfertilised eggs and embryos [27]. This should be valuable for testing the viability of eggs and embryos from species where they are refractory to culture. Briefly, it consists of treating the embryos after thawing with a non-fluorescent derivative of fluorescein (fluorescein diacetate) which, unlike fluorescein, is freely permeable. It reacts with the intracellular esterases, releasing fluorescein which is retained by cells which have intact plasma membranes, but damaged or dead cells quickly lose their fluorescence.

The ability of mouse embryos to retain fluorescence after freezing and thawing is highly correlated (0.96) with their ability to develop *in vitro* [25].

Conclusions

The mechanisms which protect cells from freezing and thawing injury have formed the basis for determining suitable methods of preserving mammalian embryos at low temperatures. In general, mammalian embryos react to the stresses of freezing and thawing in a similar fashion to other types of mammalian cells. Given the limited number of embryos that are available for experimentation, the method of preserving embryos that has been developed has an extremely high rate of success (up to 80%), much better than that achieved for many other cell types, e.g. with tissue culture cells frozen at a concentration of 10^4/ml survival may be only 10% but this is sufficient to enable the cell line to be re-established. The high survival rates have already led to the practical application of embryo storage in mammalian genetics [6, 13, 15, 49] and in the breeding of farm animals [6, 7, 48, 49].

As seen from the work of Leibo [31, 32] mammalian embryos provide a suitable model for studying some of the more fundamental aspects of low temperature biology, and it is certain that they will make further contributions to the understanding of cryobiology in the future.

References

1 Polge, C., Smith, A.U. and Parkes, A.S. (1949). *Nature* **164**, 666
2 Smith, A.U. (1953). In *Mammalian Germ Cells*, p. 217. Ciba Foundation Symposium. London: Churchill
3 Whittingham, D.G. (1973). *Bibliography of Reproduction* **21**, 273
4 Whittingham, D.G., Leibo, S.P. and Mazur, P. (1972). *Science* **178**, 411
5 Wilmut, I. (1972). *Life Sciences* **11**, 1071
6 Elliott, K. and Whelan, J. (eds.) (1977). *The Freezing of Mammalian Embryos*. Ciba Foundation Symposium 52. Amsterdam: Elsevier
7 Willadsen, S.M. (1977). In *The Freezing of Mammalian Embryos*, p. 175. Ciba Foundation Symposium 52. K. Elliott and J. Whelan (eds.). Amsterdam: Elsevier
8 Whittingham, D.G., Wood, M., Farrant, J., Lee, H. and Halsey, J.A. (1979). *Journal of Reproduction and Fertility* **56**, 11
9 Daniel, J.R. Jr (ed.). (1971). *Methods in Mammalian Embryology*, San Francisco: Freeman
10 Daniel, J.R. Jr (ed.). (1978). *Methods in Mammalian Reproduction*. New York and London: Academic Press
11 Betteridge, K.J. (ed.). (1977). *Embryo Transfer in Farm Animals*. Monograph 16. Canada Department of Agriculture
12 Gates, A.H. (1971). In *Methods in Mammalian Embryology*, p. 64. J.C. Daniel, Jr (ed.). San Francisco: Freeman
13 Whittingham, D.G., Lyon, M.F. and Glenister, P.H. (1977). *Genetical Research* **30**, 287
14 Whittingham, D.G. and Bavister, B.D. (1974). *Journal of Reproduction and Fertility* **38**, 489
15 Whittingham, D.G. (1974). *Genetics* **(Supplement)** 78, 395
16 Whittingham, D.G. (1975). In *The Early Development of Mammals*, p. 1. M. Balls and A.E. Wild (ed.). British Society for Developmental Biology Symposium 2. Cambridge University Press

17 Whittingham, D.G. (1974). *Journal of Reproduction and Fertility* **37**, 159
18 Polge, C. (1977). In *The Freezing of Mammalian Embryos,* p. 3. Ciba Foundation Symposium 52. K. Elliott and J. Whelan (eds.) Amsterdam: Elsevier
19 Mazur, P. (1970). *Science* **168**, 939
20 Wilmut, I. and Rowson, L.E.A. (1973). *Veterinary Record* **92**, 686
21 Mazur, P., Leibo, S.P., Farrant, J., Chu, E.H.Y., Hanna, M.G. Jr and Smith, L.H. (1970). In *The Frozen Cell,* p. 69. Ciba Foundation Symposium. London: Churchill
22 Whittingham, D.G. (1975). *Journal of Reproduction and Fertility* **43**, 575
23 Bank, H. and Maurer, R.R. (1974). *Experimental Cell Research* **89**, 188
24 Leibo, S.P., Mazur, P. and Jackowski, S.C. (1974). *Experimental Cell Research* **89**, 79
25 Jackowski, S.C. (1977). In *Physiological Differences between fertilized and unfertilized mouse ova: glycerol permeability and freezing sensitivity.* PhD Dissertation, University of Tennessee, USA)
26 Biggers, J.E., Borland, R.M. and Powers, R.D. (1977). In *The Freezing of Mammalian Embryos,* p. 129. Ciba Foundation Symposium 52. K. Elliott and J. Whelan (ed.). Amsterdam: Elsevier
27 Leibo, S.P. and Mazur, P. (1978). In *Methods in Mammalian Reproduction,* p. 179. J.C. Daniel (ed.). New York and London: Academic Press
28 Woolgar, A.E. and Morris, G.J. (1973). *Cryobiology* **10**, 82
29 Thorpe, P.E., Knight, S.C. and Farrant, J. (1976). *Cryobiology* **13**, 126
30 Whittingham, D.G. (1977). In *The Freezing of Mammalian Embryos,* p. 97. Ciba Foundation Symposium 52. K. Elliott and J. Whelan (eds.). Amsterdam: Elsevier
31 Leibo, S.P. (1977). In *Les Colloques de l'Institut National de la Santé et de la Recherche Médicale, Vol. 62, Cryoimmunologie,* p. 311. D. Simatos, D.M. Strong and J.M. Turc (ed.). Paris: Inserm
32 Leibo, S.P. (1977). In *The Freezing of Mammalian Embryos,* p. 69. Ciba Foundation Symposium 52. K. Elliott and J. Whelan (eds.). Amsterdam: Elsevier
33 Rapatz, G. and Luyet, B. (1965). *Biodynamica* **9**, 333
34 Mazur, P., Farrant, J., Leibo, S.P. and Chu, E.H.Y. (1969). *Cryobiology* **6**, 1
35 Leibo, S.P., Farrant, J., Mazur, P., Hanna, M.G. Jr and Smith L.H. (1970). *Cryobiology* **6**, 315
36 Farrant, J., Knight, S.C. and Morris, G.J. (1972). *Cryobiology* **9**, 516
37 Mazur, P. (1966). In *Cryobiology,* p. 214. H.T. Meryman (ed.). New York and London: Academic Press
38 Farrant, J. (1977). *Philosophical Transactions of the Royal Society, London Series B* **278**, 191
39 Mazur, P. (1963). *Journal of General Physiology* **47**, 347
40 Mazur, P. (1977). *Cryobiology* **14**, 251
41 Levin, R.L., Cravalho, E.G. and Huggins, C.E. (1977). *Cryobiology* **14**, 549
42 Farrant, J., Knight, S.C., McGann, L.E. and O'Brian, J. (1974). *Nature* **249**, 452
43 Meryman, H.T. (1967). In *Cellular Injury and Resistance in Freezing Organisms,* p. 231 E. Asahina (ed.). Sapporo, Japan: Hokkaido University
44 Rapatz, G., Luyet, B. and Mackenzie, A. (1975). *Cryobiology* **12**, 293
45 Miller, R.H. and Mazur, P. (1976). *Cryobiology* **13**, 404
46 Sherman, J.K. (1962). *Anatomical Record* **144**, 171
47 Farrant, J., Walter, C.A., Lee, H. and McGann, L.E. (1977). *Cryobiology* **14**, 273
48 Willadsen, S.M., Polge, C. and Rowson, L.E.A. (1978). In *Current Topics in Veterinary Medicine,* p. 427. J.M. Screenan (ed.). The Hague: Martinus Nujhoff
49 Polge, C. and Willadsen, S.M. (1978). *Cryobiology* **15**, 370
50 Lyon, M.F., Whittingham, D.G. and Glenister, P.H. (1977). In *The Freezing of Mammalian Embryos,* p. 273. K. Elliott and J. Whelan (eds.). Ciba Foundation Symposium 52. Amsterdam: Elsevier
51 Whittingham, D.G., and Lyon, M.F. and Glenister, P.H. (1977). *Genetical Research* **29**, 171
52 Whittingham, D.G. (1978). *Cryobiology* **15**, 245
53 McGrath, J.J., Cravalho, E.G. and Huggins, C.E. (1975). *Cryobiology* **12**, 540

CHAPTER FIVE

Cryopreservation of Red Cells and Platelets

A W Rowe   Leslie L Lenny   P Mannoni

Introduction

The ever-increasing demand for blood, especially its components, erythrocytes, leucocytes and platelets has been satisfied to some extent by technical advances both in collection and preservation. Among these technical innovations are the introduction of anticoagulant preservatives to extend the storage life of liquid blood, the introduction of integrally connected plastic bags that facilitates the separation of components and minimises the risk of contamination, the availability of refrigeration equipment to facilitate the preparation of components by differential centrifugation and finally the introduction of methods to preserve blood and its cellular components by freezing.

The difficulty in maintaining a correct balance between demand and supply has produced a need for adequate techniques for the preservation of both liquid and frozen blood. However, it was soon discovered that it was impossible to preserve each blood component separately in the liquid state with the same degree of recovery as when blood was stored as whole blood. Each cellular or plasma derivative requires specific conditions of storage depending upon its metabolism. In the frozen state preservation of plasma proteins was achieved long ago when it was demonstrated that freezing did not alter the functions of the majority of these proteins. However, the cryopreservation of blood cells is considerably more difficult. As reported in other chapters, the cellular damage induced by cooling and by freezing is often lethal to cells but research over the past several decades had led to ways by which such injury can be circumvented. The rapidly increasing knowledge concerning the properties and functions of erythrocytes, platelets, leucocytes and plasma components at low temperatures promises to stimulate even more research and improvement in techniques for blood cell cryopreservation, storage and transfusion of these cells.

Direct clinical need has been the major driving force behind the vast research effort in the field of red cell and platelet preservation. Blood components like red cells, platelets, fresh frozen plasma and granulocytes are required primarily to improve oxygen transport and oxygen delivery to tissue, to restore blood volume, to correct or prevent bleeding disorders related to platelet deficiency or abnormalities, to prevent or treat sepsis associated with granulocytopenia, and to prevent or correct a specific defect of one or several plasma proteins. These main indications emphasise the importance of the maintenance of functions and viability of blood derivatives used for the support of patients. Therefore as the knowledge of cell physiology has improved, so have techniques of preservation.

It has been more than 25 years since Mollison and co-workers [1, 2] transfused into a patient the first unit of frozen and thawed red cells. But despite the increasing demand for blood products and the developments in blood banking, the cryopreservation of red cells and especially platelets still remains a rather specialised procedure. This may be due not only to the technology involved but also to the fact that the preservation of the quality of red cells and platelets was not the first goal of most blood banks. However, the importance of the maintenance of optimal function of the cells to be transfused has now been recognised by investigators in the field. For this reason we think that storage procedures at very low temperatures will continue to improve because this approach represents the only way to preserve the biochemical and physical functions of a cell for prolonged periods. Theoretical approaches to cell freezing and the mechanisms of action of cryoprotective additives are reviewed in Chapter 1. The present chapter serves to review the current research and development of methods of red cell and platelet preservation and to discuss various prospective approaches for its further improvement.

Preservation of red cells

Historical perspective

The preservation of red cells by freezing grew out of a serendipitous event involving studies on semen preservation (see Chapter 3). It is of historical interest to recap this important discovery which has had such an impact on blood freezing. In 1948, early researchers involved in cell freezing were primarily interested in preserving spermatozoa in the frozen state for artificial insemination. Since the principal metabolic substrate of sperm is fructose, present in relatively high concentrations in seminal fluid, freezing of sperm was attempted in solutions containing considerable amounts of fructose, but these experiments resulted in loss of viability and motility. However, in the group at the National Institute for Medical Research at Mill Hill in London, Polge, Smith and Parkes [3] discovered by chance that after freezing and thawing in a solution of glycerol, the sperm were actively motile. Fortuitously, they thought that they were trying a new solution of fructose; in fact, it was a mixture of albumin and glycerol. This was discovered when, during the experiment a small amount of the solution was accidentally spilled onto a hot iron tripod. An organic chemist recognised the odour of acrolein and realised that the spilled solution contained glycerol. This was confirmed by chemical analysis which revealed albumin and a high concentration of glycerol. It was then found that the bottle containing this solution had been adjacent to the bottles containing fructose in the cold room. Albumin–glycerol solutions were used in another laboratory for histological examinations. When some labels had become unstuck in the damp atmosphere of the cold room, a technician had stuck them back on the wrong bottles. As Sloviter writes: 'Thus the relatively poor quality of labels in the post-war period in England led to an important scientific discovery: the successful preservation of cells with glycerol in a frozen state'.

This discovery involving glycerol led to attempts to preserve other cells and tissues at temperatures below freezing. In some classical experiments Smith [4, 5] showed that erythrocytes could be frozen in a solution of 15% glycerol in autologous plasma. After storage at −70°C (for a period up to 3 months) and thawing, almost all the erythrocytes were intact. Attempts to recover red cells frozen in glycerol by washing them with plasma or isotonic fluids always resulted in complete lysis of the cells. Indeed, rapid dilution of the glycerolised erythrocytes quickly decreased the osmotic pressure outside the red cells and water entered the cells much faster than glycerol moved out of the cell. The result was that the cells became swollen and burst. It was possible to prevent this osmotic lysis by slowly removing the glycerol from the glycerolised cells by dialysis using solutions of gradually decreasing glycerol concentration [6].

In order to demonstrate that erythrocytes were viable after freezing, thawing and removing of glycerol, Sloviter labelled rabbit erythrocytes with a radioactive marker and injected them into the donor rabbit after storage at −79°C for a period up to 42 days. The results of these transfusion experiments showed that the survival of the erythrocytes preserved by freezing was not significantly different from that of unfrozen erythrocytes. The next step was made when Mollison and co-workers [1, 2] gave the first transfusion of previously frozen human erythrocytes in April, 1951 at the Hammersmith Hospital. The post-transfusion survival was not different from that of unfrozen erythrocytes. This experiment was confirmed by other transfusions of frozen red cells kept at −79°C for periods up to 9 months, with good recovery and no apparent loss of viability during storage.

Other British investigators, the most notable of whom was Lovelock [7], followed the leads of Polge, Smith, Mollison, Sloviter and others by investigating the nature of freezing damage to red cells. Theoretical considerations dealing with the nature of freezing damage have been treated in detail in numerous reports [5, 8–15, see also Chapter 1].

Pioneering efforts to make practicable the transfusion of blood preserved in the frozen state have been reviewed elsewhere [1, 2, 9, 11, 15–35] and are referenced here because they were instrumental in stimulating the more recent approaches to red cell preservation. Since the pioneering studies of the English group in the early 1950s, a wide variety of procedures and preservatives have been employed for freezing red cells. The cryoprotective additives most commonly investigated have been the intracellular (permeating) compounds such as glycerol and dimethylsulphoxide (DMSO), and the extracellular (non-permeating) compounds such as polyvinylpyrrolidone (PVP), hydroxyethyl starch (HES), dextran, lactose, glucose (dextrose) and albumin.

Intensive investigation by numerous groups in the United States and Europe has led to the development of methods to preserve red blood cells in the frozen state that are suitable on thawing for transfusion to patients. Frozen blood is now recognised as a valuable and practical adjunct in the blood bank and it is the purpose of this chapter to review some of the more recent advances in the field of red cell cryopreservation.

Cryoprotective agents and freezing of red cells

Extracellular cryoprotective additives: *Rapid freezing with polymers:* One approach to the problem of freezing of blood suitable for transfusion seeks to avoid the requirement of removing the cryoprotective additive prior to transfusion. Since red cells undergo a potentially damaging transient osmotic stress associated with the addition and removal of an intracellular additive such as glycerol, this problem can be avoided by using extracellular additives which do not permeate the red cell membrane. A number of chemically unrelated compounds, including sugars and polymers, do not pass through the red cell membrane but will afford protection against freezing damage as long as cooling is rapid. Those extracellular additives most extensively studied have been sugars such as glucose, lactose, and sucrose [36–39] and polymers such as polyvinylpyrrolidone (PVP) [9, 29, 36, 40–44], hydroxyethyl starch (HES) [45–50] and dextran [51].

The most notable blood freezing process using extracellular additives and rapid heat transfer has been that pioneered by Rinfret and co-workers [29, 36 41–43]. These investigators carried out extensive studies to achieve conditions of rapid cooling that would preserve red cells suspended in various extracellular additives, including polymers. When suspended in the extracellular polymeric blood–plasma substitute, PVP, the blood (in a corrugated metal container) was frozen to −196°C in about 90 seconds, by agitation of the container in a bath of liquid nitrogen. This rapid removal of heat was enhanced by coating the container with a film of the polymer or silica powder to facilitate heat transfer and nucleate boiling in liquid nitrogen. After thawing, the recovered cells could be transfused immediately without any time-consuming post-thaw processing to remove the polymeric additive. While the advantage of having blood available for transfusion immediately upon thawing is obvious, especially for military needs, blood frozen using this rapid freeze approach must meet the following necessary criteria: a high recovery of functionally intact red blood cells should occur, because free haemoglobin and cell debris are unacceptable; second the cryoprotective additive must display physiological acceptability in large amounts by being both non-toxic and totally excretable; and finally any osmotic stress imposed on the cells by the cryoprotective additive itself must not be so great as to result in osmotic haemolysis upon infusion to the recipient.

Cryopreservation of blood using extracellular additives and rapid cooling is still under investigation. The value of a low molecular weight hydroxyethylated amylopectin (cryoHES or hydroxyethyl starch) as an effective cryoprotectant for red cells has been demonstrated [45, 46, 48–50, 52]. Plasma clearance and renal excretion studies indicate that cryoHES is eliminated from the body rapidly and thus is safe for direct transfusion to recipients of unwashed frozen blood [48].

While the potential for an acceptable extracellular cryoprotectant is still great, practical methods for preserving blood in the frozen state are at present limited to red cells protected with glycerol with associated post-thaw washing to remove glycerol, free haemoglobin and cell debris (for example, leucocytes, platelets, ghosts etc.).

Ultra-rapid (droplet) freezing with sugars: Blood which has been frozen in a large volume (unit) suitable for transfusion therapy is not practical for cross-match or other diagnostic use where only a few drops are required. The preservation of red cells in small quantities is therefore also desirable.

Because they are non-nucleated, human red cells can withstand a wide range of cooling rates, and as such, it is theoretically possible to freeze them rapidly enough in the *absence* of added protective additives to be able to recover the majority of cells intact upon thawing. Such ultra-rapid cooling rates are approached by spraying blood directly onto a moving film of liquid nitrogen on which the blood freezes immediately in the form of discrete droplets. Cells may then be recovered intact upon rapid thawing [53, 54]. The addition of some sugars–glucose and sucrose–enhances the recovery of cells from these droplet frozen pellets. A droplet freezing apparatus automatically atomises the blood sample through a syringe needle into a revolving drum which contains liquid nitrogen. The droplets are frozen almost immediately upon contact with liquid nitrogen. The thousands of discrete droplets which result are collected in a rigid plastic storage vial at the bottom of the revolving drum and are stored at −165°C or −196°C until needed.

The particular advantage of this freezing method is that only a few droplets need be thawed at any one time for use in blood testing. Because they are non-sterile, such droplet frozen cells are not suitable for transfusion to human recipients, although it has been demonstrated that these cells survive well upon transfusion to non-human primates. Huntsman and his colleagues [10, 55] have found full retention of red cell enzymes in droplet frozen blood. Maintenance of blood group characteristics upon thawing has led to the extensive use of both droplet and glycerol frozen red cells for diagnostic identification of red cell antibodies found in multiple-transfused patients [31, 53, 54, 56–62]. Subsequent to antibody identification using a diagnostic panel of droplet frozen blood, compatible blood for transfusion can be obtained from either the fresh or frozen repository. Such a 'bank' of frozen blood for both diagnostic and transfusion purposes makes possible the ability to solve complex as well as routine transfusion problems.

Intracellular cryoprotective additives: *Dimethylsulphoxide:* In the absence of electrolytes, red cells can be induced to agglomerate (reversibly) into a mass of cells which will rapidly settle to the bottom of the container. The application of the agglomeration approach for washing red cells free of cryoprotectant was first described by Huggins to remove DMSO from thawed cells [63–65].

Utilising the principle of reversible agglomeration, Huggins developed a non-centrifugal method for cryopreservation of red cells by using non-ionic diluents at low pH. These solutions cause the thawed cells to agglomerate and settle to the bottom of the freezing bag thereby permitting decantation without resorting to centrifugation. The phenomenon of reversible agglomeration is not the same as red cell agglutination (aggregation), the latter having haematological connotations. Reversible agglomeration occurs in a pH range between 5.2 and 6.1, and involves formation of a reversible complex between the 7S gammaglobulins of plasma and the lipoprotein of the red cell membrane. Reduction in ionic strength through dilution with a non-electrolyte solution causes precipitation of the globulins with coprecipitation of the attached erythrocytes. Agglomerated cells can be resuspended either by addition of electrolytes e.g. NaCl, to break the gamma-globulin bond or by raising the pH which breaks the gamma-globulin-lipoprotein bond [63–66]. Prior to reports on the adverse pharmacological effects of DMSO, Huggins abandoned this cryoprotective additive in favour of glycerol for use with his 'cytoagglomeration' process.

Glycerol: A number of effective and clinically safe techniques for cryopreservation of red cells have been developed in the past two decades. The three most commonly used today are the 'high glycerol–slow cooling' methods, with or without the 'cytoagglomeration' procedure, and the 'low glycerol–rapid cooling' method. All of these procedures use glycerol as the cryoprotective agent but differ in the concentration used, the rate of cooling, the temperature of storage, and the method by which glycerol is both introduced to and removed from the cells.

High glycerol–slow cooling preservation: A major effort to develop a practical blood preservation procedure began two decades ago and was pioneered by Tullis and his associates. This high glycerol–slow cooling approach involved treatment of red cells with concentrations of glycerol approaching 40–50%, followed by slow cooling to −80°C over a period of several hours [19. 28. 35. 67–76]. Specially engineered, continuous flow centrifugal equipment was designed to combine the glycerol with the red cells before freezing and to remove it from the cells before use for transfusion [76, 77]. Since then, the Cohn bowl used by Tullis and his colleagues has undergone a number of design changes from the original stainless steel reusable bowl to the presently used plastic disposable bowl. Briefly, this process involved the centrifugal addition and removal of glycerol with a specially designed bowl, (also known as Cohn–ADL Fractionator, ADL–Latham bowl, Abbott disposable bowl and more recently the Haemonetics disposable bowl); high intracellular concentrations of glycerol 42% (±4%); slow (uncontrolled) rates of cooling; storage at −80°C; and slow (uncontrolled) rates of rewarming. Tullis has reported that, although their techniques varied somewhat, cells stored at −80°C for periods of 6–11 years are still viable in accordance with the NIH limitation on post-transfusion survival rates [71, 78].

Since this method is the oldest procedure for freezing blood, both the process and the utilisation of frozen blood in practical blood banking have been described in numerous reports [15, 19, 28, 35, 67–76, 79, 80]. Valeri has published extensively on his studies evaluating this and other methods [15].

The principal obstacle to widespread clinical application of frozen blood described by Tullis and his colleagues [19, 72] has been the lack of satisfactory methodology for the addition and removal of the high concentrations of glycerol necessary in the original procedures. In 1972 a modification of the Tullis procedure for the glycerolisation and deglycerolisation of red cells was reported by Meryman and Hornblower [25, 81, 82] who introduced the method into the American Red Cross Blood Program.

The addition and removal of glycerol, especially at high concentrations, create transient osmotic gradients across the red cell membrane. The limits of osmotic tolerance of human red cells range from one-half to four times isotonic. In exploring the colligative properties of glycerol and red cells, Meryman and Hornblower [82–84] found that these hypotonic and hypertonic limits are not exceeded when glycerol is added to the cells slowly and with agitation, as through a tube with a controlled flow rate. Upon thawing, the tolerated limits of osmotic gradient are again not exceeded when the glycerol is removed from the cells through a stepwise dilution involving the addition of a 12% buffered NaCl solution for three minutes, followed by further dilution with 1.6% buffered NaCl solution, and a final stage of washing with saline–glucose in a cell-washing apparatus, e.g. the IBM or Haemonetics processors. This method was simplified subsequently by Meryman and Valeri [15, 84, 85] using a modified process whereby the glycerolised cells were frozen as packed cells in a 850 ml freezing bag instead of in a 2 litre bag. The smaller volume of cells allowed deglycerolisation to be accomplished by washing with an 8.5% NaCl solution followed by a glucose–saline solution.

Processing of frozen blood was introduced into the American Red Cross Blood Program in 1972 and by the end of 1976, more than 250000 units had been frozen and deglycerolised by the Red Cross alone. As of July, 1977, frozen red cells represented almost 2% of all red cell preparations transfused in the United States.

Another high glycerol–slow cooling method in current use is the 'cytoagglomeration' process developed by Huggins. Originally developed using DMSO as the cryoprotectant, this method was modified to utilise glycerol in a non-ionic environment so that the cryoprotectant could be removed by reversible agglomeration [66, 86].

Huggins' technique involves addition of an 8.6 M (79% w/v) glycerol solution containing dextrose, fructose, and EDTA to packed red cells followed by slow cooling to −80°C. After storage, a 50% aqueous glucose solution, followed by a 5% fructose solution, is added to the thawed cells contained in a large agglomeration bag, causing the red cells to agglomerate and settle. Following several washes, the cells are resuspended by adding saline as the disagglomerating electrolyte [66, 86]. Huggins modified his original method by reducing the large volume of deglycerolising solution from 6 to 3 litres and has accumulated extensive clinical experience with the use of such red cells [65, 66, 86]. Blood frozen by the Huggins procedure had been used for a short duration under active wartime conditions [87, 88]. This process also has received extensive critical evaluation by Valeri *et al* [15, 89, 90].

Low glycerol–rapid cooling preservation: A kinetic approach to the preservation of blood by freezing has involved the application of rapid and controlled techniques for the removal of heat and the use of clinically acceptable intracellular cryoprotectants in concentrations much lower than those used in the conventional approach of slow cooling. This low glycerol–rapid cooling approach was based on an observation by Luyet and Rapatz [11] that the recovery of red cells was a function of both the concentration of cryoprotectant and the rate of heat transfer. The same good recovery obtained with high glycerol–slow cooling techniques could be achieved when the glycerol concentration was lowered and the rate of cooling increased.

Reduction of the required glycerol concentration avoids the requirement for deglycerolisation by specialised mechanical devices since lower concentrations of glycerol are more easily removed from the cells. Rather than using a 40–50% solution of glycerol as employed in the Tullis procedure (including Meryman's modification) or in the Huggins process, a low glycerol concentration of about 15–18% has been used by Pert [27, 91], Krijnen [47, 92, 93], and Rowe [30, 32, 61, 62, 94–98].

The improved recovery of red cells obtained with the low glycerol process is due to the increased rate of heat removal. Rapid cooling is achieved by controlling the geometry of the red cell sample to obtain a large ratio of surface area to volume by employing a flat, stainless steel container [61, 94, 99] aluminium canister [26, 47], or more recently, a plastic bag supported between two metal plates [31, 32, 62, 100]. When blood is frozen to temperatures approaching −196°C, metal rather than plastic containers have been used to keep the glycerolised blood in a specific geometric configuration during freezing. Conventional blood containers made of polyvinyl chloride (PVC) are not suitable as containers during freezing because they tend to become brittle and crack upon exposure to the very low temperatures of liquid nitrogen. Recently, however, important advances have been made in the fabrication of blood bags from plastics capable of withstanding freezing to the temperature of liquid nitrogen. Those plastics which have been used for freezing containers have been Teflon, Teflon–Kapton laminated, and polyethylene (UCAR–Hemoflex, formerly Perflex, a biaxially oriented polyolefine, first reported by Rowe [100] as an example of a new class of plastics with durability at very low temperatures and a potential for use as containers for blood. The UCAR–Hemoflex polyolefine bag has the advantage of containing no plasticiser, i.e. diethyl hexyl phthalate (DEHP), as found in polyvinyl chloride. Because of its stability at low temperatures, this UCAR–Hemoflex bag is now used in most centres for the freezing of blood. Other plastics such as ethylene ethyl acrylate (EEA) and ethylene vinyl acetate (EVA) have shown much potential as conventional low temperature storage bags for blood (Rowe and Lenny, unpublished observations, 1975).

This highly successful approach to blood freezing is used at The New York Blood Center and has become known as the 'low glycerol–rapid cooling' technique [30, 32, 61, 62, 94–98]. The procedure involves the direct addition of glycerol solution to packed fresh red cells. The red cells containing 14% (v/v) or 17.5% (w/v) glycerol can be frozen, in a stainless steel container, aluminium bottle, or plastic UCAR–Hemoflex bag between two metal plates, by direct immersion into liquid nitrogen. Freezing is complete in 2.5 minutes, and the frozen unit is stored at −165 to −196°C. Even after 11 years of storage at these temperatures, units of blood frozen by this method have been used successfully for transfusion. Following rapid thawing by agitation of the container in warm water (42–45°C), the glycerol in the cells is readily removed by manual or automated techniques using either serial batch washing or continuous flow centrifugation thus rendering the frozen-thawed-washed cells suitable for transfusion. Rowe and his colleagues [30–32, 61, 62, 94, 95, 98, 99] point out that the principal advantage of this approach to the preservation of blood is its simplicity, versatility and adaptability to change. As technological advances have been made (e.g. new plastic containers, automated deglycerolising equipment, processing solutions etc.), they have been readily introduced into the process and used interchangeably with a variety of cryopreservation techniques.

Blood frozen and thawed by the low glycerol–rapid cooling technique has been used with considerable success for transfusion to patients in the United States [15, 30–32, 61, 62, 94, 95, 98, 99, 101–108] as well as Europe [20, 21, 23, 24, 33, 47, 92, 93, 109–113] and Asia [34, 114, 115].

Low temperature storage of frozen blood

The use of cryoprotective additives and control of the rates of cooling and warming to reduce damage to biological specimens by freezing serves no useful purpose unless the temperature of storage is maintained sufficiently low so as to limit any degradative processes that might affect biological integrity.

Red cells frozen slowly in a high concentration of glycerol (40%) using the Tullis technique have been preserved for up to 10 years at −85°C [71, 78, 116, 117], but Valeri *et al* [118] reported that glycerolised red cells frozen by the Huggins technique showed a progressive loss of viability during a one to two year period of storage at −85°C. Valeri's study [118] has been disputed by Huggins and Polesky who have since provided experimental and clinical evidence that such blood can be stored successfully in excess of 9 years.

A storage temperature of −85°C is inadequate for preserving intact the red blood cells frozen by either of the rapid cooling techniques described. The stability of the red cells preserved using the low glycerol–rapid cooling process and stored at various temperatures has been studied by Rowe *et al* [119]. Glycerolised cells rewarmed from −196°C were almost completely haemolysed within hours of storage at −20°C. After storage at −85°C, thawed cells showed a progressive increase in haemolysis which became considerably worse during the washing steps of the procedure; this contrasts with the stability of cells maintained in storage at liquid nitrogen temperatures. Studies on the stability and viability of cells stored for a period of 12 years have shown no difference with storage temperatures between −165°C and −196°C [119; Rowe and Lenny unpublished observations, 1979]. The temperature of −165°C was determined by placing thermocouples into a container of blood that was stored in the vapour phase of a liquid nitrogen refrigerator (LR-1000). All of the haemolysis data for storage temperatures of −165°C and −196°C were accumulated on human blood stored for several years before it was thawed, washed, and transfused into patients. Derrick *et al* [102–104] have made a comprehensive evaluation of the metabolic integrity of human red cells after cryopreservation. For the optimal preservation of metabolic characteristics, they conclude that blood should be frozen at least within ten (but preferably within five) days after collection. Rheological characteristics [120] and erythrocyte lipids [121] are also well preserved when the red cells are stored for many years at very low temperatures.

Although biological activity *per se* is completely arrested at these low temperatures, the loss of viability of cells noted during storage at −85°C emphasises the need to store biological material at the lowest temperature feasible e.g. −196°C. Such studies emphasise our need to learn more about the effects of low temperature *per se* on the structural characteristics of ice formation in biological systems.

Clinical use of frozen blood

The officially recognised name of cryopreserved red cells in the United States, is 'Red Blood Cells (Human) Deglycerolized' and the clinical utility of such cells has now been well documented. Those factors which determine the clinical acceptability of red cells preserved in the frozen state include: *in vivo* survival rates, mode of removal of non-viable cells (with or without haemoglobinaemia), amount of supernatant haemoglobin, total amount of extracellular potassium, presence of a significant number of red cell ghosts and remnants, concentration of remaining cryoprotective additive, stability of the red cells on resuspension, pH of the unit of blood, and sterility.

Numerous studies on the therapeutic effectiveness of blood frozen by the 'high glycerol' method have been reported by Tullis and others [8, 19, 20, 35, 70–74, 79–81, 122, 123], Meryman and colleagues [25, 34, 81, 85, 114, 122, 124, 125], Valeri [14, 15, 22, 116, 118, 122, 123, 126–136], Huggins and others [8, 20, 64–66, 85, 101, 112, 125, 137, 138] and similarly with the 'low glycerol' method by Rowe and others [8, 20, 21, 23, 24, 26, 30–34, 47, 61, 62, 92–95, 98, 99, 101, 106, 107, 110, 111, 113, 125, 126, 139–142]. Based on hundreds of thousands of transfusions over the past decade, frozen blood has now proved itself beneficial in a variety of clinical circumstances.

The utility of cryopreserved red cells can best be illustrated by looking at the actual use based on the experience of a community blood centre serving a large metropolitan area. Although the numbers used in a smaller hospital environment might change, the categories would probably remain the same. The clinical use of frozen blood in patients requiring transfusions is summarised in Table 5.1. These data were compiled by The New York Blood Center over a typical one year period based on hospital-initiated request forms that include diagnosis and reason for using frozen blood.

Clinical indications: The clinical categories for which frozen blood has been most used are listed in Table 5.1 and include:

(1) Autotransfusion—e.g. rare blood, multiple antibody problems, religious convictions, hepatitis problems. Low temperature storage of blood, now a reality, soon may make it feasible for a healthy individual to bank his own blood and later draw on his own personal reserve should the need arise. However, because of limited storage space, only patients with specific medical requirements are used as donors for autologous blood storage.
(2) Exchange transfusion in the newborn.
(3) Red cell antibody problems which require either specific rare blood types or fully typed units because the patient has single or multiple antibodies.
(4) Non-haemolytic transfusion reactions caused by antibodies to leucocytes or plasma proteins. Thawed washed blood is virtually free of leucocytes and plasma proteins.
(5) Anaemia—frozen blood has been beneficial in various types of anaemia (e.g. thalassaemia major) where multiple blood transfusions are required on a prolonged basis. Once again, non-haemolytic transfusion reactions, commonly associated with leucocyte-containing fresh blood, are virtually eliminated.
(6) Surgery—frozen cells are often used in organ transplantation to reduce the risk of leucocyte sensitisation, and in open heart surgery to reduce the risk of hepatitis.
(7) Blood storage—frozen blood can be used as a back up when the availability of fresh blood is inadequate to fulfil the requirements of hospitals. Unfortunately, in the United States this approach has not received more widespread use because of the logistical limitations of the 24-hour post-thaw outdate period imposed by the Food and Drug Administration (FDA).

As noted above and in Table 5.1, in addition to the use of frozen blood for autologous needs or for rare and multiple antibody problems, the major use of frozen blood in anaemia and renal-related problems stems from the unique advantage that these cells are virtually free from leucocytes, plasma proteins, and perhaps, even hepatitis.

Table 5.1 Clinical use of red blood cells (human) deglycerolised

Clinical indication[a]	*No. units*	*Per cent*
Autologous transfusion[b]	23	0.2
Intrauterine and exchange transfusions[c]	14	0.1
Red cell antibodies		
Single or multiple (2107 units–23%)[d]	–	–
Rare[e]	59	0.6
Chronic renal disease		
Dialysis; renal failure; kidney transplant	2117	22.5
Anaemias		
Thalassaemia major	2886	30.7
Non-specific[f]	1567	16.7
Aplastic; DiGeorge; bone marrow transplant	957	10.2
Sickle cell	202	2.1
Haemolytic; PK deficiency	30	0.3
Paroxysmal nocturnal haemoglobinuria	24	0.3
Fanconi's	11	0.1
Haemophilia	3	0.03
Refractory	2	0.02
Cancer		
Leukaemia; myelometaplasia; myelofibrosis; Hodgkins; lymphoma, DiGuglielmo's syndrome	701	7.5
Bleeding episodes		
Gastrointestinal, etc.	304	3.2
Surgery		
Other than kidney transplant–Open heart; gastrectomy, etc.	158	1.7
Burn patients	122	1.3
Inventory supplementation	10	0.1
No diagnosis[g]	146	1.6
Culture[h]	63	0.7
Total	9399	100.0%

a Clinical indication was obtained from hospital blood bank which provided patient diagnosis and reason for requesting frozen blood.
b 21 of the 23 units involved rare and other antibody problems.
c One of the 14 units involved a rare blood (U neg).
d Single and multiple antibody problems are included in all the other categories and involve 2107 units or 23% of the frozen blood used.
e Rare blood problems involved k, Yt^a, Kp^b, Ko, Co^a, Lu^b, Js^b, Vel, U, Tj^a, Bombay, Lan, Winbourne and others.
f Specific patient diagnosis not indicated.
g No diagnosis listed for the specific patient receiving blood.
h Units thawed and cultured for routine sterility testing were not used for transfusion.

Removal of leucocytes: Frozen deglycerolised red cells have been found to be particularly useful for patients sensitised to leucocyte or platelet antigens. The combination of glycerolisation, freezing, thawing, and washing destroys the majority of leucocytes and removes them during the deglycerolisation procedure [15, 23–26, 32, 34, 61, 62, 65, 81, 86, 94, 98, 101, 107, 114, 115, 122, 124, 138, 139, 143–149]. Leucocyte removal is greater than 95% efficient and the low glycerol method is somewhat more efficient than the high glycerol method. Efficiency of leucocyte removal improves with increasing age of the blood before freezing although most blood is frozen before it is 6 days old.

Removal of plasma: Since some plasma proteins may be antigenic and can cause febrile transfusion reactions in sensitised patients, frozen blood is preferred for these patients because residual plasma is virtually absent after post-thaw washing. Microaggregates found in plasma can become large enough to be embolic, producing vascular obstruction and tissue damage, but such microaggregates are essentially absent from deglycerolised blood [150, 151].

Removal of hepatitis: Post-transfusion hepatitis has been reported to be reduced substantially in patients receiving only frozen-thawed blood [15, 32, 34, 73, 81, 96, 152]. The observed reduction in the incidence of hepatitis with the use of frozen red cells may be the result of a reduction of the infectious agent due to the extensive dilution associated with deglycerolisation. In New York we have measured the dilution factor for radioactively labelled plasma protein and have determined it to be 8×10^{-5} using a Haemonetics cell washer (Rowe and Lenny, unpublished observations, 1978). The use of frozen red cells to minimise the risk of hepatitis is justified largely on the basis of empirical evidence. The validity of such observations has yet to be confirmed by more controlled clinical studies.

Frozen blood–future perspectives

Although frozen blood is now widely used with impunity, there are still problems to be solved before it can achieve its full potential. Coordinated basic and developmental research efforts are vitally necessary to solve these problems.

Problems to be overcome: *Cost:* The development of plastic containers stable at low temperatures and automated cell washers has done much to reduce the cost of freezing. Despite the increased use of frozen red cells, however, processing costs are expected to remain the same, that is, two to three times more than those for fresh unprocessed blood. Any further reduction in costs probably will not come from the processing *per se* but rather from more efficient blood utilisation and management of transfusion-related problems.

Time and manpower required for post-thaw processing: New washing methods for glycerol removal are now being developed. These will reduce both the cost and time of processing thawed blood, thus making it more readily available for routine use rather than for specialised problems, as is currently the case.

Limitations due to the short 24-hour outdating period of washed red blood cells: The single major obstacle to additional widespread use of frozen blood is due to its 24-hour outdating period [153]. This limitation is imposed in the United States by the FDA due to the possibility of contamination of the blood during processing. Studies have indicated, however, that contamination is not a problem in frozen-thawed-deglycerolised red cells. Using blood that was intentionally contaminated with pathogenic organisms, Myhre *et al* [154] and Kahn *et al* [155] have observed that the process of washing blood to deglycerolise it reduced the number of organisms by several orders of magnitude. Most pathogenic microorganisms e.g. *Staphylococcus aureus, Pseudomonas aeruginosa, Escherichia coli, Klebsiella spp* and *Enterobacter spp* grow at such a slow rate at 4°C that they do not pose any great hazard to the recipient unless present in unusually high numbers [154]. It therefore appears that the current 24-hour limit on post-thaw storage of deglycerolised blood may be too restrictive and that extension of the post-thaw shelf-life to at least 72 hours should be considered. Technological improvements in equipment for glycerolising and deglycerolising blood could further reduce problems of contamination by reducing the number of possible entry sites.

The 24-hour outdating problem will probably not be eliminated, however, until sterile 'docking devices' have been designed to permit the joining of two pieces of plastic tubing under guaranteed conditions of sterility. A prototype pull-tab connector is such a device being developed in our laboratory [156]. Other devices under development involve heat sterilisation in various forms [157, 158].

Prefreeze and post-thaw rejuvenation or supplementation of red cells with biochemical metabolites [159, 160]: Valeri and Zaroulis [15, 135, 136] have advocated biochemical modification of liquid stored

red cells by 'rejuvenation' to restore or improve oxygen transport before and after freeze-preservation. Rejuvenation solution (PIGPA) containing pyruvate, inosine, glucose, phosphate, adenine and sodium chloride are designed to restore the intracellular levels of 2, 3-diphosphoglyceric acid (2, 3-DPG) in stored blood to the level found in freshly drawn blood. Adenosine triphophate (ATP) levels are also elevated, thus permitting normal red cell viability. The rejuvenation solutions are removed during post-thaw washing. Additional metabolic and clinical studies are necessary to determine whether the benefits of rejuvenating red cells are justified from both biological and practical standpoints.

Relationship between 2, 3-DPG and the oxygen dissociation curve of haemoglobin and its significance in the patient [160] : More studies are necessary to determine the importance of normal or even supranormal levels of 2, 3-DPG in transfused cells.

Inadequate documentation concerning the removal of hepatitis virus during freezing, thawing, and washing [161] : The empirical evidence suggesting that post-transfusion hepatitis is reduced by the use of frozen blood [15, 32, 34, 81, 86, 152] requires additional support to the limited controlled clinical studies already done [73]. Recently Alter *et al* [162] administered to chimpanzees thawed-washed blood that had been inoculated with hepatitis B virus before freezing. All four of the chimpanzees showed evidence of hepatitis infection based on the appearance of antigen or the development of antibody to the virus. Freezing, thawing and deglycerolisation apparently did not eliminate the added hepatitis B virus under these conditions of processing. However, because of mandatory testing of blood for hepatitis B, virtually all transfusion hepatitis that occurs is not due to hepatitis B virus but rather to non-A or non-B types. Considerable work remains to be done on both the efficiency of detection and the removal of all forms of hepatitis virus from both fresh and frozen blood.

Inadequate information on the reduction or elimination of non-haemolytic transfusion reactions (i.e. the removal of white cells as a function of the freezing and thawing regimen): Transfusion of frozen cells is useful for patients who are sensitised to white cell or platelet antigens [25, 32, 86, 138, 139, 143, 144, 163]. It has been shown that frozen deglycerolised cells are less likely to immunise dialysis patients against histocompatibility antigens [115, 143, 144, 148, 149]. As can be seen in Table 5.1, one major use of thawed cells prepared by The New York Blood Center is for administration to dialysis patients. Opelz and Terasaki [146] have reported that the survival of kidney transplants at 12 months were better in patients receiving whole blood than in patients receiving frozen cells or no transfusions. By contrast, many investigators have evidence [34, 115, 138, 147–149] to show that the use of frozen blood results in better survivals of kidney transplants at 12 months. This controversy still requires much investigation to resolve the best therapy for dialysis patients.

Extension of post-thaw storage stability beyond 24 hours, pending elimination of potential contamination problems and maintenance of guaranteed sterility: After frozen-thawed red cells have been deglycerolised, their post-thaw stability during subsequent storage at 4°C is not the same as that of fresh blood in plasma. Deglycerolised red cells are devoid of glucose and plasma protein. Resuspension of thawed red cells in plasma is not feasible because of the increased risk of hepatitis, therefore, most resuspension solutions contain only crystalloids: NaCl, glucose and phosphate. Extension of the outdating period to 72 hours or longer will require much research to determine which resuspension media will prevent or minimise post-thaw haemolysis and best maintain the biochemical integrity of the cell.

Donor identification: There are potential hazards to the recipient, caused by loss of donor identification, especially under conditions of large-scale usage.

Basic studies: Further understanding is required of the basic mechanisms underlying the effects of freezing, storage, thawing and washing of red cells. Red cells also may serve as models for other cell systems, such as platelets.

Conclusions

Resolution of the problems such as those outlined above will extend the use of frozen red cells beyond those centres supplying frozen blood at present. Although the improved patient care realised with frozen blood more than justifies its increased cost, the cost and time of processing must be reduced to the point at which all blood banks will be using frozen blood. Frozen blood implies a feature of purity, potency, and efficacy not always obtainable in liquid stored blood. If such factors are not inherent with freezing, is there any less expensive alternative? Freeze-preservation of blood still remains the best hope for its indefinite storage. It is only a matter of time before frozen blood becomes an integral part of the logistics of every blood bank striving to provide for the most diversified and highest quality transfusion service.

Preservation of platelets

Introduction

Thrombocytopenia resulting from intensive chemotherapy of patients with leukaemia, other malignancies or who have received organ transplants requires an ever-increasing demand for platelets. Storage of platelets in the liquid state at +22°C and the modest increase in shelf-life to 72 hours has eased but not eliminated the problems of supply (especially from random donors and over long weekends and holidays). Significant benefit would be derived from a practical procedure for the cryopreservation of platelets. Long term storage would make practical the collection of autologous platelets by pheresis from leukaemia patients during remission or from patients with other forms of cancer prior to chemotherapy. Practical cryopreservation of platelets would also make possible the banking of large quantities of not only random donor platelets but also pretyped single donor platelets for use in patients who have become refractory to platelets from random donors.

The cryopreservation of platelets has posed a more formidable problem than with red cells or lymphocytes because of their extreme lability. In reviewing the problems of preserving blood platelets Morrison [164], Tullis *et al* [75], Chanutin [165], Gardner [166, 167] and others have emphasised that problems are encountered not only in the choice of a suitable container material, anticoagulants, separation and prefreeze storage of platelets, but also are compounded by the problems associated with the fragile nature and complex functioning of the platelet. Thus, it seems that the problems encountered in attempting to preserve platelets are both difficult to define and to analyse. This is largely due to the difficulty in relating the relative importance of those *in vitro* properties of the platelet that arise from the complexity of platelet function to the requirements that are necessary for *in vivo* survival and haemostatic function. In addition to the formation of the platelet plug in haemostasis, platelets have a role in initiating the intrinsic system of coagulation. They even participate in phagocytosis. One fundamental aspect of the problem of platelet cryopreservation that stands out clearly is the necessity for understanding more about the metabolism and function of platelets. This is intimately related to the understanding of mechanisms of freezing injury and cryoprotection. Since all of these aspects are still not completely understood or sufficiently developed, most approaches to the problem of platelet cryopreservation have been empirical. However, with the recent developments in the field, it is anticipated that more systematic approaches will be used to develop an efficient reliable method of platelet cryopreservation.

Cryopreservation of platelets

The inability of platelets to remain viable for any appreciable length of time in the fresh state at 4°C or at 22°C [168, 169] suggests that freezing remains the only feasible means to achieve long term preservation of this clinically important blood component. During the past two decades many investigators have reported varying degrees of success in the cryopreservation of platelets.

Lyophilised platelets: In 1956 the first reported attempts at platelet freezing were done without cryoprotective agents by Klein *et al* [170] who transfused platelets that had been concentrated and stored in plasma at −15°C for up to six weeks, achieving a transitory haemostasis in thrombocytopenic patients. Klein *et al* [171] subsequently reported that lyophilised platelet material administered to children with acute leukaemia and aplastic anaemia resulted in *some* transient haemostasis and corrected defective prothrombin consumption. This transient haemostasis may have been due to thromoplastin released from damaged platelets. Platelet thromboplastin is a potent accelerator of haemostasis and therefore its release could have resulted in the temporary improvement.

Since these initial attempts at platelet preservation, a number of investigators have attempted to store platelets at low temperatures using protective additives (e.g. DMSO, glycerol etc.) and those slow cooling methods that had previously been found to be effective for red cells and certain nucleated types of cells.

Platelets preserved with DMSO: DMSO has been used with success for the cryopreservation of red cells and tissue culture cells; it was subsequently studied as a cryoprotective additive for platelets. Iossifides *et al* [172] found that by using moderate cooling rates and a medium of 50% plasma with 15% DMSO, they could recover the clot retracting activity of platelets. They also reported some loss of clot retracting activity with platelets stored for various periods of time in liquid nitrogen.

Djerassi and Roy [173] reported high viability values of rat platelets preserved by freezing in 5% DMSO and 5% dextrose. The hypertonic concentration of dextrose served to dehydrate the platelets before freezing. Transfusion into thrombocytopenic rats elevated platelet counts to 70–80% of the control and corrected abnormal bleeding times. DMSO was consistently less effective alone than it was when combined with sugars. Subsequently, they cooled human platelets at 1°C/min to −196°C in 5% DMSO with 5% dextrose and, after having transfused them without washing to patients with leukaemia and aplastic anaemia, reported platelet recoveries approximately one-third of those observed with fresh platelets [173, 174].

Djerassi and colleagues reported some side effects from DMSO after transfusing unwashed platelets that had been suspended in 5% DMSO. Although the toxicity of DMSO is thought to be low, Perry and Yankee [175] have noted no severe side effects after transfusion of DMSO, but local venospasms and pain are not uncommon. Nausea and vomiting also occur with some frequency, and the odour of DMSO is very objectionable to many patients. The long term effects of repeated DMSO transfusions are not well known.

Using a procedure similar to Djerassi, Lundberg and Estwick [176] froze platelets in 7% DMSO with 7% dextrose but washed the platelets with plasma after thawing; they reported good platelet increments after transfusion.

Cattan *et al* [177] froze platelet .concentrates suspended in 11% DMSO using a slow (1°C/min) rate of cooling. These platelets were tested both *in vivo* and *in vitro* and were claimed to have definite activity after several months of storage.

A detailed protocol for platelet freezing was published in 1972 by Handin and Valeri [178, 179] who described a controlled rate of addition and post-thaw removal of 5% DMSO, thereby minimising any adverse effect of the DMSO on post-transfusion survival. Platelets in 5% DMSO were frozen at 1°C/min to −80°C and stored at −150°C. After thawing, the cells were diluted with 2% DMSO in ACD plasma, centrifuged, decanted and resuspended in plasma. When approximately 95% of the DMSO was removed from thawed washed platelets before transfusion, the ^{51}Cr recovery was about 60% of that of the value for fresh platelet concentrates. Valeri and his collaborators [180–184] advocated suspending the platelets in 6% DMSO and freezing at 2–3°C/min by placing the suspension in a mechanical refrigerator at −80°C. Following thawing, the platelets were washed and after transfusion they exhibited good survival. They were also capable of correcting aspirin-induced prolongation of bleeding time. Spector *et al* [185–187] from Valeri's group subsequently modified their procedure by reducing the DMSO concentration to 4% or 5% with −80°C storage and reported post-transfusion survivals approximately 50% of controls.

Careful processing of frozen platelets involving the slow addition and removal of DMSO to reduce osmotic shock has also been reported in 1973 by Kim and Baldini [188, 189] as leading to successful preservation. Using a similar procedure as above, platelets in 5% DMSO were cooled at 1°C/min to −35°C followed by direct immersion into a −79°C bath. After thawing, the cells were diluted slowly with plasma, centrifuged to remove DMSO, and resuspended in plasma. Survival after transfusion was reported to be approximately 82% of control. They also reported that the UCAR polyolefine freezing bags gave superior results over the polyvinyl chloride (PVC) bags. These investigators reported [188–192] further modifications of their procedure involving post-thaw dilution to remove DMSO. Murphy *et al* [193] used similar procedures to freeze platelets in DMSO.

Slichter and Harker [194] froze platelets in 10% DMSO in TC 199 using a cooling rate of 1°C/min to −40°C and then 4°C/min to −80°C followed by storage at −196°C. The thawed cells were transfused without washing and survivals of approximately 80% of control were reported.

Kahn in 1974–78 [195–199] has reported on another modification of the DMSO procedure that involved 4% DMSO and uncontrolled cooling to −80°C. After post-thaw dilution with saline–glucose and washing, the cells exhibited a survival rate of approximately 60% of the normal life span.

Still another variation of the basic DMSO procedure was described in 1976 by Schiffer *et al* [200] who used thawed platelets in the treatment of leukaemia. Satisfactory increments in platelet counts were reported in 21 patients. Side effects from residual DMSO in the transfused platelets were not a problem [201]. Schiffer found it especially convenient to freeze autologous platelets obtained from patients in remission [200, 201].

Platelets preserved with glycerol: The use of glycerol as a cryoprotectant was applied to the storage of platelets in 1958 by Weiss and Ballinger [202] who reported apparently successful preservation of function. Platelets suspended in 15% glycerol in saline were frozen to −79°C and, following thawing, they exhibited normal morphology. In 1959 Djerassi and Klein froze platelets in 20% glycerol at very slow rates of cooling (0.25–0.35°C/min); they also reported normal morphology. The following year Baldini *et al* [203] reduced the glycerol concentration to 9% and cooled platelets at 1°C/min. After storage at −79°C and thawing, the cells were diluted with 35% dextrose in plasma. The deglycerolised platelets demonstrated a survival of 24% of normal controls and also had a normal life span. Ballinger, Weiss and colleagues [204, 205] reported in 1962 on the use of 8% glycerol in plasma with the cooling of the platelets at 1°C/min to −20°C followed by 10°C/min to −79°C. After thawing, the cells were reinfused without washing and the reported survivals averaged 25–30% of controls with an occasional survival of 50–70% of control. In 1966, Cohen *et al* [121] and Cohen and Gardner [120, 206, 207] extended to human platelets their freezing techniques originally developed in 1962 with dog platelets. They used a stepwise addition of glycerol to a final concentration of 12% in platelet concentrates together with slow cooling (1°C/min to −30°C followed by 5°C/min to −70°C). After thawing, they diluted the platelets with 13% sodium citrate and infused the washed platelets, labelled with ^{51}Cr, into human recipients. Although the early studies with frozen dog platelets were very promising, the same cryopreservation method applied to human platelets yielded a survival of only about 30–35% of control although those that did survive exhibited a normal life span.

Sumida [208–210] reports that after freezing platelets at 2–3°C/min in 14% glycerol, storing them at −80°C and deglycerolising them twice with 13% sodium citrate, they were, with some limited success, transfused into five patients with idiopathic and iatrogenic thrombocytopenias. The *in vitro* recovery before transfusion was 46% and survival was reported to be about 5% of controls. However, complete haemostasis was seen in 15 min.

The concern over the toxicity of DMSO and the necessity of having to reduce the DMSO concentration by washing prompted Rowe and his colleagues [60, 117, 211–216] to re-explore the effect of glycerol on platelets in an attempt to re-establish glycerol as an additive of choice for the cryopreservation of platelets. They investigated the effect of glycerol concentration as a function of various freezing regimens on diverse *in vitro* platelet viability parameters [117, 212]. This stimulated a fresh approach to the freezing of platelets in glycerol because Rowe and co-workers [60, 212–215] discovered that a relatively low concentration of glycerol (5%), together with glucose (4%) and coupled with moderate cooling rates (approximately 30°C/min) could result in a high (70–80%) recovery of platelets that could be assessed as viable by numerous *in vitro* parameters and also subsequently by survival *in vivo*. Use of a low concentration of glycerol allowed for the direct transfusion of the thawed platelets without washing. This glycerol–glucose moderate cooling rate protocol first described by Rowe in 1971 [212] was reported in detail in 1976 [215]. Basically the technique involves suspending an acidified platelet concentrate in 5% glycerol with 4% glucose and 30% plasma and reconcentrating in a volume of 10–12 ml. The suspension of glycerolised platelets in the polyolefine bag placed between two metal plates is cooled at a controlled rate of 30°C/min to −80°C and then transferred directly into liquid nitrogen for storage. Achieving the proper cooling rate is critical and still somewhat cumbersome in this procedure. After thawing, the highly concentrated glycerolised platelets were diluted with about 50 ml of autologous acidified plasma and incubated at room temperature for 0.5–4 hours. Approximately 90% of the platelets were recovered intact. The thawed platelet suspensions, that had been labelled with ^{51}Cr prior to freezing, were infused directly without washing and exhibited a survival of about 80% in comparison with unfrozen control platelets stored at 22°C for 24 hours. The transfused cryopreserved platelets had a half-life of approximately 3.8 days. In addition, Rowe reported that the infused platelets demonstrated their haemostatic capacity in their ability to shorten aspirin-prolonged bleeding time [60, 213, 214].

Although this glycerol–glucose moderate cooling rate method reported by Rowe and his colleagues is still somewhat cumbersome to carry out, especially the control of cooling rate, it does represent a novel approach and a departure from the conventional cooling rates of 1°C/min used by most other investigators. Valeri, who admittedly departed radically from the published protocol by using uncontrolled rates of cooling, did not get good survivals in one experiment. Meryman has indicated that Kahn in his laboratory (unpublished) followed the glycerol–glucose moderate cooling rate procedure [217] and duplicated the recovery and survival results of Rowe and colleagues although some results were variable.

Using the identical glycerol–glucose procedure of Rowe and colleagues, Sumida [218] reported that after transfusion of such cryopreserved platelets stored at −196°C there was a significant elevation of platelet counts, good life span of ^{51}Cr-labelled platelets and that they were haemostatically effective in the patients.

Herve *et al* [219, 220] have reported on a variation of the glycerol–glucose method. They suspended the platelets in 3% glycerol with 3% dextrose and cooled at a rate of 10°C/min to −3°C followed by 35°C/min to −150°C. After thawing, the platelets were infused without further processing and these authors reported a survival half-life of 56 hours compared to a control value of 72 hours for fresh platelets. When evaluated in patients who were bleeding, the haemostatic effectiveness of the infused platelets was measured by observing stoppage of bleeding and absence of relapse for at least three days. In 75% of 65 observations of thrombocytopenic patients the efficiency of frozen-thawed platelets appeared convincing in that bleeding was controlled for at least three days.

Platelets preserved with dimethyl acetamide (DMAC): In addition to the conventional cryoprotective additives, DMSO and glycerol, other compounds such as ethylene glycol [221] and dimethyl acetamide (DMAC) have been studied for their cryoprotective capacity.

Very rapid cooling rates of more than 100°C per second in liquid helium damaged unprotected rabbit platelets according to Pfisterer *et al* [222, 223]. Using high cooling rates, they also found that 10% glycerol with 5% dextrose was not suitable for the preservation of rabbit platelets, but results obtained using 5% DMAC with 5% dextrose were encouraging. They concluded that DMAC–dextrose was superior to glucose–dextrose for cryoprotection. Djerassi *et al* [224] found that DMAC was a good cryoprotective agent for rat platelets. In 1971 they reported that human platelets frozen at 3°C/min with 5% DMAC and 5% dextrose could be transfused without washing into thrombocytopenic patients with acute leukaemia and that no side effects were noted. These transfusions of DMAC preserved platelets were associated with temporary increases of their platelet counts. The toxicity of DMAC remains a problem, however.

Platelets preserved with polymers: In addition to the permeating additives, glycerol, DMSO, etc., non-permeating polymeric additives have been studied for their cryoprotective potential with platelets. Doebbler *et al* [9] observed that gross destruction of platelets occurs when whole blood treated with polyvinylpyrrolidone (PVP) is rapidly cooled in liquid nitrogen. Platelets suspended in solutions of dextran sulphate, dextran or PVP and frozen at a rate of 20°C/min were found in 1971 by Reuter and Gross [225] to remain functionally intact. In 1972, Wybran *et al* [226] investigated a combination of 5% DMSO and 5% PVP as a cryoprotectant for cooling platelets at 1–2°C/min to −40°C before transfer to −196°C. Such cryopreserved platelets were reported to have a survival of 30% of controls.

In addition to using DMSO in concentrations of 2–14%, Raymond *et al* [227] explored several concentrations of polyethylene glycol

(PEG) 6000, hydroxyethyl starch (HES) and dextran. DMSO and PEG 6000 were superior to other additives tested, with the optimal concentration being 5%. In this study the assay involved the perfusion of an isolated rabbit kidney and examination for vascular integrity and presence of platelet thrombi.

Spencer *et al* [228, 229] have reported on the use of hydroxyethyl starch (HES) as a cryoprotective agent for monkey platelets. Platelets suspended in 6% HES and cooled at a rate of 37°C/min to −80°C were recovered from storage at −140°C and found to retain good *in vitro* function with an *in vivo* survival of 33–55% in rhesus monkeys after 24 hours. These investigators suggested that a system utilising HES may be applicable for use with human platelets. Choudhury and Gunstone [230] have reported that they obtained a higher recovery of cryopreserved platelets treated with 4% HES than with 5% DMSO after cooling at 1°C/min and storage in liquid nitrogen. They emphasised that the final concentration of HES (4%) was critical in that variations of 5–7% and above were detrimental to platelet function. A preliminary clinical trial involving transfusion of HES-preserved platelets in five thrombocytopenic patients confirmed their haemostatic effectiveness.

Cryopreservation of platelets—future perspectives:

From the foregoing it is obvious that there have been numerous approaches to the problem of cryopreservation of platelets. Most methods have been developed empirically and, although some involving DMSO and glycerol have been used clinically, there is not yet one simple method suitable for routine use in a blood bank. We suggest that the ideal cryopreservation procedure would involve:

(1) a clinically acceptable cryoprotectant such as glycerol or perhaps HES;
(2) a rigidly controlled, preferably automated, cooling rate to minimise technical error;
(3) a minimum of post-thaw processing; and
(4) sufficient post-thaw storage stability to permit ample time to transport the thawed platelets from the blood bank to the patient.

Current methods for preserving platelets in the frozen state have largely resulted in inadequate *in vitro* and low *in vivo* survival. While freezing adversely affects the majority of platelets, it does seem possible to preserve platelet integrity in a portion of the platelet population. While results so far obtained are encouraging, we cannot overcome the inherent problems associated with the preservation of platelets until we have a greater understanding of the system as a whole, i.e. the collection, processing, freezing, storage and post-thaw processing. The inability to preserve completely viable platelets has been due largely to a lack of basic cryobiological knowledge of factors related to the freeze-thaw survival of platelets; however, some recent advances have been made in this area.

Implicating lysosomes as targets of freezing injury, Rowe and his group [117, 212] postulated that loss of platelet function from cryo-injury may be due to damage to lysosomes resulting in release of hydrolytic lysosomal enzymes into the platelet cytoplasm, bringing about irreversible cell changes. Using the release of lysosomal enzymes as indicators of cryoinjury at the subcellular level, they found a direct relationship existed between freeze-thaw injury to platelets and release of the hydrolytic enzymes, β-glucuronidase, acid phosphatase and cathepsin. An optimal cooling rate at various glycerol concentrations (1–10%) was about 30°C/min, while the optimal glycerol concentration at all cooling rates was 0.55 M. Under the optimal cooling conditions, the activation, i.e. release of β-glucuronidase, indicated that 90–95% of platelet lysosomes were intact. Greiff *et al* [231, 232] have used aminopeptidase activity as a measure of the intactness of platelets preserved by freezing in DMSO.

Most investigators have employed a slow rate of cooling of the order of 1–3°C/min for the cryopreservation of platelets in DMSO and glycerol. The finding by Rowe and his colleagues [60, 212–216] that platelets suspended in 5% glycerol with 4% glucose and 30% plasma could withstand faster cooling rates of the order of 30°C/min has opened up new areas of investigation. Leibo and Mazur [233] have shown that optimal cooling rate can vary widely from one cell type to the other. Mazur [234] has discriminated between two types of freezing injury, namely injury due to intracellular freezing at excessive cooling rates and injury resulting from extracellular ice that concentrates the extracellular solution, i.e. the so-called solution effect. Farrant *et al* [235] believe that the cell damage related to intracellular freezing may rather correlate with the total amount of ice formed per cell rather than with the size of the individual crystals. They also suggest that the cell injury that occurs during rewarming is primarily osmotic in nature. Meryman [236] has speculated on osmotic mechanisms of freezing injury in platelets. Meryman *et al* [12] contend that cell damage from the concentration of the extracellular solute is not a direct result of solute concentration *per se* but is associated with the cell dehydration and volume reduction that occurs in response to the increased extracellular osmolarity.

Perhaps the unique cellular structure of platelets allows for several optima in cooling rates depending upon the cryoprotective additive type and concentration. Platelet cryopreservation offers a unique challenge and test of prevailing theories of freezing injury namely:

(1) at slow cooling rates the cell becomes dehydrated beyond its limit of tolerance;
(2) as cooling rate is increased less time is available for efflux of water from the cell, thus avoiding excessive dehydration; and
(3) with still further increase in cooling rate, the cell interior may not be sufficiently dehydrated so that intracellular ice forms (see Chapter 1).

It is possible that cryoprotectants may influence the diffusion of water. The optimal cooling rate with the usual concentration of DMSO appears to be between 1 and 2°C/min whereas with a comparable concentration of glycerol the optimal cooling rate is about 30°C/min. It is possible that DMSO may affect the permeability of the platelet membrane to water diffusion. Rowe and associates [211] have reported on differences in human platelet permeability to glycerol and DMSO. A change in platelet volume takes place during the addition of cryoprotective additives which could account for the activation of lysosomal enzymes and an altered metabolism. While the uptake of glycerol and DMSO is linear, glycerol causes shrinkage of the platelet. This suggests a partial barrier to the entry of glycerol into platelets and may account for the difficulties other investigators have experienced using glycerol as the cryoprotective additive. Bakry and McGill [237] have studied the effect of glycerol on the volume, shape and ultrastructure of human platelets. Crowley *et al* [238] described changes in shape and structure of freeze preserved platelets based on electron micrographs.

Rowe and Peterson [216] found that a 6–8% concentration of either glycerol or DMSO was optimal for preserving platelets based on the *in vitro* measurement of uptake of ^{14}C-serotonin, spontaneous release of preadded ^{14}C-serotonin, activation of platelet factor III as measured by Stypven time, and ADP-induced aggregation. Undeutsch *et al* [239] have also investigated the effect of glycerol on platelet function while Schiffer *et al* [240] have reported on the effect of DMSO on platelet function. Odink and collaborators [241–246] have also studied many *in vitro* parameters that could influence platelets during cryopreservation.

The effects of cryoprotective agents and freezing on the glycolysis and adenine nucleotide metabolism have been studied by Kim and Baldini [190, 191], who found that the increase in glycolytic activity which occurred before freezing in platelets exposed to glycerol, DMSO and sugars may be interpreted as a reaction to cellular damage; that freezing and thawing produced abrupt depression of platelet glycolysis and reduction of ATP; and finally that the least change was observed in platelets treated with DMSO only.

Kahn [195–197] has recently reviewed the biochemical changes that have been reported in platelets especially in relation to cryopreservation. He also reported on his own studies involving the alteration of the platelet membrane glycoprotein pattern due to freezing.

Rowe and Lenny [247, 248] have studied the ^{51}Cr labelling of human platelets cryopreserved by the glycerol–glucose method and the DMSO method. Freezing and thawing by either method caused elution of some free and protein bound ^{51}Cr from platelets. Chromium loss was unrelated to cell loss and presumably was due to selective membrane leakage induced by freezing and osmotic stress.

The mechanisms of freezing injury in platelets still remain an enigma. 'Colligative' cryoprotections as used in red cell freezing with high glycerol concentration does not appear to be a feasible approach with platelets. The major obstacle to this approach with platelets is the addition and removal of the necessary high concentrations of glycerol required, as well as the inherent cellular toxicity of the additive at high concentration. Clearly, future approaches to resolving the problems in cryopreservation of platelets will have to encompass mechanisms involving not only the effect of cryoprotectant during freezing but also the effect on the physiological function of the platelet *per se*.

The requirement of low concentrations of cryoprotectant for platelets combined with cooling at a controlled rate indicates that kinetic approaches will direct the future for developing successful platelet cryopreservation procedures. The 'low glycerol–rapid cooling' method used for red cells is an example of such a kinetic approach to cryopreservation. Greater understanding of the underlying mechanisms involved with control of cooling rates and the effect on platelets is a necessary requirement to make the kinetic approach of any cryopreservative procedure a practical reality. The relatively low concentrations of a cryoprotective additive, especially glycerol–glucose, are very attractive from a practical point of view since such cryopreserved platelets can be used for transfusion with little or no post-thaw processing.

It is difficult to reconcile the various *in vitro* changes which have been noted by many investigators using the best procedures for preserving platelets in either the liquid or frozen state. It appears that not only the method of collection, the nature of the collecting vessel or bag, the storage temperature in the liquid state, the pH of the collecting media but also the cryoprotective additive type and concentration, rates of cooling, temperature of storage, and effects of washing are all important for the cryopreservation of platelets. Variations in these parameters unquestionably affect the results. The clinical demands for preserved platelets have emphasised empiricism in approaches to preserving platelets based on the success experienced with preserving red cells. Future approaches to preserving platelets indicate that only through more basic study of the biochemistry and physiology of platelets can we hope to achieve full preservation of platelet viability.

Advances in the understanding of the mechanism of freezing injury and the recent approach using moderately rapid rates of cooling with glycerol for platelet storage encourage hope for the early development of a simple practical procedure for the cryopreservation of platelets.

References

1 Mollison, P.L. and Sloviter, H.A. (1951). *Lancet* **ii**, 862

2 Mollison, P.L., Sloviter, H.A. and Chaplin, H. (1952). *Lancet* **ii**, 501

3 Polge, C., Smith, A.U. and Parkes, A.S. (1949). *Nature* **164**, 666

4 Smith, A.U. (1950). *Lancet* **ii**, 910

5 Smith, A.U. (1961). In *Biological Effects of Freezing and Supercooling*. Baltimore: The Williams and Wilkins Company

6 Chaplin, H. and Veall, N. (1953). *Lancet* **i**, 218

7 Lovelock, J.E. (1957). *Proceedings of the Royal Society, Series B Biological Sciences* **147**, 426

8 Callahan, A.B. (1967). In *International Working Conference on the Freeze-Preservation of Blood*. Washington, DC: Office of Naval Research Report DR-143

9 Doebbler, G.F., Rowe, A.W. and Rinfret, A.P. (1966). In *Cryobiology*, p. 407. H.T. Meryman (ed.). New York: Academic Press

10 Hurn, B.A.L. (1968). In *Storage of Blood*. New York: Academic Press

11 Luyet, B. and Rapatz, G. (1970). *Cryobiology* **6**, 425

12 Meryman, H.T., Williams, R.W. and Douglas, M.S.J. (1977). *Cryobiology* **14**, 287

13 Turner, A.R. (1970). In *Frozen Blood–A Review of the Literature 1949–1968*. New York: Science Publishers, Inc.

14 Valeri, C.R. (1970). In *Critical Reviews in Clinical Laboratory Science*, p. 381. Cleveland: CRC Press

15 Valeri, C.R. (1976). *Blood Banking and the Use of Frozen Blood Products*. Cleveland: CRC Press

16 Chaplin, H. (1978). *New England Journal of Medicine* **298**, 679

17 Chaplin, H. and Mollison, P.L. (1953). *Lancet* **i**, 215

18 Crosby, W. (1967). *Military Medicine* **132**, 119

19 Haynes, L., Tullis, J., Pyle, H., Sproul, M., Wallach, S. and Turnville, W.C. (1960). *Journal of the American Medical Association* **173**, 1657

20 Högmann, C.F. and Åkerblom, O (1970). In *Modern Problems of Blood Preservation*, p. 212. W. Spielmann and S. Seidl (eds.). Stuttgart: Fisher Verlag

21 Jenkins, W.J. and Blagdon, J. (1971). *Journal of Clinical Pathology* **24**, 685

22 Keitt, A.S. (1972). In *Hematology*, p. 1301. W.J. Williams, E. Beutler, A.J. Erslev and R.W. Rundles (eds.). New York: McGraw-Hill

23 Mannoni, P. and Beaujean, F. (1976). *Les Collogues de l'INSERM/Cryoimmunology* **62**, 121

24 Mannoni, P., Beaujean, F. and LeForestier, C. (1975). *Revue Française de Transfusion et Immuno-Hematologie* **18**, 425
25 Meryman, H.T. (1977). In *Clinical and Practical Aspects of the Use of Frozen Blood,* p. 1. Washington, DC: AABB
26 Pepper, D.S. (1976). *Clinical Haematology* **5**, 53
27 Pert, J.H. and Schork, P.K. (1970). In *Modern Problems of Blood Preservation,* p. 168. W. Spielmann and S. Seidl (eds.) Stuttgart: Fischer-Verlag
28 Pyle, H.M. (1964). *Cryobiology* **1**, 57
29 Rinfret, A.P. (1968). *Cryosurgery* **1**, 82
30 Rowe, A.W. (1967). In *International Working Conference on the Freeze-Preservation of Blood,* p. 44. A.B. Callahan (ed.). Washington, DC: ONR Report DR-143
31 Rowe, A.W. (1971). *Mechanical Engineering* **93**, 37
32 Rowe, A.W. (1973). In *Red Cells Freezing,* p. 55. Washington, DC: AABB
33 Seidl, S. (1977). *Infusionstherapie* **4**, 88
34 Sumida, S. (1974). In *Transfusion of Blood Preserved by Freezing.* G. Thieme (ed.). Stuttgart: Lippincott Co.
35 Tullis, J.L. and Pyle, H.M. (1960). *Vox Sanguinis* **5**, 70
36 Rinfret, A.P. (1960). *Annals of the New York Academy of Sciences* **85**, 576
37 Strumia, M.M., Colwell, L.S. and Strumia, P.V. (1958). *Science* **128**, 1002
38 Strumia, M., Colwell, L. and Strumia, P. (1962). In *Proceedings of the 8th Congress of the International Society for Blood Transfusion, Tokyo, 1960,* p. 433
39 Strumia, M.M., Eusebi, A.J. and Strumia, P.V. (1968). *Journal of Laboratory and Clinical Medicine* **71**, 138
40 Bricka, M. and Bessis, M. (1955). *Comptes Rendus des Séances de la Société de Biologie et des Ses Filiales* **149**, 875
41 Doebbler, G.F., Sakaida, R.R., Cowley, C.W. and Rinfret, A.P. (1966). *Transfusion* **6**, 104
42 Rinfret, A.P. (1963). *Federation Proceedings* **22**, 94
43 Rinfret, A.P., Doebbler, G.F., Rowe, A.W., Cowley, C.W. and Sakaida, R.R. (1963). *Progress in Refrigeration Science and Technology,* p. 1551
44 Robson, C.R. (1967). In *International Working Conference on the Freeze-Preservation of Blood,* p. 53. A.B. Callahan (ed.). Washington, DC: ONR Report DR-143
45 Bear, S. (1973). *Transfusion* **13**, 2
46 Knorpp, C.T., Starkweather, W.H., Spencer, H.H. and Weatherbee, L. (1971). *Cryobiology* **8**, 511
47 Krijnen, H.W., Kuivenhoven, A.C. and deWit, J.J.F.M. (1970). In *Modern Problems of Blood Preservation,* p. 176. W. Speilmann and S. Seidl (eds.). Stuttgart: Fischer-Verlag
48 Mishler, J.M., Parry, E.S. and Petrie, A. (1978). *British Journal of Haematology* **40**, 231
49 Parry, E.S., Mishler, J.M. Bushrod, J. and Whitcher, H. (In press)
50 Weatherbee, L., Allen, E.D., Spencer, H.H., Lindenauer, S.M. and Permoad, P.A. (1975). *Cryobiology* **12**, 119
51 Pribor, D.B. and Pribor, H.C. (1973). *Cryobiology* **10**, 93
52 Lionetti, F.J. and Hunt, S.M. (1975). *Cryobiology* **12**, 100
53 Rowe, A.W. and Allen, F.H. (1965). *Transfusion* **5**, 379
54 Rowe, A.W., Borel, H. and Allen, F.H. (1972). *Vox Sanguinis* **22**, 188
55 Huntsman, R., Hurn, B., Ikin, E., Lehmann, H. and Liddell, J. (1967). *British Medical Journal* **4**, 458
56 Behzad, O. and Lee, C.L. (1977). *Transfusion* **17**, 650
57 Bronson, W. and McGinniss, M. (1962). *Blood* **20**, 478
58 Cunningham, R.K. and Johnson, R.D. (1978). *Cryobiology* **15**, 670
59 Gibbs, M., McCord, E., Collins, W., Schrider, C. and Akeroyd, J. (1962). *Transfusion* **2**, 100
60 Rowe, A.W., Dayian, G., Reich, L.M., Turc, J.M and Mayer, K. (1974). In *Platelet Preservation and Transfusion. Proceedings of the International Society for Blood Transfusion, Symposium* **32**, 51
61 Rowe, A.W., Derrick, J.B., Miles, W., Allen, F.A. and Kellner, A. (1970). In *Modern Problems of Blood Preservation,* p. 184. W. Spielmann and S. Seidl (eds.). Stuttgart: Fischer-Verlag

62 Rowe, A.W. and Lenny, L.L. (1978). *Problemy Techniki W Medycynie* **9**, 1
63 Huggins, C.E. (1963). *Surgery* **54**, 191
64 Huggins, C.E. (1965). *Journal of the American Medical Association* **193**, 941
65 Huggins, C.E. (1966). *Monographs in Surgical Sciences* **3**, 133
66 Huggins, C.E. (1970). In *Modern Problems of Blood Preservation,* p. 138. W. Spielmann and S. Seidl (eds.). Stuttgart: Fischer-Verlag
67 Sproul, M.T. (1965). In *Proc. Comm. Blood and Transfusion Problems.* National Academy of Sciences–National Research Council
68 Sproul, M.T., Grove-Rasmussen, M., Jago, S. and Pyle, H.M. (1965). In *Proceedings of X Congress of the International Society for Blood Transfusion,* Karger, Basel, p. 731
69 Tullis, J.L. (1963). *Transfusion* **3**, 155
70 Tullis, J.L. (1964). In *The Red Blood Cell–A Comprehensive Treatise,* p. 491. C. Bishop and D.M. Surgenor (eds.). New York: Academic Press
71 Tullis, J.L., Gibson, J.G., Sproul, M.T., Tinch, R.J. and Baudanza, P. (1970). In *Modern Problems of Blood Preservation,* p. 161. W. Spielmann and S. Seidl (eds.). Stuttgart: Fischer-Verlag
72 Tullis, J., Haynes, L., Pyle, H., Wallach, S., Pennell, R., Sproul, M. and Khoubesserian, A. (1960). *Archives of Surgery* **81**, 169
73 Tullis, J., Hinman, J., Sproul, M. and Nickerson, R. (1970). *Journal of the American Medical Association* **213**, 719
74 Tullis, J., Ketchel, M., Pyle, H., Pennell, R., Gibson, J., Tinch, R. and Driscoll, S. (1958). *Journal of the American Medical Association* **163**, 399
75 Tullis, J.L., Surgenor, D.M. and Baudanza, P. (1959). *Blood* **14**, 456
76 Tullis, J.L., Surgenor, D.M., Tinch, R.J., D'Hont, M., Gilchrist, F.L., Driscoll, S.G. and Batchelor, W.G. (1956). *Science* **124**, 792
77 Ketchel, M., Tullis, J., Tinch, R., Driscoll, S. and Surgenor, D. (1958). *Journal of the American Medical Association* **168**, 404
78 Valeri, C.R. and Runck, A. (1969). *Transfusion* **9**, 5
79 O'Brien, T., Haynes, L., Hering, A., Tullis, J. and Watkins, E. (1961). *Surgery* **49**, 109
80 O'Brien, T. and Watkins, E. (1960). *Journal of Thoracic and Cardiovascular Surgery* **40**, 611
81 Meryman, H.T. (1973). In *Red Cell Freezing,* p. 73. Washington, DC: AABB
82 Meryman, H.T. and Hornblower, M. (1972). *Transfusion* **12**, 145
83 Meryman, H.T. (1971). *Cryobiology* **8**, 173
84 Meryman, H.T. and Hornblower, M. (1977). *Transfusion* **17**, 438
85 Hornblower, M. and Meryman, H.T. (1977). *Transfusion* **17**, 417
86 Huggins, C.E. (1973). In *Red Cell Freezing,* p. 31. Washington, DC: AABB
87 Moss, G. (1969). In *Hematologic Problems in Surgery,* p. 93. H. Laufman and R.B. Erichson (eds.). Philadelphia: Saunders
88 Moss, G., Valeri, C. and Brodin, C. (1968). *New England Journal of Medicine* **278**, 747
89 Almond, D.V. and Valeri, C.R. (1967). *Transfusion* **7**, 95
90 Valeri, C.R., Runck, A. and Sampson, W. (1969). *Transfusion* **9**, 120
91 Pert, J.H., Schork, P.K. and Moore, R. (1963). *Clinical Research* **11**, 197
92 Krijnen, H.W., deWit, J.J.F.M., Kuivenhoven, A.C.J., Loos, J.A. and Prins, H.K. (1964). *Vox Sanguinis* **9**, 559
93 Krijnen, H.W., deWit, J.J.F.M., Kuivenhoven, A.C.J. and Reyden G. van D. (1975). *Proceedings of the 10th Congress of the International Society for Blood Transfusion, Stockholm,* p. 683
94 Rowe, A.W., Derrick, J.B., Miles, W., Allen, F.H. and Kellner, A. (1971). *Bibliotheca Haematologica* **38**, 320
95 Rowe, A.W., Eyster, E., Allen, F.H. and Kellner, A. (1966). *Transfusion* **6**, 521
96 Rowe, A.W., Lenny, L.L., Dayian, G. and Mayer, K. (1973). *Cryobiology* **10**, 522
97 Rowe, A.W., Lenny, L.L., Dayian, G. and Mayer, K. (1973). *Transfusion* **13**, 355
98 Rowe, A.W., Miles, W., Derrick, J.B., Allen, F.H., Eyster, E., Schoor, J.B. and Kellner, A. (1970). *New York Medicine* **26**, 246
99 Rowe, A.W., Eyster, E. and Kellner, A. (1968). *Cryobiology* **5**, 119
100 Rowe, A.W. (1967). In *Plastics in the Medical Sciences,* p. 86. J. Frados (ed.). New York: Regional Technical Conference of the Society of Plastics Engineers, Inc.

101 Bryant, L.R. (1977). In *Clinical and Practical Aspects of the Use of Frozen Blood,* p. 77. Washington, DC: AABB Monograph
102 Derrick, J.B., Lind, M. and Rowe, A.W. (1969). *Cryobiology* **6**, 260
103 Derrick, J.B., Lind, M. and Rowe, A.W. (1969). *Transfusion* 9, 317
104 Derrick, J.B., McConn, R., Lind-Sorovacu, M. and Rowe, A.W. (1972). *Transfusion* **12**, 400
105 Frozen Preservation of Red Blood Cells (1977). In *Technical Manual,* p. 47. Washington, DC: AABB
106 Gerst, P.H., Horowitz, H.I., Rowe, A.W., Allen, F.H. and Kellner, A. (1967). *New York State Journal of Medicine* **67**, 830
107 Jones, C., Horowitz, H.I., Spielvogel, A.R. and Rowe, A.W. (1968). *Transfusion* 8, 323
108 Puno, C., Rowe, A.W., Allen, F.H., Spielvogel, A.R. and Horowitz, H.I. (1967). *Transfusion* 7, 391
109 Åkerblom, O. and Högman, C. (1974). *Transfusion* **14**, 16
110 Combrisson, A. (1972). *Revue du Practicien* **22**, 1479
111 Combrisson, A., Laforest, J., Champcommunal, A. and Pannetier, C. (1968). *Revue Française de Transfusion* 9, 7
112 Eygonnet, J.P., Saint-Blancard, J., Fort, V., Berroche, L. and Maupin, B. (1968). *Revue Française de Transfusion* 9, 385
113 Seidl, S. (1971). *Bibliotheca Haematologica* **38**, 190
114 Sumida, S. (1969). *Cryobiology* **6**, 580
115 Sumida, S., Sumida, M., Miyata, K., Sakai, T. and Uchida, H. (1975). *Low Temperature Medicine* **1**, 227
116 Valeri, C.R. (1966). *Military Medicine* **131**, 705
117 Dayian, G. and Rowe, A.W. (1970). *Cryobiology* **6**, 579
118 Valeri, C.R., Szymanski, I.O. and Runck, A.H. (1970). *Transfusion* **10**, 102
119 Rowe, A.W., Lenny, L.L. and Hirsch, R.L. (1971). *Proceedings of the XIII Congress on Refrigeration, I.I.R.,* p. 908. New York: Pergamon Press
120 Cohen, P. and Gardner, F.H. (1960). *Clinical Research* 8, 207
121 Cohen, P., Pringle, J.C. and Gardner, F.H. (1958). *Clinical Research* **6**, 199
122 Kevy, S.V., Jacobson, M, and Button, L. (1976). In *Clinical Uses of Frozen Thawed Red Blood Cells,* p. 89. J.P. Griep (ed.). New York: Alan R. Liss, Inc.
123 Szymanski, I.O. and Carrington, E.J. (1977). *Transfusion* **17**, 431
124 Telishi, M., Krmpotic, E. and Moss, G. (1975). *Transfusion* **15**, 481
125 Wajcman, H., Saint-Blancard, J., Fort, V. and Fabre, G. (1971). *Revue Française de Transfusion* **14**, 247
126 Contreras, T.J. and Valeri, C.R. (1976). *Transfusion* **16**, 539
127 Umlas, J. and Gootblatt, S. (1976). *Transfusion* **16**, 636
128 Valeri, C.R. (1966). *Transfusion* **6**, 112
129 Valeri, C.R. (1966). *New England Journal of Medicine* **275**, 365 and 425
130 Valeri, C.R. (1970). In *Modern Problems of Blood Preservation,* p. 125. W. Spielmann and S. Seidl (eds.). Stuttgart: Fischer-Verlag
131 Valeri, C.R. (1971). *New England Journal of Medicine* **284**, 81
132 Valeri, C.R. (1974). *Transfusion* **14**, 1
133 Valeri, C.R. (1975). *Transfusion* **15**, 195
134 Valeri, C.R., Runck, A. and Brodine, C. (1969). *Journal of the American Medical Association* **208**, 489
135 Valeri, C.R. and Zaroulis, C.G. (1972). In *Progress in Transfusion and Transplantation,* p. 343. P.J. Schmidt (ed.). Washington, DC: AABB
136 Valeri, C.R. and Zaroulis, C.G. (1972). *New England Journal of Medicine* **287**, 1307
137 Fabre, G., Fort, V., Ginesie, J., Hainaut, J., Monteux, J., Saint-Blancard, J. and Wajcman, H. (1971). *Revue Française de Transfusion* **14**, 235
138 Fuller, T.C., Delmonico, F.L., Cosimi, B., Huggins, C.E., King, M. and Russell, P.S. (1978). *Annals of Surgery* **187**, 211
139 Haber, J., Erlandson, E., Miles, W. and Rowe, A.W. (1968). *Transfusion* 8, 324
140 Runck, A.H. and Valeri, C.R. (1969). *Transfusion* 9, 297
141 Runck, A.H. and Valeri, C.R. (1972). *Transfusion* **12**, 237
142 Runck, A.H., Valeri, C.R. and Sampson, W.T. (1968). *Transfusion* 8, 9

143 Crowley, J.P., Skrabut, E.M. and Valeri, C.R. (1974). *Vox Sanguinis* **26**, 513
144 Crowley, J.P. and Valeri, C.R. (1974). *Transfusion* **14**, 590
145 Miles, W., Issitt, P.D., Adebahr, M., Allen, F.H. and Rowe, A.W. (1968). *Transfusion* **8**, 324
146 Opelz, G. and Terasaki, P.I. (1974). *Lancet* **ii**, 696
147 Perkins, H.A. (1974). *Transfusion* **14**, 509
148 Polesky, H.F. (1976). In *Clinical Uses of Frozen Thawed Red Blood Cells,* p. 141. J.A. Griep (ed.). New York: Alan R. Liss, Inc.
149 Polesky, H.F. (1977). In *Clinical and Practical Aspects of the Uses of Frozen Blood,* p. 113. Washington, DC: AABB Monograph
150 Kreuger, A. (1976). *Vox Sanguinis* **30**, 239
151 Kreuger, A. and Blombäck, M. (1974). *Haemostasis* **3**, 329
152 Contreras, T.J. and Valeri, C.R. (1976). *Transfusion* **16**, 594
153 Barker, L.L. and Crouch, M.L. (1975). In *Frozen Red Cell Outdating,* p. 7. P.B. Sherer (ed.). DHEW Publ. No. (NIH) 76-1004
154 Myhre, B.A., Nakasako, Y.Y. and Schott, R. (1977). *Transfusion* **17**, 454
155 Kahn, R.A., Meryman, H.T., Syring, R.L. and Flinton, L.J. (1976). *Transfusion* **16**, 215
156 Rowe, A.W., Marshall, E.T., Lenny, L.L. and Graham, J.R. (1975). In *Frozen Red Cell Outdating,* p. 53. P.B. Sherer (ed.). DHEW Publ. No. (NIH) 76-1004
157 Berkman, R.M., Arnett, J.C., Eleland, E.L. and Wardle, M.D. (1975). In *Frozen Red Cell Outdating,* p. 17. P.B. Sherer (ed.). DHEW Publ. No. (NIH) 76-1004
158 Tenczar, F.J. (1975). In *Frozen Red Cell Outdating,* p. 61. P.B. Sherer (ed.). DHEW Publ. No. (NIH) 76-1004
159 Beutler, E. and Duron, O. (1965). *Transfusion* **5**, 17
160 Beutler, E., Meul, A. and Wood, L.A. (1969). *Transfusion* **9**, 109
161 Werch, J., Gray, R.E., Hersh, T. and Melnick, J.L. (1971). *Journal of the American Medical Association* **218**, 93
162 Alter, H.J., Taber, E., Meryman, H.T., Hoofnagle, J.H., Kahn, R.A., Holland, P.V., Gerety, R.J. and Barker, L.F. (1978). *New England Journal of Medicine* **298**, 637
163 Schechter, H.P., Soehnlen, F. and McFarland, W. (1972). *New England Journal of Medicine* **287**, 1169
164 Morrison, F.S. (1968). *Cryobiology* **5**, 29
165 Chanutin, A. (1970). In *Platelet Biochemistry–A Review.* Washington, DC: Department of Defense Report No. DADA 1769-G-9301
166 Gardner, F.H. (1968). *Cryobiology* **5**, 43
167 Gardner, F.H. (1972). In *Hematology,* p. 1305. W.J. Williams, E. Beutler, A.J. Erslev and R.W. Rundles (eds.). New York: McGraw-Hill
168 Murphy, S. and Gardner, F.H. (1969). *New England Journal of Medicine* **280**, 1094
169 Murphy, S. and Gardner, F.H. (1971). *Journal of Clinical Investigation* **50**, 370
170 Klein, E., Farber, S., Djerassi, I., Toch, R., Freeman, F. and Arnold, P. (1956). *Journal of Pediatrics* **49**, 517
171 Klein, E., Toch, R., Farber, S., Freeman, G. and Fiorentino, R. (1956). *Blood* **11**, 693
172 Iossifides, I., Geisler, P., Eichman, M.F. and Tocantins, L.M. (1963). *Transfusion* **3**, 167
173 Djerassi, I. and Roy, A. (1963). *Blood* **22**, 703
174 Djerassi, I., Farber, S., Roy, A. and Cavins, J. (1966). *Transfusion* **6**, 572
175 Perry, S. and Yankee, R.A. (1971). In *The Circulating Platelet,* p. 541. S. Johnson (ed.). New York: Academic Press
176 Lundberg, A. and Estwick, N. (1969). *Cryobiology* **6**, 570
177 Cattan, A., Schwarzenberg, L., Hayat, M., Seman, G., Ouazan, A. and Pradel-Balade, O. (1968). *Revue Française d'Etudes Cliniques et Biologiques* **13**, 500
178 Handin, R.I. and Valeri, C.R. (1971). *New England Journal of Medicine* **285**, 538
179 Handin, R.I. and Valeri, C.R. (1972). *Blood* **40**, 509
180 Valeri, C.R. (1974). *New England Journal of Medicine* **290**, 353
181 Valeri, C.R. (1974). In *Platelet Preservation and Transfusion. Proceedings of the International Society for Blood Transfusion,* Symposium 41
182 Valeri, C.R. and Feingold, H. (1974). In *Platelets: Production, Function, Transfusion and Storage,* p. 377. M.G. Baldini and S. Ebbe (eds.). New York: Grune and Stratton

183 Valeri, C.R., Feingold, H. and Marchionni, L.D. (1974). *Blood* **43**, 131
184 Valeri, C.R., Feingold, H. and Marchionni, L.D. (1974). *Transfusion* **14**, 331
185 Spector, J.I., Flor, W.J. and Valeri, C.R. (1977). *Blood* **50** (Suppl. 1), 253
186 Spector, J.I., Skrabut, E.M. and Valeri, C.R. (1977). *Transfusion* **17**, 99
187 Spector, J.I., Yarmala, J.A., Marchionni, L.D., Emerson, C.P. and Valeri, C.R. (1977). *Transfusion* **17**, 8
188 Baldini, M.G. and Kim, B.K. (1974). *Proceedings of the International Society of Blood Transfusion,* Symposium, 32
189 Kim, B.K. and Baldini, M.G. (1973). *Proceedings of the Society for Experimental Biology and Medicine* **142**, 345
190 Kim, B.K. and Baldini, M.G. (1974). *Proceedings of the Society for Experimental Biology and Medicine* **145**, 830
191 Kim, B.K. and Baldini, M.G. (1974). *Transfusion* **14**, 130
192 Kim, B.K., Tanove, K. and Baldini, M.G. (1976). *Vox Sanguinis* **30**, 401
193 Murphy, S., Sayer, S.N., Ardou, N.L. and Gardner, F.H. (1974). *Transfusion* **14**, 139
194 Slichter, S.J. and Harker, L.A. (1972). *Clinical Research* **20**, 571
195 Kahn, R.A. (1974). *Cryobiology* **11**, 564
196 Kahn, R.A. (1974). In *Platelets: Production, Function, Transfusion and Storage,* p. 355. M.G. Baldini and S. Ebbe (eds.). New York: Grune and Stratton
197 Kahn, R.A. (1977). *Cryobiology* **14**, 700
198 Kahn, R.A. (1978). In *The Blood Platelet in Transfusion Therapy* p. 167. New York: Alan R. Liss, Inc.
199 Kahn, R.A. and Golterman, J. (1977). *Cryobiology* **14**, 699
200 Schiffer, C.A., Buckholz, D.H., Aisner, J., Wolff, J.H. and Wiernik, P.H. (1976). *Transfusion* **16**, 321
201 Schiffer, C.A., Aisner, J. and Wiernik, P.H. (1978). *New England Journal of Medicine* **299**, 7
202 Weiss, A.J. and Ballinger, W.F. (1958). *Annals of Surgery* **148**, 360
203 Baldini, M., Costea, N. and Dameshek, W. (1960). *Blood* **16**, 1669
204 Ballinger, W.F., Weiss, A.J., Jackson, L.G. and Barr, M.A. (1962). *Journal of the American Medical Association* **179**, 148
205 Weiss, A.J., Ballinger, W.F., Jackson, L.G. and Barr, M.A. (1961). *Clinical Research* **9**, 168
206 Cohen, P. and Gardner, F.H. (1959). *Journal of Clinical Investigation* **38**, 995
207 Cohen, P. and Gardner, F.H. (1966). *New England Journal of Medicine* **274**, 1400
208 Sumida, S. (1971). *Cryobiology* **8**, 393
209 Sumida, S. (1974). In *Platelet Preservation and Transfusion. Proceedings of the International Society for Blood Transfusion,* Symposium 52
210 Sumida, S. and Sumida, M. (1974). *Cryobiology* **11**, 564
211 Dayian, G., Chin, S.N. and Rowe, A.W. (1971). *Cryobiology* **8**, 376
212 Dayian, G., Chin, S.N. and Rowe, A.W. (1971). *Cryobiology* **8**, 393
213 Dayian, G., Reich, L.M., Mayer, K., Turc, J.M. and Rowe, A.W. (1974). *Transfusion* **14**, 511
214 Dayian, G., Reich, L.M., Mayer, K., Turc, J.M. and Rowe, A.W. (1974). *Cryobiology* **11**, 563
215 Dayian, G. and Rowe, A.W. (1976). *Cryobiology* **13**, 1
216 Rowe, A.W. and Peterson, J. (1971). *Cryobiology* **8**, 397
217 Meryman, H.T. and Burton, J.L. (1978). In *The Blood Platelet in Transfusion Therapy,* p. 153. New York: Alan R. Liss, Inc.
218 Sumida, S. (1977). *Cryobiology* **14**, 700
219 Herve, P., Masse, M., Potron, G., Kieffer, Y., Carbillet, J.P., Bidet, A., Bosset, J.F. and Peters, A. (1978). In *Cell Separation and Cryobiology,* p. 239. G. Rainer (ed.). Stuttgart: Schattauer Verlag
220 Herve, P., Masse, M., Potron, G., Kieffer, Y., Carbillet, J.P. and Bidet, A. (1977). *Revue Française de Transfusion* **20**, 603
221 Kahn, R.A. and Flinton, L.J. (1973). *Cryobiology* **10**, 148
222 Pfisterer, H., Michlmayr, G. and Weber, F. (1968). *Cryobiology* **4**, 257
223 Pfisterer, H., Weber, F. and Michlmayr, G. (1969). *Cryobiology* **5**, 379
224 Djerassi, I., Roy, A., Kim, J. and Cavins, J. (1971). *Transfusion* **11**, 72
225 Reuter, H. and Gross, R. (1971). *Bibliotheca Haematologica* **38**, 376

226 Wybran, J., Stacquez, C. and Govaerts, A.E. (1972). *Transfusion* **12**, 413
227 Raymond, S.L., Pert, J.H. and Dodds, W.J. (1975). *Transfusion* **15**, 219
228 Spencer, H.H., Starkweather, W.H. and Knorpp, C.T. (1969). *Cryobiology* **6**, 285
229 Spencer, H.H., Starkweather, W.H. and Knorpp, C.T. (1969). *Clinical Research* **17**, 344
230 Choudhury, C. and Gunstone, M.J. (1978). *Cryobiology* **15**, 493
231 Greiff, D., Brooker, D. and Mackey, S. (1969). *Cryobiology* **6**, 194
232 Greiff, D. and Mackey, S. (1970). *Cryobiology* **7**, 9
233 Leibo, S.P. and Mazur, P. (1971). *Cryobiology* **8**, 447
234 Mazur, P. (1977). *Cryobiology* **14**, 251
235 Farrant, J., Walter, C.A., Lee, H. and McGann, L. (1977). *Cryobiology* **14**, 273
236 Meryman, H.T. (1974). In *Platelets: Production, Function, Transfusion and Storage*, p. 339. M.G. Baldini and S. Ebbe (eds.). New York: Grune and Stratton
237 Bakry, M. and McGill, M. (1977). *Cryobiology* **14**, 699
238 Crowley, J.P., Rene, A. and Valeri, C.R. (1974). *Blood* **44**, 599
239 Undeutsch, J., Reuter, H. and Gross, R. (1975). *Thrombosis Research* **6**, 459
240 Schiffer, G.A., Whitaker, C.L., Schmukler, M., Aisner, J. and Hilbert, S.L. (1976). *Thrombosis and Haemostasis* **36**, 221
241 Odink, J. (1976). *Thrombosis and Haemostasis* **36**, 182
242 Odink, J. and Brand, A. (1977). *Cryobiology* **14**, 519
243 Odink, J. and Brand, A. (1977). *Transfusion* **17**, 203
244 Odink, J., deWitt, J.J., Janssen, C.L. and Prins, H.K. (1978). *Transfusion* **18**, 21
245 Odink, J. and Sprokholt, R. (1976). *Thrombosis and Haemostasis* **36**, 192
246 Odink, J., Sprokholt, R. and Offerijns, F.G.J. (1974). *Cryobiology* **11**, 564
247 Rowe, A.W. and Lenny, L.L. (1977). *Cryobiology* **14**, 699
248 Rowe, A.W., Lenny, L.L. and Reich, L.M. (1978). *Cryobiology* **15**, 698

CHAPTER SIX

Preservation of Leucocytes

Stella C Knight

Introduction

The successful and by now routine recovery of functional lymphocytes following storage in liquid nitrogen contrasts sharply with the extreme sensitivity of granulocytes to damage during procedures that are aimed at but have not yet reached the goal of preservation. Despite the difference between the results of preservation techniques for the two types of cell, there is evidence that both require, for their protection, a protocol during which some cell water is removed. Lymphocytes shrink as a result of the extracellular hypertonic solution when the freezing technique allows ice to form outside the cells but not inside them. Cells as grossly shrunken during freezing as that shown in Figure 6.1 are then recovered functionally intact [1]. When 10% of granulocytes are shrunken during freezing following a particular cooling technique as in Figure 6.2, a similar proportion of these cells will actively ingest particles by phagocytosis on thawing. Alternatively, when a large amount of ice is formed within cells during freezing and no shrinkage is apparent (Figures 6.3 and 6.4) recovery of function is not obtained.

The mechanism by which cell shrinkage during freezing also allows recovery of living cells is not understood completely (see Chapter 1). However, whether protection is due to the prevention of intracellular ice formation, by some effect on the constituents of the cell membranes or via osmotic stresses on the cells during thawing, the protection produced by the complex and interacting conditions of freezing and thawing appears to be related to the amount of shrinkage achieved. Variables such as sample volume, amount and type of cryoprotective agent, and the cooling and rewarming procedures are all inter-related and can be manipulated to provide many combinations of circumstances which will allow a suitable degree of shrinkage for successful storage of lymphocytes.

However, a high proportion of granulocytes examined in the frozen state are found to have ice inside them after the use of cooling techniques which would allow adequate shrinkage of lymphocytes (Figure 6.3). In this chapter some of the variables of preservation procedures applied to lymphocytes and attempted with granulocytes are discussed. But systems allowing good recovery of stored cells can not be worked out until methods of assessing this recovery are available.

Figures 6.1–4 Electron micrographs of human peripheral blood leucocytes which have been cooled in 5% dimethylsulphoxide (0.2 ml samples in plastic vials) by holding them in air in a refrigerator set at −26°C before plunging them to −196°C. The samples were then freeze-substituted at −80°C and prepared for electron microscopy as previously described [1]. The bar lines represent 3 μm

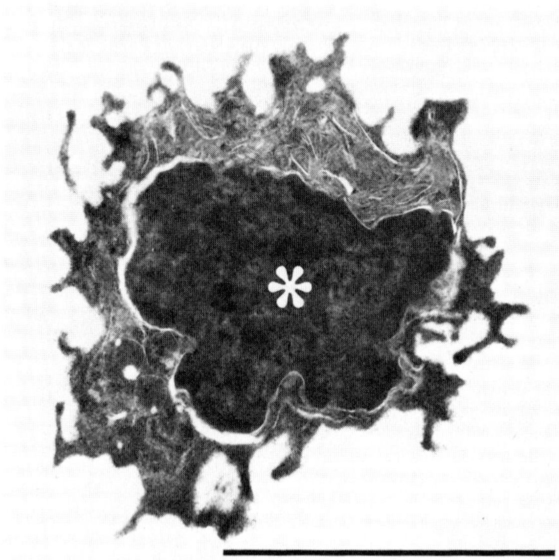

Figure 6.1 A lymphocyte which was held for 60 minutes in the refrigerator followed by rapid cooling to −196°C. Compared with non-frozen cells this cell is very shrunken with numerous villous-like protrusions and has extremely condensed nuclear material (white asterisk). Cells with this dehydrated appearance survive on thawing [1]

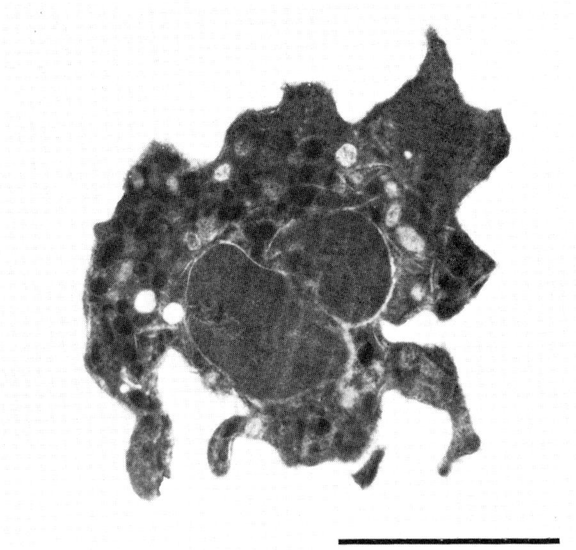

Figure 6.2 A granulocyte which was held for 60 minutes in the refrigerator set at −26°C followed by rapid cooling to −196°C. Approximately 10% of the cells had this shrunken appearance following freeze-substitution. A similar proportion of the granulocytes retained the ability to phagocytose particles and reduce the dye nitro-blue tetrazolium to formazan after this cooling procedure

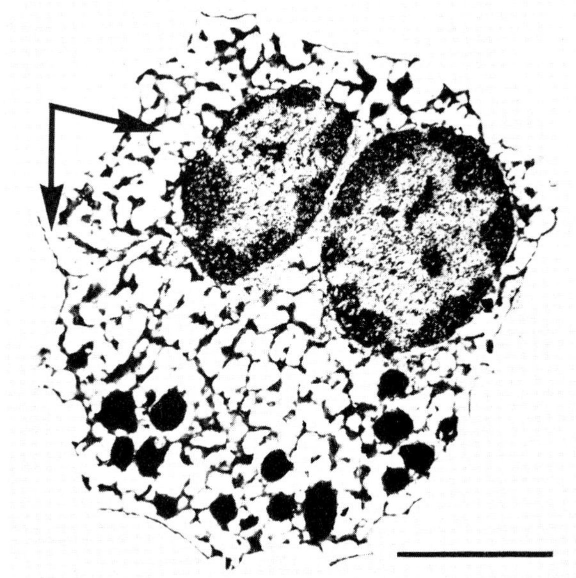

Figure 6.3 This granulocyte was held for 60 minutes in the refrigerator followed by rapid cooling to −196°C. It contains intracellular ice cavities following the formation of ice within the cell during the cooling procedure (arrows). Ice cavities were found in 90% of granulocytes

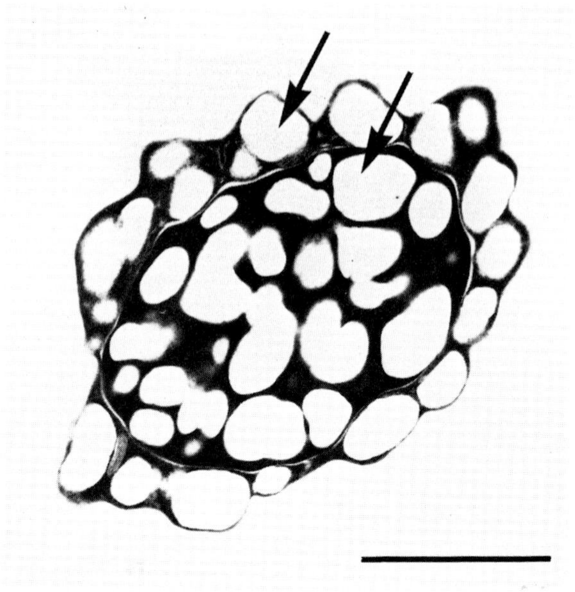

Figure 6.4 This lymphocyte was held for only 15 minutes in the refrigerator followed by rapid cooling to −196°C. This time was insufficient for the sample to reach −26°C and no shrinkage occurred. Large cavities where ice was present after the cooling procedure are seen within the cytoplasm and the nucleus of the cell (arrows). These cells did not survive the cooling and rewarming procedures

Assessing recovery of stored leucocytes

Two aspects of assaying the recovery of cells following storage cause confusion and inaccuracy. First, attempts are often made to use the recovery obtained by one assay (e.g. dye exclusion) in the estimate of recovery by another assay (e.g. DNA synthesis). Second, when assaying the functions of a whole population of cells confusion can occur between changes in the number of cells recovered and changes in the activity of individual cells. In relation to the former question, there has been much discussion concerning the most appropriate method of assessing 'viability' of leucocytes following storage. Counts of cells under phase contrast microscopy, dye exclusion tests and the use of fluorescent dyes have all had their proponents. Even using these direct counting methods, however, there are differences in 'recovery' measured, confirming that the only test for the integrity of a particular function after freezing and thawing is a test for that function itself.

This can be illustrated by comparing the levels of recovery of granulocytes obtained when different assay systems are used. For example, using one procedure, 80% of granulocytes excluded dye following freezing although only 5% were capable of phagocytosis [2]. Although an improvement of storage procedures can result in a much higher recovery of granulocytes able to phagocytose particles [3], the majority of cells able to perform this function may be unable to reduce the dye nitro-blue tetrazolium to formazan. As the recovery of granulocytes after freezing and thawing has been extremely poor when judged by their activities in the more demanding *in vitro* tests, it is unfortunately premature to use frozen and thawed granulocytes for transfusions into leucopenic patients. The main emphasis in granulocyte freezing continues to be the manipulation of the freezing conditions in order to improve recovery in the *in vitro* assays. In the final analysis however, it will be the performance of the granulocytes *in vivo* that will be the yardstick for assessing competence of the preserved cells for clinical use.

Many assays do not record the functions of individual cells but the net effect of a population of cells in the sample. In some systems, particularly those measuring the growth of cells, the response does not increase linearly with numbers of cells present. When a response of a sample is dependent on the numbers of cells in this way the relationship between cell concentration and the function has to be considered. If it is not taken into account false estimates of recovery can result. The responsiveness of lymphocytes to stimulation *in vitro* with a mitogen as measured by the uptake of ^{3}H-thymidine into DNA can be used to illustrate this point. In fact, in this type of assay both major problems in assaying cells are encountered. Conventionally, post-thaw cells are cultured at a single concentration for the stimulation assay using an assessment of recovery from a different assay—namely dye exclusion, to set up the cell concentration. Functional survival is then estimated from the uptake of ^{3}H-thymidine without reference to the relationship between cell numbers and responsiveness.

The full consequences of this situation can be seen from the data in Figure 6.5. Increasing the concentration of non-frozen lymphocytes in culture has a marked effect on the ^{3}H-thymidine uptake into stimulated cultures which ranged from no response at all to a maximal response at a cell concentration one log unit higher. At still higher cell concentrations the response fell (Figure 6.5, open circles). The potential problems of assaying recovery of cells with this function following storage are emphasised in this example by the deliberate use of sub-optimal cooling conditions. After thawing and dilution out of the cryoprotective agent only half of the cells excluded the dye trypan

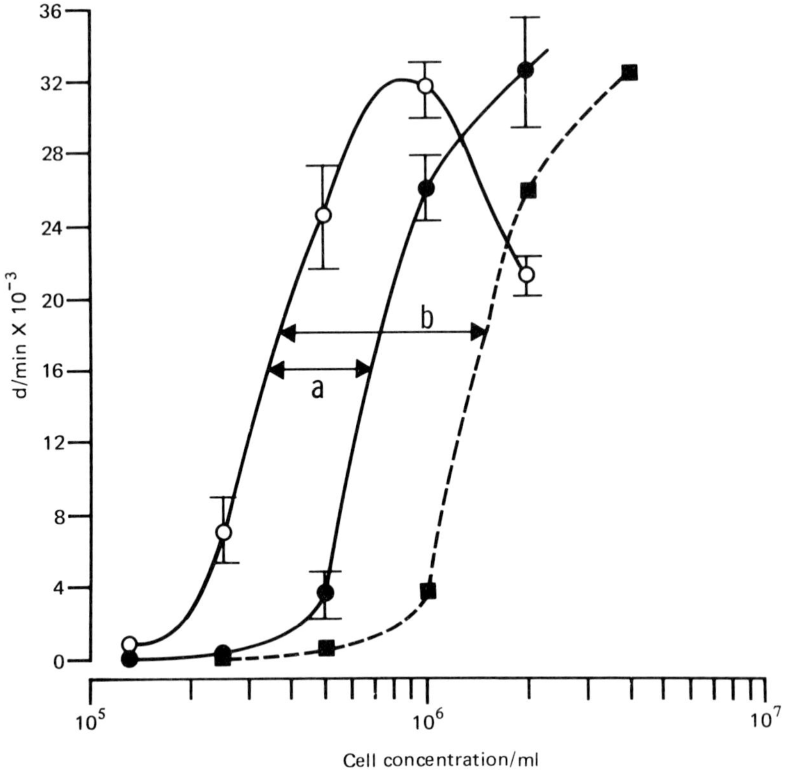

Figure 6.5 Uptake of ^{3}H-thymidine on the fourth day of culture with the mitogen phytohaemagglutinin by fresh or frozen (0.4°C/min, 5% DMSO) human peripheral blood lymphocytes (○, fresh cells; ●, dye excluding frozen cells; ■, frozen cells assuming 100% survival of cell numbers)

blue and the cells were cultured using a range of concentrations of dye excluding cells (Figure 6.5, closed circles). The response curve with increasing concentrations of dye excluding thawed cells paralleled that of the non-frozen cells in that the curve had a similar slope and maximal response. However, more thawed cells than unfrozen cells were required to obtain this response. There is some evidence that the similar response curves show that functional individual thawed cells respond in the same way as non-frozen cells to stimulation [4]. The shift in the response curve thus indicates that fewer of the cells are responding. The extent of this loss of cell numbers by freezing and thawing can be assessed from the lateral shift of the response curve on the cell concentration axis (Figure 6.5, a). This emphasises again the problem of trying to extrapolate from recovery assessed by one assay (dye exclusion) to that in a second (DNA synthesis). The population of dye excluding cells after thawing contained only about half the expected number of responsive cells. To assess the total recovery of responsive cells, the cell concentration curve was set up using the counts of cells before freezing (Figure 6.5, dotted line) rather than by making an adjustment for cell loss based on the count of dye excluding cells. The lateral shift (Figure 6.5, b) now gives the total loss of cells responsive in the culture (i.e. 80% loss of cells in this example). From this same data, the consequences of comparing the uptake of ^{3}H-thymidine on stimulation of the cells before and after freezing using only one concentration of dye excluding cells in culture can be seen. At 2×10^6 cells/ml the uptake of ^{3}H-thymidine of the frozen cells is higher than that of the non-frozen cells (145%); at 1.3×10^6 cells/ml it is the same (100%) and at 5×10^5 cells/ml it is much less (16%). The apparent recovery will thus vary according to the cell concentration chosen for the assay. By contrast the curve shift method of assessing survival gives a value for recovery which is an inherent property of the sample rather than being determined by assay variables [4].

The use of different cooling conditions causes the recovery of different numbers of responding cells. Also, the relationship between the proportion of cells excluding dye and the proportion of responding cells recovered will vary with the severity of the freezing stress. A further complication is that the original position of the responses to stimulation on the cell concentration axis may differ with different stimuli in non-frozen cells. Simple loss of total cell numbers during freezing can therefore result in apparent selective effects when single concentrations of cells only are studied in the assay [4]. Reports of selection during freezing [5–7] may therefore require reassessment using curve shift methods before they can be substantiated.

A second example of the problems that result from assaying the mean recovery of a sample of cells can be seen in the investigation of possible changes to cell surface receptors for ligands after freezing. A preliminary report suggested that up to 80% of the receptors for the mitogen concanavalin A on the surface of human lymphocytes were lost following the use of routine preservation techniques [8]. This estimate was obtained from a study of the binding capacity of the whole population of cells in a sample with tritiated concanavalin A. However, when the receptors for fluoresceinated concanavalin A on individual cells were examined using a fluorescence activated cell sorter it was found that a high proportion of cells which retained the size characteristics of normal lymphocytes following freezing also had normal numbers of receptors for the concanavalin A [9]. Just as there may be a population of cells recovered from liquid nitrogen which is unaltered in terms of mitogen responsiveness there appears also to be a population of cells which are unchanged in their density of available cell surfaces receptors for concanavalin A.

Two rules of thumb may aid the assessment of recovery of a property of cells after freezing and thawing. The recovery of a particular function can only be assessed by study of that function itself and distinction should be sought between simple changes in numbers of responsive cells and changes in the efficiency of individual cells themselves.

Freezing of Lymphocytes

Current techniques for preserving lymphocytes are still based on those which appeared in the early descriptions of successful recovery of lymphocytes [10–12]. These techniques usually employ about 10% dimethylsulphoxide DMSO (or more rarely 10–15% glycerol) with 10–20% serum in a medium isotonic with the cells. Samples are then frozen using the 'traditional' cooling rate of 1°C/min [13]. Below −30°C or −40°C the cooling rate is often increased and the cells are finally stored in liquid nitrogen. Using this type of approach some cells retaining virtually all the properties of non-frozen cells can be recovered. Some of the slight variations in this technique and descriptions of some of the functions recovered may be found in other publications[5, 14–16]

An alternative to the use of continuous reduction in temperature is to put the sample with cryoprotective additive (usually a small volume sample to allow rapid heat exchange) into a constant temperature bath set at a subzero temperature which will allow extracellular ice formation. Time at this intermediate holding temperature will then protect the cells against damage on a subsequent plunge into liquid nitrogen and thawing and dilution [17]. For human lymphocytes with 5% DMSO, being at −30°C for just a few minutes will protect them against damage during a subsequent plunge into liquid nitrogen, thawing and dilution out of the DMSO [18]. The continuous slow cooling technique (cooling rate) and the use of intermediate holding temperature (two-step) presumably represent two methods of achieving the same end – namely cell shrinkage – and very many other combinations of cooling rates and holding temperatures can also produce the requisite shrinkage.

As with the storage of other cell types, the cooling conditions required to protect cells are interrelated with many variables including the amount and type of cryoprotective additive, volume of sample, type of vessel and methods of thawing, diluting and post-thaw handling. As these factors interact it is impossible to isolate their effects, so that in any cooling 'recipe' a variation in one of these factors may influence the requirements for the others. For the sake of clarity some of these factors will be considered individually and an indication of the way they influence the cooling conditions will be given.

Effects of different types and amounts of cryoprotective additives

Some cryoprotective additives are toxic to lymphocytes; this can generally be minimised by adding them at lower temperatures. Although toxicity of DMSO (10% v/v) appears to be minimal for human lymphocytes over short periods of time (e.g. 30 min) at room temperature, mouse lymphocytes are more sensitive [19]. The toxicity in the latter case can be reduced by lowering the temperature, amount of DMSO or time for which DMSO is present before freezing. This last factor, however, may have to be balanced against the time required before the additive will exert its maximum protective effect; this has been suggested to be about 20 minutes with 5% DMSO at 0°C for mouse lymphocytes [19] although other workers found that time had no effect [16]. Two studies on the possible differential effects of cryobiological techniques on subpopulations of mouse lymphocytes have suggested that T lymphocytes may be more sensitive than B lymphocytes to the osmotic stresses of addition and removal of DMSO [20, 21].

The relationship between the amount of cryoprotective additive present during cooling and the cooling conditions required for preservation (see Chapter 1) is also seen with lymphocytes. This relationship is demonstrated in Figure 6.6 which shows that the responsiveness of human lymphocytes to the mitogen, pokeweed, is recovered after freezing using slower cooling rates when the amount of DMSO in the sample is increased [5, 19].

In addition to DMSO or glycerol [10], the non-penetrating polymer polyvinylpyrrolidone (PVP) has also been used as a cryoprotectant for lymphocytes [22]. In the presence of serum only (100%) a high proportion of human lymphocytes has been recovered using a two-step method for preservation as demonstrated in Figure 6.7 [23].

Effect of sample volume and type of freezing vial

With larger samples the rate of heat exchange with the environment will be slower. Thus in two-step methods, for example, a large sample will

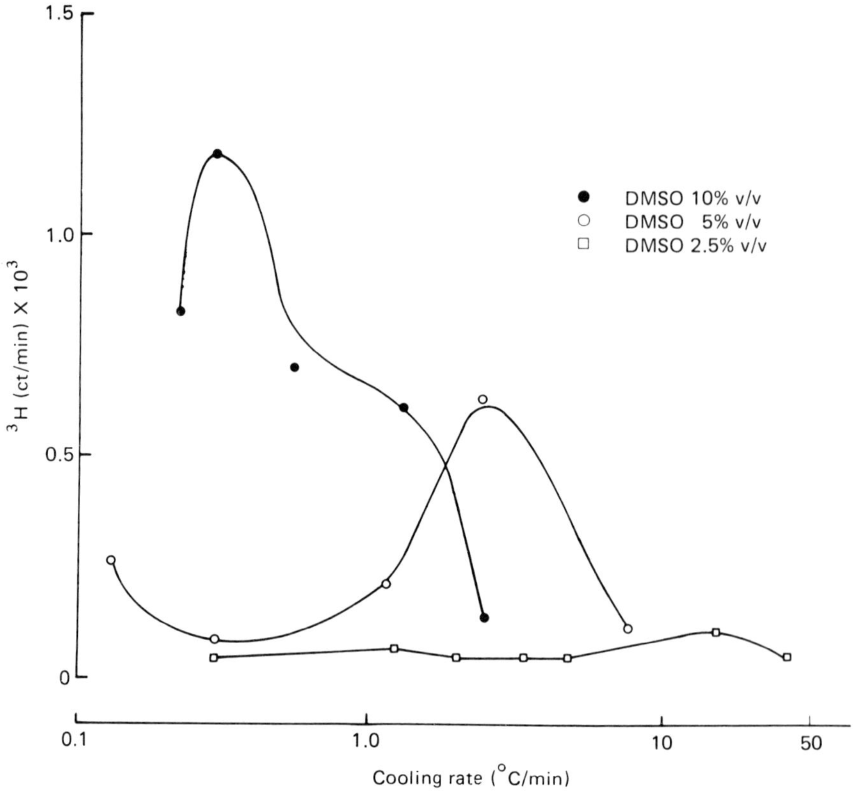

Figure 6.6 Uptake of ^{3}H-thymidine on the fourth day of culture with pokeweed mitogen of thawed human lymphocytes after cooling at different rates to −60°C with different amounts of DMSO [5]

take more time to reach the holding temperature. Similar effects are seen when plastic vials are used rather than glass ones. With human peripheral lymphocytes in a 0.25 ml sample in 5% DMSO in a 0.5 dram glass vial (Johnson and Jorgenson), 15 minutes at −26°C in an unstirred glycerol bath will protect against damage after plunging the sample into liquid nitrogen, whereas approximately 30 minutes is required for the protection of the same sample in a small polypropylene tube (Sterilin 12 × 35 mm).

The thawing of larger volume samples or those in plastic vessels will also be slower. With lymphocytes as with other cell types slow thawing is usually detrimental [19], perhaps because it will allow time for the ice crystals in the cells to grow (Chapter 1).

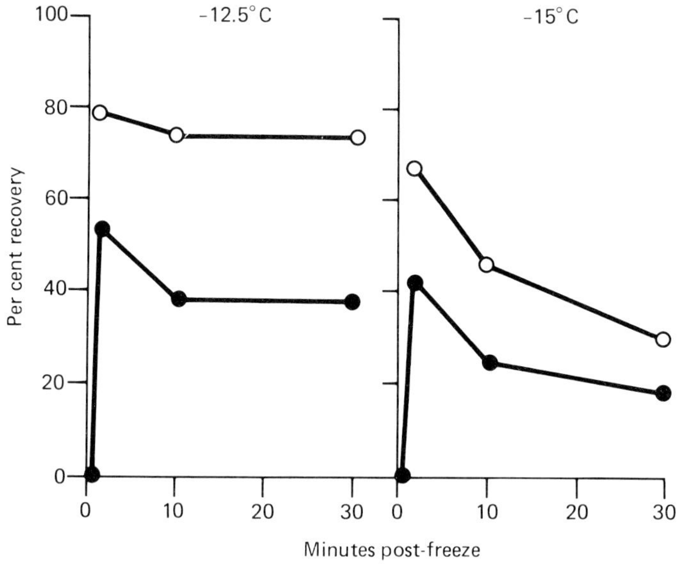

Figure 6.7 Percentage recovery of lymphocytes responding to phytohaemagglutinin (estimated from the cell concentration response curves) after freezing samples (0.2 ml in glass vials) in 100% autologous serum for different periods at −12.5°C or −15°C and either thawing directly (○) or cooling to −196°C before thawing (●)

Handling on thawing and dilution

Following most storage procedures using DMSO as a cryoprotective agent, rapid thawing to 37°C and slow dilution by dropwise addition of warm medium containing some protein seems to give optimal recovery [14, 19, 24]. Removal of DMSO from cells at low temperatures or very rapidly can result in damage to lymphocytes, presumably due to the swelling of cells because of the entry of water before the DMSO leaves. Thus, although DMSO is more toxic to lymphocytes at higher temperatures, that effect is probably less important than the osmotic problems resulting from dilution at low temperatures. The use of room temperature (20–22°C) for dilution out of DMSO provides a possible compromise between the toxic and osmotic effects of the DMSO. Following thawing and dilution, mouse lymphocytes were shown to be more susceptible than non-frozen cells to injury by mechanical trauma [19]. The marked effects that post-thaw handling technique can have on the recovery of functional lymphocytes is demonstrated in Figure 6.8 where recovery of mouse lymphocytes at suboptimal cooling rates can be much increased when post-thaw handling procedures are improved.

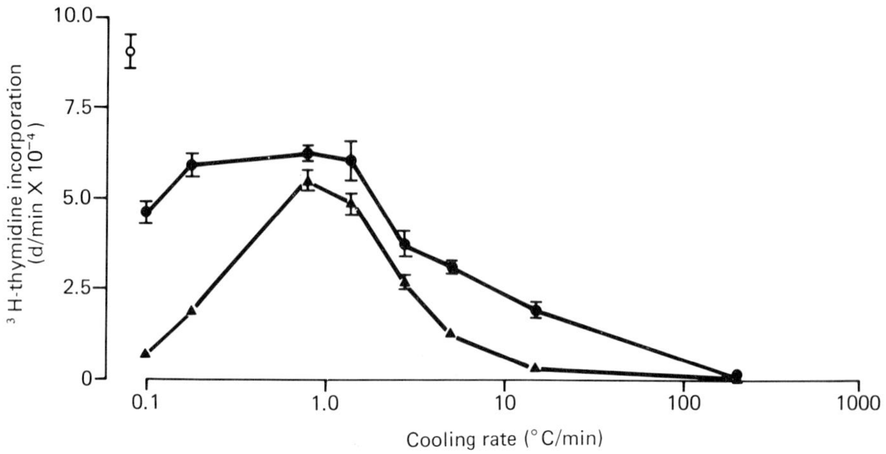

Figure 6.8 The effect of the post-thaw handling procedures on the uptake of ^{3}H-thymidine in response to stimulation with concanavalin A of mouse lymphocytes cooled at a range of cooling rates in DMSO (10% v/v) [19]. ●, optimal dilution procedure (2.5 minutes for a 10-fold dilution at 25°C with fetal calf serum (10%) in medium followed by centrifugation at 60 g; ▲, suboptimal dilution procedure (rapid dilution at 20°C in fetal calf serum (10%) and centrifugation at 370 g)

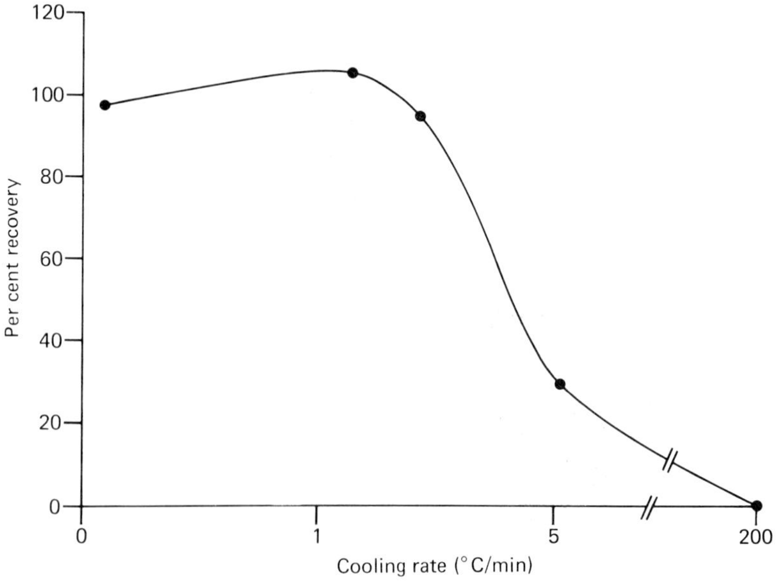

Figure 6.9 Percentage recovery estimated by the curve shift method (see section on assessing recovery) of human lymphocytes responding to phytohaemagglutinin on the third day of culture following cooling at different rates to −60°C in 0.5 ml volumes in plastic vials in 10% DMSO before plunging them into liquid nitrogen

Recovery of lymphocytes using cooling rate or two-step technique

Using the curve shift method of assessing the recovery of human peripheral blood lymphocytes responding to mitogenic stimulation (see section on assessing recovery) the recovery of cells frozen in 10% DMSO to −60°C at different cooling rates before being plunged into liquid nitrogen is shown in Figure 6.9. Figure 6.10 gives the recoveries of cells stored by a two-step method in 5% DMSO. For both techniques

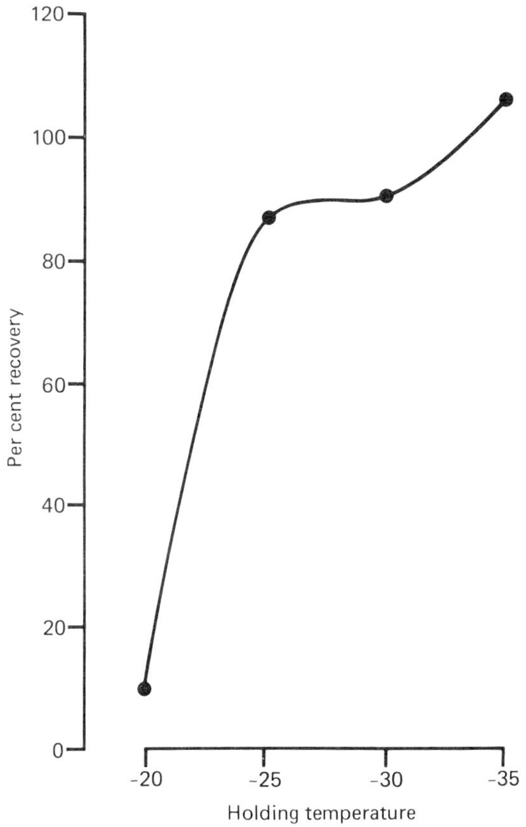

Figure 6.10 Percentage recovery estimated by the curve shift method (see section on assessing viability) of human lymphocytes responding to phytohaemagglutinin on the third day of culture after holding the samples (0.2 ml in 5% DMSO in plastic vials) for 30 minutes at different holding temperatures before plunging them into liquid nitrogen

complete recovery of this functional population was found. In another assay dependent on cell concentrations in culture the recovery of killer T cells was found by curve shift analysis to vary between 15 and 65% using a single set of 'standardised' freezing conditions [25].

When cooling conditions are damaging to the cells, some information may be gained by thawing the sample directly from an intermediate temperature, before the cells would normally be plunged into liquid nitrogen. Damage at this stage may be due to the presence of ice in the cells already or alternatively could result from too much exposure to hypertonic solutions. An example of this is seen from the 70% damage caused after 30 min at $-15°C$ when human lymphocytes are frozen with serum alone (Figure 6.7). This low survival can be improved by reducing the time of the hold or by using a higher holding temperature (Figure 6.7). In a cooling rate experiment, reducing the range of temperature over which cooling is controlled or increasing the rate of cooling might bring about an improvement. An alternative approach might be to increase the amount of cryoprotective additive present. Alternatively, if complete recovery is obtained following controlled cooling or the first step of a two-stage procedure, but damage occurs in cells plunged into liquid nitrogen, it is likely that insufficient shrinkage has occurred to the cells and that time will be required at a lower temperature to produce sufficient shrinkage for protection.

Freezing of Granulocytes

No entirely satisfactory routine method for the storage of human granulocytes is available; the recovery estimated from a variety of functional tests following many different cooling schedules is disappointing (see references 26, 27). An example of the recovery of granulocytes frozen in 5% DMSO over a range of cooling rates is shown in Figure 6.11. The best recovery was 10% obtained at 0.3°C/min. If the usual relationship between amount of cryoprotective additive and optimal cooling conditions (e.g. Figure 6.6) holds true for granulocytes, then rates slower than 0.3°C/min would be required for optimal recovery with a higher concentration of DMSO. Using the same assay to assess any recovery of granulocytes a two-step method with 0.2 ml samples containing 10% DMSO was done. This gave recovery which never exceeded 30% with a holding temperature between -30 and $-40°C$ for 2 hours before the sample was cooled in liquid nitrogen. Recovery within this range is not considered adequate for clinical transfusions [28, 29]. The loss of functional cells occurs early during freezing as

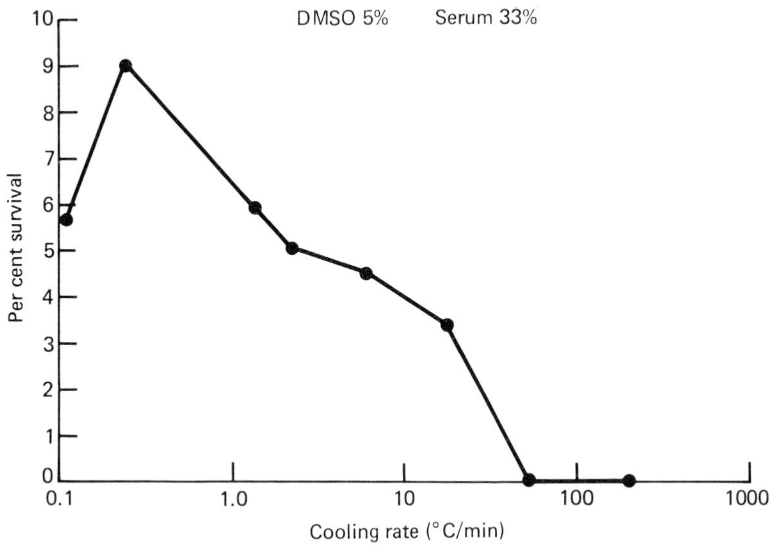

Figure 6.11 Samples of human leucocytes (0.2 ml in 5% DMSO in 0.5 dram glass vials, Johnson and Jorgenson) were cooled at different rates to −60°C before being plunged to −196°C. Recovery is the percentage of cells counted which after a two hour incubation had phagocytosed yeast particles (boiled for 20 minutes and added to 1-2% nitro-blue tetrazolium in saline) and reduced the dye on them to insoluble blue-black formazan

shown in Figure 6.12 where samples were frozen and cooled to different temperatures. The majority of thawed cells were already injured when rewarmed from conditions before those where protective shrinkage might have occurred. Some protection of cells against damage on subsequent plunging into liquid nitrogen and thawing was seen at temperatures between −20 and −30°C. From the work of Rapatz and Luyet [30] and as demonstrated in Figure 6.3, many granulocytes may have ice crystals which form within the cells at comparatively high subzero temperatures — perhaps even at −5°C — so preventing shrinkage of the cells. Thus, more successful storage of granulocytes may require conditions which prevent the formation of ice inside the cells.

A second problem with granulocytes is their extreme sensitivity to osmotic stresses [31, 32]. Thus, only 11% of granulocytes were able to reduce the dye nitro-blue tetrazolium after 10 min exposure to 0.5 M sodium chloride, although lymphocytes were not damaged by this

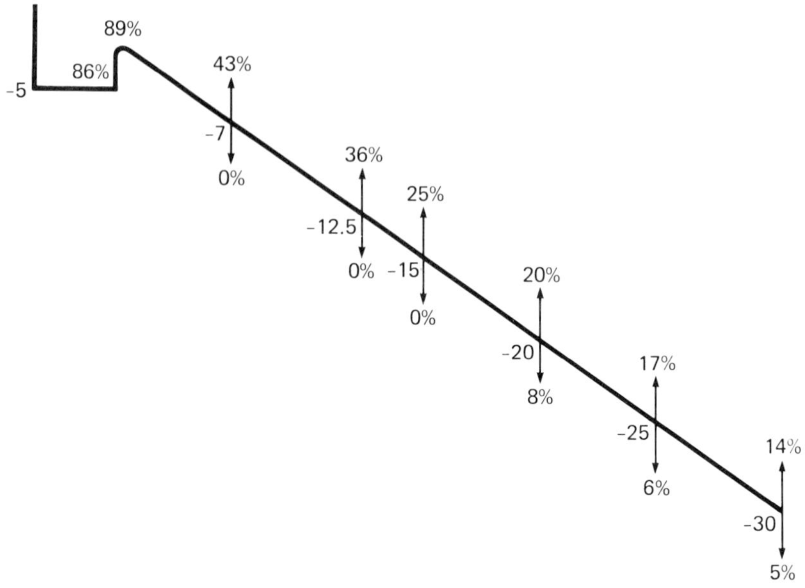

Figure 6.12 Samples of granulocytes (0.2 ml, 5% DMSO) were frozen at −5°C and then cooled at a rate of 0.3°C/min. Percentage recovery (see legend Figure 6.11) was assessed in samples thawed either directly from different temperatures during cooling or after plunging the sample to −196°C from different temperatures during cooling

treatment. Crowley and his colleagues [33] observed that after freezing and thawing every granulocyte had an outpouching of the cell membrane which suggested an osmotic disturbance of the cells.

Thus, although a proportion of granulocytes can be recovered intact from liquid nitrogen, complete recovery of more demanding functions such as bacterial killing has not been obtained. There are at least two major problems which may be related. First, intracellular ice probably occurs at relatively high subzero temperatures so preventing shrinkage of the cells and second, granulocytes are exquisitely sensitive to osmotic stresses.

The injurious effects of cooling procedures on granulocytes may be peculiar to human cells as a higher proportion of dog granulocytes can be recovered from liquid nitrogen [34].

Conclusions

Many successful methods for long term storage of lymphocytes are available ranging from cooling rate to two-step techniques. Complete recovery of lymphocyte function can be obtained using assays as demanding as lymphocyte stimulation in culture. Further studies using more reliable assay techniques are required to confirm whether storage procedures exert selective effects on lymphocyte subpopulations.

In contrast, although some functioning granulocytes can be found after storage in liquid nitrogen, a satisfactory method for recovering a high proportion of human granulocytes suitable for transfusion into leucopenic patients has not been achieved. Thus, while more ambitious ideas such as the recovery of functional cells following freeze-drying are now being pursued for lymphocytes [35], most granulocytes still stubbornly refuse to recover following attempts at the more modest goal of preservation in liquid nitrogen.

Acknowledgements

Dr. C.A. Walter kindly provided the electron microscope pictures of the freeze-substituted leucocytes.

References

1 Walter, C.A., Knight, S.C. and Farrant, J. (1975). *Cryobiology* **12**, 103
2 Cavins, J.A., Djerassi, I., Roy, A.J. and Klein, E. (1965). *Cryobiology* **2**, 129
3 Cavins, J.A., Djerassi, I., Aglieu, E. and Roy, A.J. (1968). *Cryobiology* **5**, 60
4 Knight, S.C. and Farrant, J. (1978). *Journal of Immunological Methods* **22**, 63
5 Farrant, J., Knight, S.C. and Morris, G.J. (1972). *Cryobiology* **9**, 516
6 Knight, S.C., Farrant, J. and Morris, G.J. (1973). In *Cryopreservation of Normal and Neoplastic cells,* p. 133. R.S. Weiner, R.K. Oldham and L. Schwarzenberg (ed.). Paris: INSERM
7 Jewett, M.A.S., Gapta, S., Hansen, J.A., Cunningham-Rundles, S., Siegal, F.P., Good, R.A. and Dupont, B. (1976). *Clinical and Experimental Immunology* **25**, 449
8 Bryan, D.J., Fuller, T.C. and Huggins, C.E. (1976). *Cryobiology* **13**, 655
9 Bryan, D.J., Knight, S.C. Edwards, A.J. and Farrant, J. (1979). *Cryobiology* **16**, 1
10 Atkins, L. (1962). *Nature* **195**, 610
11 Ashwood-Smith, M.J. (1964). *Blood* **23**, 494
12 Pegg, P.J. (1965). *British Journal of Haematology* **11**, 586
13 Polge, C., Smith, A.U. and Parkes, A.S. (1949). *Nature* **164**, 666
14 Eijsvoogel, V.P. du Bois, M.J.G.J., Wal, R., Huismans, D.R. and Raat-Koning, L. (1973). In *Cryopreservation of Normal and Neoplastic Cells,* p. 101. R.S. Weiner, R.K. Oldham and L. Schwarzenberg (ed.). Paris: INSERM
15 Strong, D.W. and Sell, K.W. (1977). In *Cryoimmunology,* p. 81. D. Simatos, D.M. Strong and J. Turc (ed.). Paris: INSERM
16 Birkeland, S.A. (1976). *Cryobiology* **13**, 442
17 Luyet, B. and Keane, J. (1955). *Biodynamica* 7, 281

18 Farrant, J., Knight, S.C., McGann, L.E. and O'Brien, J. (1974). *Nature* **249**, 452
19 Thorpe, P.E., Knight, S.C. and Farrant, J. (1976). *Cryobiology* **13**, 126
20 Strong, D.M., Ahmed, A., Sell, K.W. and Greiff, D. (1974). *Cryobiology* **11**, 127
21 Thorpe, P.E. (1976). PhD Thesis, Council for National Academic Awards, UK
22 Thomas, D., Edwards, D.C. and Damjanovic, V. (1976). *Cryobiology* **13**, 191
23 Knight, S.C., Farrant, J. and McGann, L.E. (1977). *Cryobiology* **14**, 112
24 Weiner, R.S. (1976). *Journal of Immunological Methods* **10**, 49
25 Lambalgen, R., Farrant, J. and Bradley, B.A. (1979). *Journal of Immunological Methods* **27**, 327
26 Knight, S.C., O'Brien, J. and Farrant, J. (1976). In *Cryoimmunology*, p. 139. D. Simatos, D.M. Strong and J. Turc (ed.). Paris: INSERM
27 Roy, A.J. (1978). *Cryobiology* **15**, 232
28 Lionetti, F.J., Hunt, S.M., Gore, J.M. and Curby, W.A. (1975). *Cryobiology* **12**, 181
29 Graham-Pole, J., Davie, M. and Willoughby, M.L.N. (1977). *Journal of Clinical Pathology* **30**, 758
30 Rapatz, G. and Luyet, B. (1971). *Biodynamica* **11**, 69
31 Rapatz, G. and Luyet, B. (1971). *Biodynamica* **11**, 53
32 Skeel, R.T., Yankee, R.A., Spivak, W.A., Novikovs, L. and Henderson, E.S. (1969). *Journal of Laboratory and Clinical Medicine* **73**, 327
33 Crowley, J.P., Rene, A. and Valeri, C.R. (1974). *Cryobiology* **11**, 395
34 French, J.E., Flor, W.J., Grissom, M.P., Parker, J.L., Sajko, G. and Ewald, W.G. (1977). *Cryobiology* **14**, 1
35 Thomas, D. and Zola, H. (1977). *Cryobiology* **14**, 45

CHAPTER SEVEN

Bone Marrow Stem Cells

U W Schaefer

Introduction

Progress in transplantation biology has renewed the interest in the transplantation and preservation of bone marrow. Although the graft versus host disease (GVH) is still the main obstacle in allogeneic bone marrow transplantation, there is hope that immunological incompatibility reactions might be better controlled in the future. When autologous bone marrow is transplanted graft rejection and GVH reaction do not occur. Therefore the use of cryopreserved haemopoietic stem cells as an autograft in the management of malignant diseases is an attractive concept. This approach would permit the administration of considerably larger doses of radiation or cytotoxic chemotherapy. By retransfusion of autologous marrow cells the toxic limitations of the aggressive therapy might be overcome. In allogeneic situations the need for marrow preservation is less urgent but the possibility of storing typed marrow of blood bank donors or cadavers could provide therapeutic chances if no related donor is available. When cell separation techniques are applied to remove from the graft immunocompetent lymphocytes or tumour cells a reliable preservation method would permit the storage of different marrow fractions until the results of specific *in vitro* assays are known.

Since the initial observation of Barnes and Loutit many authors have shown in a variety of experimental systems that haemopoietic cells can be cryopreserved. In 1955 Barnes and Loutit demonstrated that the injection of frozen-thawed isologous spleen cells led to the recovery of lethally irradiated mice [1]. Following the preservation method devised by Polge *et al* [2] for spermatozoa and erythrocytes, Barnes and Loutit used 15% glycerol in serum as the cryoprotective medium. The cells were frozen in ampoules at a freezing rate of 1°C/min to −15°C, then the temperature was allowed to fall to −70°C at not more than 10°C/min. Before use the ampoules were thawed rapidly in a water bath at 37°C.

This technique of cryopreservation has been extended by many others. In 1961 Ashwood-Smith [3] was able to preserve mouse bone marrow by freezing in dimethylsulphoxide (DMSO), found to protect red cells by Lovelock and Bishop two years before [4]. In 1962 Persidsky and Richards reported that polyvinylpyrrolidone (PVP) protects mouse bone marrow against freezing damage [5]. Van Putten found in 1965 that the combination of intra- and extracellular protectants can be very effective [6]. Using 10% glycerol + 10% PVP 72–87% of the transplantation potential of mouse bone marrow could be recovered.

Viability tests

When dealing with bone marrow preservation the major question is whether the stem cell population has survived the freezing damage. Haemopoietic stem cells capable of self-replication and differentiation are responsible for the reconstitution of haemopoietic tissues after bone marrow transplantation. The best way to evaluate the efficacy of marrow preservation methods is therefore to test the repopulating potential of the stored cells after transplantation into lethally irradiated recipients. Viability tests on the base of morphology, dye exclusion, supravital staining, thymidine incorporation etc. do not reflect the capacity of the cells for unlimited proliferation.

In mice the spleen colony technique as described by Till and McCulloch [7] is an easy and reliable quantitative stem cell assay. It is based on the observation that intravenous injection of haemopoietic cells into irradiated mice results in the formation of distinct cell colonies in the spleen (CFU-S). Repopulation studies after lethal total body irradiation and the CFU-S technique proved to be useful guides for choosing cryoprotective agents, rates of freezing and thawing and storage temperature.

Mice

A number of authors have reported on surprisingly good stem cell recovery in cryopreserved bone marrow of mice. Under well controlled conditions survival rates of 65 – 100% are obtained with glycerol, DMSO or PVP [3, 8–11]. Extensive investigations by Lewis *et al* [12] and Leibo *et al* [10] have shown that the rates of freezing and thawing are critical. When a suboptimal freezing rate was used slow warming was more deleterious than rapid warming. Generally, a slow cooling rate of 1–2°C/min, a short heat of fusion period, a very low storage temperature and rapid warming have been found most suitable for marrow preservation.

Lewis *et al* who reported a mean CFU-S recovery of 94.54%, used 12% glycerol + 4% calf serum as the protective medium [11]. The cooling rate was 2°C/min and the storage temperature −100°C. The cells were thawed rapidly in a water bath of 40°C. After thawing the cells were resuspended in 35% glucose, then increasing volumes of 6% dextran were added. Leibo *et al* also deglycerolised the cells after thawing by a stepwise dilution in Tyrode-glucose solution [10]. These authors found a CFU-S recovery of 65% when the cells were frozen in 1.25 M glycerol at a cooling rate of 1.7°C/min.

Van Putten performed repopulation studies in lethally irradiated mice and compared different cryoprotective media and different resuspension techniques in the post-thawing phase [6]. After thawing, the cells were handled in one of three different ways. They were stepwise slowly diluted according to the Drášil method [13], diluted rapidly after addition of hypertonic glucose according to the Sloviter technique [14] or they were quickly diluted in Tyrode without special precautions. With DMSO and glycerol the slow dilution method was better than the other two techniques. Using PVP or a combination of PVP and glycerol a stem cell recovery of 70–100% has been achieved without special dilution.

Phan The Tran *et al* have studied the ability of a variety of compounds to protect mouse bone marrow cells against damage by freezing and thawing [15]. Dextran, bovine serum albumin and several polyhydric alcohols (*iso*-erythritol, *D*-mannitol, *D*-sorbitol, *D*-ribitol) had protective effects. Another interesting cryoprotective agent is hydroxyethyl starch (HES) [16–18]. Persidsky and Ellett investigated this compound in rat bone marrow [19]. Using the ^{14}C-1-glycine incorporation as the viability assay they found a 22% recovery of incorporation after cryopreservation in HES.

Larger animals

Storage techniques which were found to be effective for mouse bone marrow were later tested for their efficiency in dogs, monkeys and man. As with mice the transplantation potential of fresh and cryopreserved marrow in larger animals can be quantitatively compared in lethally irradiated recipients. Several groups performed transplantations in dogs [20–25] or monkeys [6, 26–28] grafting frozen-thawed marrow cells. Mannick *et al* injected into lethally irradiated dogs autologous bone marrow which had been frozen in 15% glycerol [20]. After thawing, these authors reduced the concentration of glycerol according to the Sloviter technique by addition of three volumes of dextrose solution. Grafting with 2.4–4.0 $\times$ 10^9 marrow cells per recipient they achieved haemopoietic reconstitutions whereas fewer cells were not effective. A quantitative evaluation of the storage efficiency, however, is not possible because the authors did not mention the weight of the recipients nor the minimal effective dose of fresh marrow. In parallel *in vitro* experiments, DNA synthesis was determined by measuring the incorporation of ^{14}C-formate into the thymine isolated for DNA. Between 40 and 50% of the original ability to synthesise DNA was preserved in the frozen samples but as shown in mice by Lewis *et al* [29] and pointed out earlier there is no evidence that DNA synthesis reflects the number or viability of the cells responsible for repopulation. Cavins *et al* grafted into lethally irradiated dogs autologous bone marrow which had been frozen in 5–15% DMSO [21]. They injected the marrow cells immediately after thawing without any precautions. Haemopoietic recovery could be demonstrated if 1.1–5.2 $\times$ 10^9 marrow cells per recipient (4.5–13.5 kg body weight) were injected. When 1.1–2.0 $\times$ 10^9 cells were given, three out of nine dogs did not show takes. Similar studies using allogeneic marrow have been published by Storb *et al* [22]. These authors preserved the marrow in 10% DMSO and injected the thawed cells without any dilution procedure. In four cases donor and recipient were litter mates, in six cases they were unrelated. The cell dose administered varied from 9.8 $\times$ 10^9 to 30.0 $\times$ 10^9 per recipient (7–12 kg b.wt.) In all dogs marrow takes could be achieved. Those animals grafted with HLA incompatible cells from unrelated donors rejected the transplant later. Assuming a mean recipient weight of 10 kg, 9.8–19.8 $\times$ 10^8 cells per kg b.wt. were grafted in the litter mate combinations and 16.7–30.0 $\times$ 10^8 per kg in the unrelated animals.

In all of these investigations no comparison with the minimal efficient cell dose of fresh marrow was attempted, and therefore quantitative conclusions of the stem cell survival cannot be drawn. According to other authors the minimal cell dose for reproducible haemopoietic reconstitution of lethally irradiated dogs is 1–5 $\times$ 10^7 per kg b.wt. for autologous and 2–3 $\times$ 10^8 per kg for for allogeneic bone marrow [23, 24, 30, 31]. Taking these numbers into account the doses of thawed cells which have been applied in the studies already mentioned exceeded the minimum dose of fresh marrow several times. Recently Gorin *et al* [23] and Appelbaum *et al* [24] performed dose titration studies in irradiated dogs comparing fresh and frozen marrow cells directly. The minimum dose of fresh bone marrow for autologous grafting was between 0.1 and 0.25×10^8 nucleated cells per kg b.wt. All animals receiving greater than 0.5×10^8 nucleated bone marrow cells per kg, stored up to 5 months, survived. The marrow became less able to repopulate the host with cryopreservation times of more than 9 months. The authors used a freezing rate of 1°C/min from room temperature to −50°C. The marrow was protected by 10% DMSO + 5% fetal calf serum or autologous plasma and kept in the vapour phase of liquid nitrogen. After thawing in a water bath at 37°C the cells were infused immediately without attempts to remove the DMSO *in vitro*.

Cryopreserved mononuclear cells from the peripheral blood have also been shown to repopulate dogs after pretreatment with irradiation or chemotherapy. Haemopoietic reconstitution could be obtained with autologous [32–35] as well as with allogeneic [36] cells frozen in 10% DMSO. However, a direct quantitative analysis of the transplantation potential after storage was not performed in these studies. Sarpel *et al* found in parallel agar culture experiments a 92% ± 12% recovery of the colony forming capacity [35].

In rhesus monkeys van Putten [6, 26], Buckner *et al* [27] and Merritt *et al* [28] investigated the transplantation potential of cryopreserved bone marrow. In these studies the marrow was transfused immediately after thawing without special precautions. Buckner *et al* preserved autologous marrow in 10% DMSO and injected $0.2–0.7 \times 10^9$ cells per kg b.wt. [27]. A haemopoietic reconstitution was observed in all animals. Merritt *et al* also used DMSO. They achieved takes of autologous marrow after administration of $0.5–4.0 \times 10^8$ cells per kg b.wt. [28]. Only the data of van Putten permit a quantitative analysis [6, 26]. This author made a comparison with the minimal efficient dose of fresh marrow which is 4×10^7 per kg for autologous and 4×10^8 per kg for Rh-incompatible allogeneic combinations. Only a few takes could be demonstrated by van Putten with 10% PVP or 10% PVP + 10% DMSO and none using 15% glycerol or 15% DMSO. The best protective medium (10% PVP + 10% glycerol) preserved 50% of the stem cell capacity in the autologous situation. With allogeneic marrow van Putten could not observe reproducible takes although he increased the cell dose up to 16×10^8 per kg b.wt. In view of these results van Putten concluded that there are species differences and extrapolation to man of the results obtained with bone marrow storage in experimental animals should be done with caution.

Man

In the past, several groups have tried the effect of reinfusion of preserved autologous bone marrow in patients who suffered from malignant diseases and had to be treated with high doses of radiation or cytotoxic chemotherapy [37–44]. In some instances there were indications of a therapeutic benefit but overall the early clinical trials were disappointing or not very conclusive. In the autologous situation it is very difficult to prove takes of a frozen transplant because genetic markers are lacking. Under these circumstances it is often impossible to differentiate between a take and a spontaneous recovery of the autochthonous cell population. As in the animal experiments a variety of so-called viability tests has been applied to measure *in vitro* the storage efficiency of different techniques. As already mentioned morphology, dye exclusion, or DNA synthesis do not necessarily reflect stem cell viability. Malinin *et al* investigated the cytomorphology of human marrow which had been stored in glycerol for 3–9 years at temperatures between −69 and −79°C or −120 and −196°C [45]. Most of the cells kept at very low temperatures seemed morphologically intact whereas storage at higher temperatures resulted in severe structural alterations. Lochte *et al* [46], Pyle and Boyer [47], Sultan and Soulas [48] and Mosimann *et al* [49] used determination of DNA synthesis as a viability assay. The recovery rates were variable and low. Mosimann *et al* measured the ^{3}H-thymidine and ^{3}H-leucine incorporation. They compared the protective effects of glycerol, DMSO and PVP using a stepwise dilution technique after thawing. The highest incorporation rates of 50% of the fresh controls were found, when the marrow was frozen in 15% glycerol or in the combination of 10% glycerol and 10% PVP. In the case of 10% DMSO or 10% PVP the rates were 40% or 10% of the controls. Comparisons of rapid and stepwise dilution techniques revealed that the slow dilution method was superior when glycerol, PVP + glycerol or DMSO were used. Collmann and Boll [50] and Adamson and Storb [51] examined the proliferative potential of human bone marrow in suspension cultures after cryopreservation in DMSO. They demonstrated that the erythropoietic cell class was morphologically and functionally preserved.

Much better information on the stem cell viability of stored bone marrow can be obtained *in vitro* by the agar culture technique first described by Pluznik and Sachs [52] and by Bradley and Metcalf [53]. Both groups found that under controlled culture conditions haemopoietic cells of mice show clonal growth. The culture technique was modified later and adapted to the different species. It is accepted that the number of colony forming units in culture (CFU-C) is a quantitative parameter of the stem cell activity although only committed stem cells are measured.

Using the agar technique as modified by Pike and Robinson [54], Berthier and Marcille examined the proliferative capacity of human marrow frozen in 15% PVP [55]. The CFU-C recovery varied from 11.6 to 38.7%. Some years later the same group used DMSO as the protective compound and applied the stepwise dilution technique described by us. Under these conditions they never measured a CFU-C recovery less than 73.5%. Rapid dilution resulted in significantly lower recovery values [56]. Several reports have appeared since 1972 describing a high CFU-C recovery of 70–100% after freezing of bone marrow or peripheral mononuclear cells from normal human individuals [33, 57–62]. Most investigators used 10% DMSO, a slow cooling rate, storage at a very low temperature, rapid thawing and after thawing special dilution procedures.

CFU-C recovery after the cryopreservation of bone marrow from leukaemia patients has been studied by a number of groups [60, 62–67]. Recovery rates of 50–100% have been reported. Ragab *et al* investigated the CFU–C potential of bone marrow in patients suffering from acute lymphocytic leukaemia in remission [63]. They compared the protective effects of DMSO and glycerol using a storage temperature of $-196°C$ and a rapid dilution after thawing. Correcting the CFU-C numbers according to the recovery of mononuclear cells, a mean CFU-C recovery of 51.8% ± 7.5% (20–93%) after one week of storage was measured. After a storage time of 12 weeks the mean value decreased to 30.7% ± 12.1%. DMSO was superior to glycerol which protected only 2.3% ± 0.8% of the CFU-C potential. Balkwill *et al* studied the marrow of patients with acute myeloid leukaemia. They compared rapid and slow dilution after thawing of the cells frozen in DMSO [64]. Significantly better viability could be obtained by use of the slow dilution technique as measured by the agar method, the trypan blue test and the ^{3}H-thymidine incorporation. An increasing number of patients with leukaemia or solid tumours have received frozen bone marrow autografts. A critical review has been published recently by Graze and Gale [68]. The renewed interest follows numerous reports of successful allogeneic marrow transplantations in acute leukaemia, aplastic anaemia and combined immunodeficiency [69–71]. Theoretically the use of autologous marrow has advantages over allogeneic cells, because reactions due to tissue incompatibility will not occur. In addition compatible related donors can only be found for a relatively small percentage of patients.

The result of an autotransplantation in leukaemia is not only dependent on the storage efficiency but also on the number of residual leukaemic cells in the graft as well as on the effectiveness of the cytotoxic treatment before transplantation. After tansplantation recurrence of leukaemia can happen due to residual leukaemic cells in the preserved graft or by the regrowth of cells surviving the cytotoxic regimen administered before transplantation.

In acute leukaemia several groups have grafted cryopreserved autologous marrow after pretreatment with chemotherapy [72, 73]. In most cases the contribution of marrow transfusion, however, is unknown since bone marrow recovery after chemotherapy may occur without cell grafting. Dicke *et al,* who pretreated the patients with a supralethal regimen including total body irradiation, could demonstrate that preserved remission marrow is able to restore haemopoiesis and to induce remissions of several months [74, 75]. In 10 out of 14 patients signs of a take were observed and 5 out of 10 patients achieved complete haemopoietic recovery. Remission duration ranged from 2 months to more than 9 months. In two patients leukaemia recurred, the others died from infections or drug toxicity. Dicke *et al* used the preservation technique described by us. They prepared CFU-C enriched marrow fractions obtained by the albumin gradient technique [91–93]. The fractions were stored in 10% DMSO for periods from 5 months to 30 months. The number of CFU-C per 10^5 cells transplanted varied from 1 to 30, whereas the grafted cell number was $8 \times 10^6 - 3 \times 10^8$ per kg b.wt. The lowest number of CFU-C successfully grafted was 5000 per kg b.wt.

In chronic myeloid leukaemia attempts have been made to treat the blastic crisis with intensive cytotoxic therapy followed by reinfusion of marrow or peripheral blood cells harvested during the chronic phase of the disease [68, 76, 77]. The Seattle group applied a supralethal chemoradiotherapy before transplantation [76]. The marrow cells were frozen in 10% DMSO and stored for 8–32 months. The thawed cells were injected without removal of DMSO. The number of nucleated cells grafted varied from 15×10^9 to 313×10^9. Two patients failed to achieve marrow repopulation, three patients had a partial marrow recovery. Two patients achieved prompt and complete re-establishment of the chronic phase of CML, one died on day 72 with infection and one developed blastic transformation within 4 months.

Also, in other malignancies autologous bone marrow transplantation has been applied as a supportive measure after high dose radio- or chemotherapy. Early trials provided evidence that stored autologous marrow could be injected without significant side effects, but the therapeutic effect was difficult to analyse. Adequate viability assays of the stem cell population were not available and the cytoreductive therapy was often suboptimal. More recently, several groups initiated studies to reinvestigate the value of autologous bone marrow transplantation in patients with solid tumours or lymphomas using better technology and better cytoreductive regimens. A more rapid recovery of the haemopoietic system after transplantation has been claimed, but often a spontaneous reconstitution could not be excluded [78–81].

Comparative studies from our laboratory

Our own group has extensively studied different aspects of marrow preservation in mice, monkeys and in man using such quantitative viability assays as the spleen colony technique of Till and McCulloch (CFU-S technique), the agar culture technique (CFU-C technique) and transplantation experiments after lethal irradiation [82–88]. We could demonstrate that marrow viability is dependent not only on the cryoprotective medium, the rates of feezing and thawing, the cell concentration during freezing and the storage temperature but also on the resuspension technique after thawing. In all three species we observed that the speed of dilution used to adapt the thawed suspension to the final cell concentration had a marked influence on the stem cell survival.

Previously, others have studied the removal of the cryoprotective agents from thawed cells by dilution [13], by resuspension in dextrose [14], centrifugation [89] or—with red cells— by reversible agglomeration [90]. In mice, several groups reported a high stem cell recovery after freezing. In this species the thawed marrow has usually to be diluted in order to get suitable cell concentrations for injection, whereas in larger animals immediate infusion is possible and was applied by many investigators in the past. We compared systematically rapid and slow stepwise dilution. The cells were diluted tenfold in Hanks' solution (HBSS) either in one step or slowly in several steps over 45–60 min. The HBSS and cells were kept at room temperature. In 1972 we demonstrated that haemopoietic stem cells of mice, monkeys and human beings survive significantly better when intracellular protective agents are removed after thawing by a slow stepwise dilution technique.

In mice we tested the cryoprotective media 10% PVP + 10% glycerol + 20% calf serum (CAS), 10% DMSO + 20% CAS, 10% PVP + 20% CAS, 17.5–20% HES + 20% CAS (Figure 7.1). After rapid dilution we measured a mean CFU-S recovery of 47.2% ± 5.5% for PVP–glycerol, 70.9% ± 4.9% for DMSO, 74.4% ± 6.3% for PVP and 88.7% ± 3.9% for HES. After stepwise dilution the recovery rates were significantly better

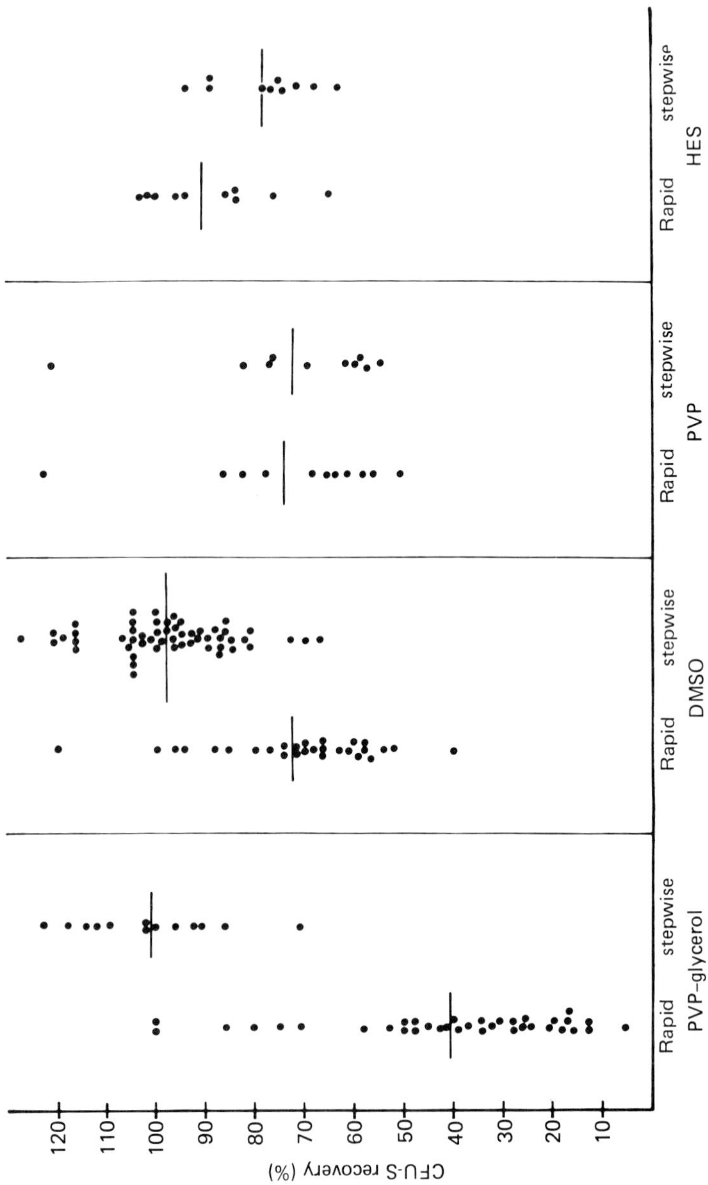

Figure 7.1 CFU-S recovery after freezing in various protective combinations comparing rapid and slow stepwise dilution after thawing

for PVP–glycerol (100% ± 4.9%) and for DMSO (98% ± 2.5%). With PVP and HES the stepwise dilution did not improve the CFU-S recovery (71.5% ± 6.3% and 77.8% ± 3.2%). It should be noted that after the suboptimal rapid dilution CFU-S were better preserved by DMSO than by the PVP–glycerol mixture. The low CFU-S recovery could not be overcome when serum was added to HBSS or the dilution was performed in pure serum. On the other hand the protective effect of the freezing media could be increased by adding 20% serum before freezing. When erythrocyte pure bone marrow was frozen at different cell concentrations optimal CFU-S recovery was measured at a range from 10×10^6 to 400×10^6 nucleated cells per ml. Preservation in DMSO at suboptimal cell concentrations of $0.5–1.0 \times 10^6$ per ml resulted in low stem cell survival. The poor recovery could be improved, however, when 90% serum, or 20% haemolysate or 1×10^9 intact erythrocytes per ml were added to the DMSO. The cryoprotective effect of serum is known; with haemolysate or red cells repair of latent cell damage by nutritional factors might play a role.

In monkeys we performed repopulation studies after lethal irradiation. The same storage technique which was found to be optimal in mice was applied. Using stepwise dilution with the PVP–glycerol mixture as well as with DMSO, reproducible takes of stored allogeneic marrow could be demonstrated. So far only autologous engraftment with 50% loss of the transplantation potential had been described in this species [6]. A high CFU-C recovery in parallel agar culture experiments and the fact that the takes could be achieved by injecting 4×10^8 nucleated cells per kg b.wt. (the minimal effective dose of fresh allogeneic marrow) indicate a very good storage efficiency. Reproducible takes were also seen when purified CFU-C rich marrow fractions obtained by the albumin gradient technique of Dicke [91–93] were grafted.

In man we tested the protective effect of DMSO and HES using the agar culture technique to assess viability. After rapid dilution 51.5% ± 6.6%, after slow stepwise dilution 98.3% ± 4.4% of the CFU-C were recovered when the cells were frozen in 10% DMSO + 20% CAS. Protection in 17.5% HES + 20% CAS resulted in a CFU-C recovery of 68.7% ± 13.0% or 56.9% ± 14.3% respectively when rapid and slow dilution were compared. Bone marrow of patients in complete remission of acute myeloid leukaemia could be preserved as effectively as normal marrow. Using DMSO and stepwise dilution after thawing we measured a CFU-C recovery of 103% ± 7.3%.

It has to be pointed out that in all our comparative CFU-S and CFU-C studies we have never applied correction factors for lost or dead cells. Applying the optimal storage procedure there was no significant loss of CFU in a tested suspension volume. The recovery of nucleated cells varied depending on the degree of contamination by mature peripheral blood cells. If marrow from exsanguinated donor mice was frozen almost 100% of the nucleated cells per ml suspension could be retrieved. In the transplantation experiments, however, when large amounts of marrow cells had to be processed and the contamination of peripheral blood was high the loss of nucleated cells ranged from 0 to 50%.

The details of one clinical autologous bone marrow transplantation we have performed are published elsewhere [88, 94]. The patient suffering from the second relapse of an acute myeloid leukaemia was treated with a supralethal chemoradiotherapy before grafting. The marrow was harvested during the first remission after one year of maintenance chemotherapy, six weeks after the last therapy course. The interval from cryopreservation to the first relapse was six months. The marrow was frozen in 10% DMSO and kept in liquid nitrogen for 15 months. The cell dose grafted was 2×10^8 nucleated cells per kg b.wt. containing approximately 3×10^4 CFU-C per kg b.wt. Post-transplantation, a rapid take and a very good repopulation of the marrow space were observed. The haemopoiesis, however, was not sufficient to prevent lethal infections. The patient died five months after transplantation. The post-mortem histology revealed an almost normal marrow cellularity. All cell classes were present. There was no evidence of a leukaemic relapse.

The clinical course indicated that the transplantation potential of frozen remission marrow might be impaired in spite of a prompt take and a normal proliferation of the thawed cells in the agar culture. It is possible that the prolonged aggressive chemotherapy which our patient received before the marrow was cryopreserved upset the self-reproduction and differentiation potential of the marrow stem cells. Observations of Hellman *et al* [95] in mice and Lohrmann *et al* [96] in man demonstrate that the haemopoiesis can be disturbed significantly and for a long time when repeated courses of chemotherapy are administered.

Conclusions

Reviewing the available data it can be stated that pluripotent haemopoietic stem cells survive storage at low temperature to provide haemopoietic reconstitution in animals and in man. The results from several investigators prove a high viability after storage periods of many months [23, 45, 88, 97]. There is indication that stem cells can survive for years when the marrow is kept at a very low temperature in the liquid phase of nitrogen [67, 75, 97], whereas preservation at −30°C to −100°C might result in deterioration during storage [24, 45]. Further experience may determine the extent to which the agar culture technique provides a quantitative parameter of the transplantation potential in frozen-thawed marrow. It has to be considered that cryoprotective agents can influence the *in vitro* proliferation and differentiation pattern by inhibition or stimulation [98–104]. If the DMSO concentration is decreased after thawing to less than 1% it is not toxic in the culture systems as shown by us and other authors [67, 98]. But besides toxicity there is another reason to apply dilution procedures after thawing. Cryoprotective agents of low molecular weight such as glycerol and DMSO which penetrate the cell membrane have a high osmolality and provide a severe osmotic stress to the cell. The osmotic stress depends on the speed of penetration. Glycerol penetrates the cell membrane more slowly than DMSO. Therefore rapid dilution by a medium of physiological osmolality or by immediate intravenous injection provides a long lasting osmotic gradient at the cell membrane when glycerol is used. Although DMSO penetrates more rapidly it is apparently better to mitigate the osmotic gradient by a stepwise dilution procedure. The extracellular protectants PVP and HES do not cause significant osmotic problems. They can be washed out by rapid dilution. PVP was found by us to be inhibitory *in vitro,* possibly due to a partial binding to the cell membrane [105]. The slow stepwise removal of DMSO described by us has been tested later also by others and appeared to be superior to a rapid dilution [56, 106].

Thus the available evidence suggests that storage of bone marrow under well controlled conditions provides reproducible stem cell recovery to permit clinical application. Rapidly increasing experience with combinations of chemotherapy and radiation and with development of technology to separate tumour cells from normal marrow by immunological or physical means may hasten the use of autologous bone marrow transplantation as a therapeutic alternative in cancer treatment.

References

1 Barnes, D.W.H. and Loutit, J.F. (1955). *Journal of the National Cancer Institute* **15**, 901
2 Polge, C., Smith, A.U. and Parkes, A.S. (1949). *Nature* **164**, 666
3 Ashwood-Smith, M.J. (1961). *Nature* **190**, 1204
4 Lovelock, J.E. and Bishop, M.W.H. (1959). *Nature* **183**, 1394
5 Persidsky, M. and Richards, V. (1962). *Nature* **196**, 585

6 van Putten, L.M. (1965). *European Journal of Cancer* **1**, 15
7 Till, J.E. and McCulloch, E.A. (1961). *Radiation Research* **14**, 213
8 Lewis, J.P. and Trobaugh, F.E. (1964). *Annals of the New York Academy of Sciences* **114**, 677
9 Schwarzenberg, L., Amiel, J.L., Tenenbaum, R. and Mathe, G. (1963). *Revue Française d'Etudes Cliniques et Biologiques* **8**, 783
10 Leibo, S.P., Farrant, J., Mazur, P., Hanna, M.G. and Smith, L.H. (1970). *Cryobiology* **6**, 315
11 Lewis, J.P., Passovoy, M. and Trobaugh, F.E. (1966). *Cryobiology* **3**, 47
12 Lewis, J.P., Passovoy, M., Conti, S.A., McFate, P.A. and Trobaugh, F.E. (1967). *Transfusion* **7**, 17
13 Drášil, V. (1960). *Folia biologica* **6**, 359
14 Sloviter, H.A. (1956). *American Journal of the Medical Sciences* **231**, 437
15 Phan The Tran and Bender, M.A. (1960). *Comptes Rendus des Séances de la Société de Biologie* **154**, 1090
16 Knorpp, C.T., Merchant, W.R., Gikas, P.W., Spencer, H.H. and Thompson, N.W. (1967). *Science* **157**, 1312
17 Weatherbee, L., Spencer, H.H., Knorpp, C.T., Lindenauer, S.M., Gikas, P.W. and Thompson, N.W. (1974). *Transfusion* **14**, 109
18 Ashwood-Smith, M.J., Warby, C., Connor, K.W. and Becker, G. (1972). *Cryobiology* **9**, 441
19 Persidsky, M.D. and Ellett, M.H. (1971). *Cryobiology* **8**, 586
20 Mannick, J.A., Lochte, H.L., Thomas, E.D. and Ferrebee, J.W. (1960). *Blood* **15**, 517
21 Cavins, J.A., Kasakura, S., Thomas, E.D. and Ferrebee, J.W. (1962). *Blood* **20**, 730
22 Storb, R., Epstein, R.B., Le Blond, R.F., Rudolph, R.H. and Thomas, E.D. (1969). *Blood* **33**, 918
23 Gorin, N.C., Herzig, G., Bull, M.I. and Graw, R.G. (1978). *Blood* **51**, 257
24 Appelbaum, F.R., Herzig, G.P., Graw, R.G., Bull, M.I., Bowles, C., Gorin, N.C. and Deisseroth, A.B. (1978). *Transplantation* **26**, 245
25 Davies, N.J., Paterson, A.H.G. and English, D. (1978). *Experimental Hematology* **6**, Suppl. 3, 22
26 van Putten, L.M., van Bekkum, D.W. and Dicke, K.A. (1969). In *Annual Report 1968. Radiobiological Institute TNO,* p. 62. Rijswijk
27 Buckner, C.D., Storb, R., Dillingham, L.A. and Thomas, E.D. (1970). *Cryobiology* **7**, 136
28 Merritt, C.B., Darrow, C.C., Vaal, L., Herzig, G.P. and Rogentine, G.N. (1973). *Transplantation* **15**, 154
29 Lewis, J.P., Farnes, M.P., Albala, M. and Trobaugh, F.E. (1964). *Annals of the New York Academy of Sciences* **114**, 701
30 Sullivan, R.D., Stecher, G. and Sternberg, S.S. (1959). *Journal of the National Cancer Institute* **23**, 367
31 Hager, E.B., Mannick, J.A., Thomas, E.D. and Ferrebee, J.W. (1961). *Radiation Research* **14**, 192
32 Bruch, C., Herbst, E., Calvo, W., Huget, P., Flad, H.D. and Fliedner, T.M. (1973). In *The Cryopreservation of Normal and Neoplastic Cells,* p. 51. R.S. Weiner, R.K. Oldham and L. Schwarzenberg. (ed.). Paris: Editions INSERM
33 Debelak-Fehir, K.M., Bennett, B.T. and Epstein, R.B. (1974). In *Abstracts of the 5th International Congress of the Transplantation Society, Jerusalem, Israel, August 25-30,* p. 200
34 Nothdurft, W., Bruch, C., Fliedner, T.M. and Ruber, E. (1977). *Scandinavian Journal of Haematology* **19**, 470
35 Sarpel, S., Harvath, L. and Epstein, R. (1978). *Experimental Hematology* **6**, Suppl. 3, 22
36 Nothdurft, W., Fliedner, T.M., Calvo, W., Flad, H.D., Huget, R., Körbling, M., Krumbacher-von Loringhofen, K., Ross, W.M., Schnappauf, H.P. and Steinbach, I. (1978). *Scandinavian Journal of Haematology* **21**, 115
37 Thomas, E.D., Lochte, H.L., Lu, W.C. and Ferrebee, J.W. (1957). *New England Journal of Medicine* **257**, 491
38 Buckner, C.D., Rudolph, R.H., Fefer, A., Clift, R.A., Epstein, R.B., Funk, D.D., Neiman, P.E., Slichter, S.J., Storb, R. and Thomas, E.D. (1972). *Cancer* **29**, 357

39 Polese, E., Massenti, S. and Rossi-Torelli, M. (1970). *Minerva medica* **61**, 4159
40 Lawrik, S. (1970). *Folia haematologica (Lpz)* **94**, 261
41 Mannheimer, E. (1970). *Wiener Zeitschrift für Innere Medizin* **51**, 297
42 Pegg, D.E., Humble, J.G. and Newton, K.A. (1962). *British Journal of Cancer* **16**, 417
43 Karrer, K. and Mannheimer, E. (1973). *Wiener Zeitschrift für Innere Medizin* **54**, 51
44 Buckner, C.D., Clift, R.A., Fefer, A., Neiman, P.E., Storb, R. and Thomas, E.D. (1974). *Experimental Hematology* **2**, 138
45 Malinin, T.I., Pegg, D.E., Perry, V.P. and Brodine, C.E. (1970). *Cryobiology* **7**, 65
46 Lochte, H.L., Ferrebee, J.W. and Thomas, E.D. (1959). *Journal of Laboratory and Clinical Medicine* **53**, 117
47 Pyle, H.M. and Boyer, H.F. (1964). *Annals of the New York Academy of Sciences* **114**, 686
48 Sultan, C. and Soulas, S. (1970). *Pathologie-Biologie* **18**, 1097
49 Mosimann, W., Furlan, M., Beck, E.A. and Bucher, U. (1972). *Schweizerische Medizinische Wochenschrift* **102**, 1600
50 Collmann, H. and Boll, I. (1972). *Blut* **25**, 265
51 Adamson, J.W. and Storb, R. (1972). *Transplantation* **14**, 490
52 Pluznik, D.H. and Sachs, L. (1965). *Journal of Cellular and Comparative Physiology* **66**, 319
53 Bradley, T.R. and Metcalf, D. (1966). *Australian Journal of Experimental Biology and Medical Science* **44**, 287
54 Pike, B.L. and Robinson, W.A. (1970). *Journal of Cellular Physiology* **76**, 77
55 Berthier, R. and Marcille, G. (1972). In *In Vitro Culture of Hemopoietic Cells,* p. 377. D.W. van Bekkum and K.A. Dicke (ed.). Rijswijk: Radiobiological Institute TNO
56 Grande, M., Berthier, R. and Hollard, D. (1977). *Experimental Hematology* **5**, 436
57 Gray, J.L. and Robinson, W.A. (1973). *Journal of Laboratory and Clinical Medicine* **81**, 317
58 Debelak-Fehir, K.M., Catchatourian, R. and Epstein, R.B. (1974). *Blood* **44**, 954
59 Weiner, R.S. (1977). *Experimental Hematology* **5**, Suppl. 2,104
60 Netzel, B., Haas, R.J., Janka, G.E. and Thierfelder, S. (1978). In *Cell-Separation and Cryobiology,* p. 255. H. Rainer, H. Borberg, J.M. Mishler and U.W. Schaefer (ed.). Stuttgart and Schattauer: New York
61 Körbling, M., Fliedner, T.M., Pflieger, H., Vassileva, D. and Heimpel, H. (1978). *Experimental Hematology* **6**, Suppl. 3, 95
62 Wells, J.R. and Graze, P.R. (1977). *Blood* **50**, Suppl. 1, 316
63 Ragab, A.H., Gilkerson, E. and Choi, S.C. (1974). *Cancer Research* **34**, 942
64 Balkwill, F., Pindar, A. and Crowther, D. (1974). *Nature* **251**, 741
65 Meyers, P.A., Clarkson, B., Murphy, M.J. Zaroulis, C.G. and Arlin, Z. (1977). *Blood* **50**, Suppl. 1, 130
66 Ottenbreit, M.J. and Inoue, S. (1977). *Blood* **50**, Suppl. 1, 157
67 Goldman, J.M., Th'ng, K.H., Park, D.S., Spiers, A.S.D., Lowenthal, R.M. and Ruutu, T. (1978). *British Journal of Haematology* **40**, 185
68 Graze, P.R. and Gale, R.P. (1978). *Transplantation Proceedings* **10**, 177
69 Thomas, E.D., Buckner, C.D., Fefer, A., Neiman, P.E. and Storb, R. (1978). *Advances in Cancer Research* **27**, 269
70 Storb, R., Thomas, E.D., Weiden, P.L., Buckner, C.D., Clift, R.A., Fefer, A., Goodell, B.W., Johnson, F.L., Neiman, P.E., Sanders, J.E. and Singer, J. (1978). *Transplantation Proceedings* **10**, 135
71 O'Reilly, R.J., Pahwa, R., Dupont, B. and Good, R.A. (1978). *Transplantation Proceedings* **10**, 187
72 Gorin, N.C., Najman, A., David, R., Hirsch-Marie, F., Stachowiak, J. and Duhamel, G. (1978). *Experimental Hematology* **6**, Suppl. 3, 8
73 Hellriegel, K.P., Hirschmann, W.D., Gauwerky, C., Gerecke, D., Borberg, H. and Gross, R. (1978). In *Cell-Separation and Cryobiology,* p. 335. H. Rainer, H. Borberg, J.M. Mishler and U.W. Schaefer (ed.). Stuttgart and Schattauer: New York
74 Dicke, K.A., McCredie, K.B., Spitzer, G., Zander, A., Peters, L., Verma, D.S., Stewart, D., Keating, M. and Stevens, E.E. (1978). *Transplantation* **26**, 169
75 Dicke, K.A., Zander, A., McCredie, K.B., Spitzer, G., Verma, D.S. and Vellekoop, L. (1978). *Experimental Hematology* **6**, Suppl. 3, 8

76 Buckner, C.D., Stewart, P., Clift, R.A., Fefer, A., Neiman, P.E., Singer, J., Storb, R. and Thomas, E.D. (1978). *Experimental Hematology* **6**, 96
77 Goldman, J.M. (1978). *Seminars in Hematology* **15**, 420
78 Tobias, J.S., Weiner, R.S., Griffiths, C.T., Richman, C.M., Parker, L.M. and Yankee, R.A. (1977). *European Journal of Cancer* **13**, 269
79 Graze, P.R., Wells, J.R., Gale, R.P. and Cline, M.J. (1977). *Blood* **50**, Suppl. 1, 314
80 Dicke, K.A., Lotzova, E., Spitzer, G. and McCredie, K.B. (1978). *Seminars in Hematology* **15**, 263
81 Appelbaum, F.R., Herzig, G.P., Ziegler, J.L., Graw, R.G., Levine, A.S. and Deisseroth, A.B. (1978). *Blood* **52**, 85
82 Schaefer, U.W., Dicke, K.A. and Klein, J.C. (1972). In *In Vitro Culture of Hemopoietic Cells,* p. 187. D.W. van Bekkum and K.A. Dicke (ed.). Rijswijk: Radiobiological Institute TNO
83 Schaefer, U.W., Dicke, K.A. and van Bekkum, D.W. (1972). *Revue Européenne d'Etudes Cliniques et Biologiques* **17**, 483
84 Schaefer, U.W. and Dicke, K.A. (1972). *Verhandlungen der Deutschen Gesellschaft fur Innere Medizin* **78**, 1601
85 Schaefer, U.W. and Dicke, K.A. (1973). In *The Cryopreservation of Normal and Neoplastic Cells,* p. 63. R.S. Weiner, R.K. Oldham and L. Schwarzenberg (ed.). Paris: INSERM
86 Schaefer, U.W., Schmidt, C.G., Dicke, K.A., van Bekkum, D.W. and Schmitt, G. (1975). *Zeitschrift für Krebsforschung* **83**, 285
87 Schaefer, U.W., Öhl, S. and Nowrousian, M.R. (1977). *Verhandlungen der Deutschen Gesellschaft für Innere Medizin* **83**, 1230
88 Schaefer, U.W., Nowrousian, M.R., Ohl, S. and Schmidt, C.G. (1978). In *Cell-Separation and Cryobiology,* p. 243. H. Rainer, H. Borberg, J.M. Mishler and U.W. Schaefer (ed.). Stuttgart and Schattauer: New York
89 Tullis, J.L., Tinch, R.J. and Latham, A. (1971). *Bibliotheca haematologica* **38**, Part 2, 338
90 Huggins, C.E. (1963). *Science* **139**, 504
91 Dicke, K.A., van Hooft, J.I.M. and van Bekkum, D.W. (1968). *Transplantation* **6**, 562
92 Dicke, K.A., Tridente, G. and van Bekkum, D.W. (1969). *Transplantation* **8**, 422
93 Dicke, K.A. (1970). Thesis, Rijksuniversiteit Leiden
94 Schaefer, U.W. and coworkers of the Bone Marrow Transplantation Team Essen (1977). *Experimental Hematology* **5**, Suppl. 2, 101
95 Hellman, S., Botnick, L.E., Hannon, E.C. and Vigneulle, R.M. (1978). *Proceedings of the National Academy of Sciences* **75**, 490
96 Lohrmann, H.P., Schreml, W., Lang, M., Betzler, M., Fliedner, T.M. and Heimpel, H. (1978). *British Journal of Haematology* **40**, 369
97 O'Grady, L.F. and Lewis, J.P. (1972). *Transfusion* **12**, 312
98 Abrahams, S., Till, J.E., McCulloch, E.A. and Siminovitch, L. (1968). *Cell and Tissue Kinetics* **1**, 255
99 Chang, C. and Simon, E. (1968). *Proceedings of the Society for Experimental Biology and Medicine* **128**, 60
100 Strong, D.M., Ahmed, A.A., Sell, K.W. and Greiff, D. (1972). *Cryobiology* **9**, 450
101 Malinin, T.I. and Perry, V.P. (1967). *Cryobiology* **4**, 90
102 Forejt, J. and Puza, V. (1968). *Folia biologica* **14**, 156
103 Friend, C., Scher, W., Holland, J.G. and Sato, T. (1971). *Proceedings of the National Academy of Sciences* **68**, 378
104 Preisler, H.D., Christoff, G. and Taylor, E. (1976). *Blood* **47**, 363
105 Rowe, A.W. (1966). *Cryobiology* **3**, 12
106 Holden, H.T., Oldham, R.K., Ortaldo, J.R. and Herberman, R.B. (1977). *Journal of the National Cancer Institute* **58**, 611

CHAPTER EIGHT

Protozoa and Helminth Parasites of Man and Animals

E R James

Introduction

The earliest recorded experiments which tested the effect of freezing on a living animal were probably those performed with the vinegar eel *Turbatrix aceti* by Henry Power [1] in 1661. According to his records, these small nematode worms endured overnight exposure to a keen frost or several hours of 'artificial freezing . . . by applying snow and salt . . . without any manifest injury being done to them'. The class Nematoda, as well as containing many free-living species such as *T. aceti*, also includes a large number of organisms which are important as parasites of man, animals and plants. Several of these parasitic species may spend a considerable time during the course of their development as free-living larval stages and thus may be exposed to extremes of temperature and to desiccation. It is perhaps not surprising therefore that many of these species can, like *T. aceti*, survive being frozen and can be stored for long periods at low temperatures.

Until relatively recently, helminths had been the subject of few experimental studies and most of the work pioneering the use of low temperatures for the preservation of parasites had been performed with protozoa; the early studies, beginning with those of Laveran and Mesnil in 1904 [2] with trypanosomes, have been well documented by Smith [3]. The trypanosomes in particular have received considerable attention, largely because during an infection these organisms spontaneously generate a succession of antigenic variants, and a stable method for conserving these individual variants for laboratory study was required. Cryopreservation of trypanosomes is now a well established routine technique [4] and the term 'stabilate' [5] (referring to the discrete population or clone of organisms with specific biological characteristics preserved on a unique occasion) has come to be used to describe the frozen samples of these and other protozoa. The advantages of stabilation have been considerable in providing a uniformity and coordination in research not previously possible, and reference banks of stored material are now available to workers throughout the world.

The reviews of Diamond [6], Muhlpfordt [7], Dalgliesh [8] and Miyata [194] listed some 80 species of parasitic protozoa which had been successfully cryopreserved up to 1975. It is possible to add a further 52 species from previously unquoted sources, or from reports published since these reviews, and from the records of the WHO Collaborating Centre for Trypanosomiasis and Cryopreservation of Protozoa in Professor Lumsden's Department of Medical Protozoology, London School of Hygiene and Tropical Medicine, which currently holds in excess of 1040 stabilated cultures (including 22 species of trypanosomes, 12 of *Leishmania,* 6 of malaria and 12 of other protozoa). This cryobank was originally modelled on the EATRO cryobank (see section on trypanosomes) established in the early 1960s by Lumsden, Cunningham and others at Tororo, Uganda. Many similar centres for parasitic and non-parasitic protozoa exist and some of these are listed in the World Directory of Collections of Cultures of Microorganisms [9], soon to be revised. Many laboratories do, however, maintain stocks of frozen parasites, the contents of which are not widely known and it is therefore impossible, and probably pointless also, to update these lists [6–8, 194] with those species which now have also been successfully cryopreserved.

The work on malaria by Coggeshall [10] suggested that the *in vitro* maintenance of parasites by cryopreservation might be used eventually to replace the costly and laborious methods of animal passage; this has now occurred for many of the species and strains of parasitic protozoa, and in the process the distribution of these organisms between laboratories has consequently also become considerably easier. Similar facilities are available on a small scale to helminthologists working with the infective larvae of domestic animal nematodes, and several pharmaceutical companies now routinely use cryopreservation as a more convenient method for storing live parasite material for screening new anthelmintics.

Cryopreservation is now also being considered as a method of increasing the shelf-life of certain organisms for possible use as anti-parasite vaccines [11–14], since under normal laboratory conditions they may remain viable for only a few days or even only a few hours. Partial protection against schistosomiasis has already been achieved in mice [15, 189] and in sheep [16] with radiation-attenuated organisms cryopreserved in liquid nitrogen and work with similar vaccines against some filarial worms and malaria is also in progress.

The application of low temperatures for the selection and destruction of parasites has also been reported. It was suggested by Joyner [17] that specific freezing schedules might be used to remove *Trichomonas foetus*, a causative agent of cattle venereal disease, from the bull spermatozoa used in artificial insemination, but following the conflicting results obtained by other workers (see [3] for discussion) this technique has not been adopted. Similar applications have been suggested for malaria [18] where selective destruction of early or late intraerythrocytic stages could be used to improve the synchronicity of development.

From studies of the effects of low temperatures on those parasites transmitted in food, and particularly the helminths, refrigeration has become mandatory in several countries as an aid to the control of diseases such as trichinosis [19], beef tapeworm and anisakiasis.

Possibly because parasite cryopreservation has generally been directed towards simplifying the maintenance of these organisms, experimental studies have been largely empirical and few attempts have been made to relate the survival of organisms to the actual conditions which occur during preservation. Concepts derived through the extensive studies of other systems such as red cells and lymphocytes can be applied to the freezing and thawing of parasites; however, with the exception of the cooling rate which has often received considerable attention, most published reports generally provide insufficient detail of the experimental procedures used to make this possible.

Cooling rates

The many accounts which now exist regarding the cryopreservation of parasites often appear to be contradictory and confusing. In an attempt to resolve this confusion it is probably most convenient to deal separately with the different groups of parasites and to relate these studies principally to the cooling rates which have been employed. Factors such as the differences which may occur between species or stages will also be mentioned along with the cooling rates within the headings of these parasite groups, while the other physical divisions of the process of cryopreservation, such as storage temperature, warming rate and cryoprotectants are probably best dealt with in general, since for these very much less work has been reported. Inevitably in dealing with the parasites as separate groups with regard to cooling rate there will be some repetition, but where this occurs it should serve to emphasise the similarities and outline the differences between the groups.

Protozoa

The protozoa differ from the helminths in two important respects with regard to cryopreservation. First, the many specialised cells such as nerve, muscle, gut, tegument, etc. in helminths probably comprise within a single organism a range of requirements for successful cryopreservation which is only found among all of the many different types of mammalian cells and for all the different species of protozoa. The conditions leading to survival are therefore possibly less critical for the protozoa as variation will occur to a greater extent between individuals than within them. Second, in contrast to almost all the helminths, most protozoa have the ability to reproduce asexually within their vertebrate hosts, and thus even if the survival is initially extremely small, high levels of parasitaemia can be attained within a relatively short time following infection with cryopreserved organisms.

Most of the important groups of parasitic protozoa have been cryopreserved with some degree of success and the only exception to this appears to be the group of intestinal ciliates including organisms such as *Balantidium coli.*

Malaria and Babesia

In 1939 Coggeshall [10] demonstrated that blood containing *Plasmodium knowlesi* and *P. inui* taken from rhesus monkeys could be stored at $-79°C$ for 70 days and still produce infections in clean monkeys after thawing and injection. Samples of defibrinated or citrated blood (1 ml) in celluloid tubes (120×20 mm) were cooled rapidly by immersion in an alcohol-solid CO_2 mixture in a Horsfall cabinet [20]. This simple technique has subsequently been used by many other workers [21–26], occasionally with minor modifications such as the use of thinner walled tubes [20, 25, 26] in order to increase the cooling rate still further and to reduce the degree of red cell lysis. Several different species of malaria have been successfully preserved in this way.

Fairly good levels of survival appear to have been achieved with the duck malaria *P. cathemerium* and a detectable parasitaemia occurred 4 days after the injection of 7×10^8 cryopreserved parasites [24]. It was also suggested in this study that the parasitised red cells may have survived cryopreservation better than the non-parasitised cells.

The early studies with human malarias were not so successful and in a comparison of normal and cryopreserved *P. vivax* [27] the latent periods for infection in recipients were 40 hours and 10 days respectively.

The conditions producing successful recovery of the erythrocytic stages of malaria and *Babesia* might be expected to depend to a large extent on those conditions giving good survival of the red cells, although there may not necessarily be a correlation between the preservation of red cells and parasites [18]. Optimum survival of erythrocytes occurs with relatively rapid rates of cooling and many authors have continued to use rapid cooling schedules for both species [18, 28–36, 190], although since the advent of cryoprotectants the use of slower rates has proved successful for *Plasmodium* [37–42] and for *Babesia* [11, 40, 41, 43, 44]. Both glycerol and dimethylsulphoxide (DMSO) apparently enhance the degree of survival of mammalian malarias, and latent development periods were considerably reduced; for example in rhesus monkeys infected with cryopreserved *P. knowlesi*, from 7–11 days without DMSO to 72 hours with DMSO.

A partial explanation for the success of both slow and rapid cooling rates may well be the existence of different optimum schedules for the different intraerythrocytic stages. Saunders *et al* [27] suggested that the younger stages appeared to be more resistant to rapid cooling and thawing than the larger trophozoites, and observations by Wilson *et al* [18] indicated that rapid cooling to −196°C preserved erythrocytes and ring forms of the parasites well but almost totally destroyed the large trophozoites and schizonts, which in turn were preserved by using slow or two-step cooling schedules.

Techniques for freezing the sporozoite stage of malaria appear to have been adapted from those used for the erythrocytic stages. Thus those reports describing the successful freezing of sporozoites in suspension [36, 38, 45, 46], or in the tissues of mosquitoes [47, 48], employed fast cooling rates. Slow cooling (2°C/min to −40°C before plunging into liquid nitrogen) of a *P. berghei* sporozoite suspension [49] did not yield any viable organisms. The recent study by Leef *et al* [191] indicated that the optimum cooling rates for *P. berghei* sporozoites varied according to the cryoprotectants used; the optimum rates with DMSO and glycerol were 1°C/min and 70°C/min respectively, whereas with serum, polyvinylpyrrolidone and hydroxyethyl starch, optimum survival occured between 20 and 60°C/min. Sporozoites of *Babesia bovis* derived from ticks were cryopreserved successfully using an unspecified rapid cooling rate [50], whereas a similar schedule used with *B. major* gave no survival and only slow cooling at approximately 1°C/min was reported [51] as successful.

The effect of different cooling rates on survival has been studied more extensively with the erythrocytic stages of *Babesia.* Dalgliesh [52] and coworkers [53, 54] used a range of cooling rates between 0.73 and 3070°C/min with glycerol or DMSO and found that the least depression of infectivity for the bovine parasite *B. bigemina* occured at 82°C/min [53] and with the mouse parasite *B. rodhaini* at 265°C/min [54] (see Tables 8.1 and 8.2).

Table 8.1 Differences in the infectivity for cattle of *Babesia bigemina* in blood containing 2 M DMSO after cooling to −196°C at eight cooling rates

	Experiment 1				*Experiment 2*				*Experiment 3*				
Cooling rate (°C/min)	3070	212	39	0.73	737	82	24	3.1	737	212	82	39	3.1
Mean pre-patent period (days)	8.57	4.14	4.00	4.71	6.43	4.75	5.50	7.13	7.00	5.25	4.63	5.25	7.00

Shorter mean prepatent periods indicate higher infectivity
From Dalgliesh and Mellors [53], by permission of the authors and publishers

Table 8.2 Infectivity of *B. rodhaini* for mice after freezing to −79°C and thawing using three cooling rates with DMSO

Cooling rate (°C/min) 0 to −40°C	2785	265	2.5
Mean prepatent period (days) for inoculation of 1.82 × 10^6 parasites	4.25	2.33	4.50

From Dalgliesh *et al* [54], by permission of the authors and publishers

These results together with those for sporozoites suggest that a difference in the cooling rate requirements for different species may exist. Differences may also occur between strains of organisms, since Warhurst [40] observed that 3.4% of a chloroquine-resistant strain of *P. berghei* survived cooling at 2°C/min and rapid warming while for a normal chloroquine-susceptible strain the survival was only 0.9%.

Trypanosomes

Of all the parasites, trypanosomes have been the most extensively studied with regard to rates of cooling. The technique of Laveran and Mesnil [2], which yielded viable and infective *Trypanosoma lewisi* after 15–75 minutes' exposure at −191°C, was to plunge the samples directly into liquid air. When cryoprotectants became available, many authors, as for malaria, continued to use rapid cooling schedules such as allowing droplets of citrated or heparinized blood to fall into liquid nitrogen [55–57], or immersing tubes containing suspensions of the organisms into alcohol–dry ice mixtures at −79°C [58, 59]. However, even before cryoprotectants had been discovered, it had been noticed [26] that the survival of *T. cruzi* could be improved by as much as 10 times if the samples were first cooled slowly to −15°C before subsequent rapid cooling to −70°C (see section on stepped cooling). Polge and Soltys [60] using a similar but slower two-step schedule for *T. brucei* (2°C/min to −20°C followed by 5°C/min to −79°C), found that the proportion surviving increased if glycerol (5–20%) was present. The same slow cooling schedule with DMSO gave good survival for *T. congolense* and *T. rhodesiense* [61].

The preservation of trypanosomes using a variety of cooling rates in combination with different cryoprotectants has been studied comprehensively by Lumsden, Cunningham and others at the East African

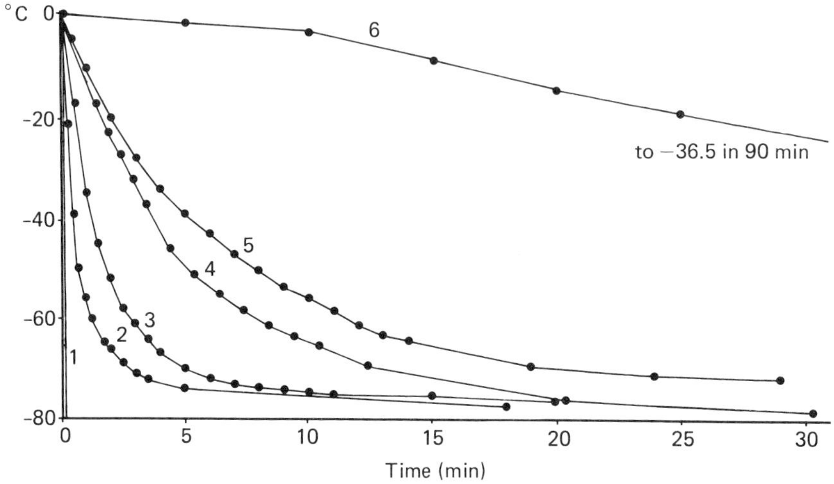

Figure 8.1 Observed cooling curves (1 to 6) for trypanosome samples in Table 8.3. (From Webber *et al* [62], by permission of the authors and publishers)

Table 8.3 Results of trypanosome counts and infectivity estimates for various samples, the cooling curves of which are shown in Figure 8.1

Sample no.	*Rate of cooling*	*Log numbers of trypanosomes per ml*	*Log numbers of ID_{50} per ml*
1	not cooled below 0°C	8.1	7.3
2		7.9	7.4
3	1	No trypanosomes seen	< 4.5
4			< 4.5
5	2	No trypanosomes seen	< 5.8
6			< 4.5
7	3	7.2	6.2
8		7.0	6.5
9	4	7.8	8.0
10		7.8	7.3
11	5	7.9	7.3
12		8.0	7.3
13	6	7.9	7.5
14		7.8	7.7

From Webber *et al* [62], by permission of the authors and publishers

Trypanosomiasis Research Organization (EATRO) between 1959 and 1968 (see Figure 8.1 and Table 8.3), and it is interesting to follow the development of a technique producing optimum recovery of trypanosomes, in the EATRO annual reports over this period.

Cunningham *et al* [63] found that the rate of cooling was especially important for a critical temperature region between −25 and −35°C, and that fast cooling did not appear to affect the survival of the organisms so long as they had been cooled slowly through this zone. In most recent studies slow cooling rates varying from 0.5 to 27.5°C/min have been employed [63–74], but the more successful of these reported the use of rates in the range of 1–2°C/min [66, 67, 69, 72, 74, 75].

Slow cooling is now part of the routine method for stabilation of trypanosomes; this important technique, which is in use in many laboratories throughout the world, was first outlined by Cunningham in 1960 [76] and has also been described elsewhere [77, 78]. Essentially it involves running infected heparinized blood containing 7% glycerol into capillary tubes (75 × 1.5 mm i.d.), sealing both ends with a flame and placing the capillary tubes into a screw cap glass tube containing 10 ml of methanol which is then inserted into an insulating jacket (25 mm thick polystyrene foam) and deposited in a solid CO_2 or mechanical refrigerator giving −79°C or below. The heat capacity of the tube with the methanol and capillaries is approximately 210 joules and a cooling rate of about 1°C/min is produced over the range of +10 to −40°C.

It must be emphasised however, that whereas this technique has been devised by careful experimentation with trypanosomes and can consistently produce a level of survival which differs little from unfrozen controls (see Table 8.3), it may yield far from maximum recovery for other protozoa; indeed when used for *Babesia* only 0.16–0.65% of the organisms were infective following thawing [44].

Leishmania

There have been reports of successful cryopreservation of *Leishmania spp* using fast cooling schedules [26, 79, 80] similar to those cited in the early studies with malaria and trypanosomes, but continuous or stepped slow cooling schedules of around 1–2°C/min at least to −30°C have also been used successfully [38, 66, 67]. However, until the study of Callow and Farrant [81] with *Leishmania tropica*, the effect of

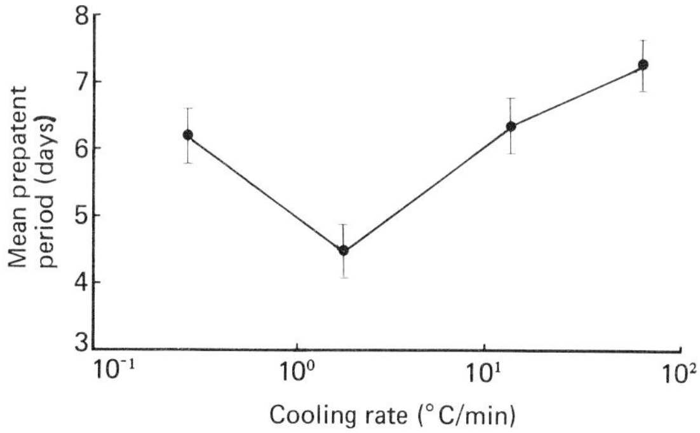

Figure 8.2 Survival of *Leishmania tropica* frozen to −196°C in 1.5 M DMSO at four cooling rates. Shorter prepatent periods following inoculation of cultures with thawed material indicate greater survival. Optimal survival was at 1.9°C/min. The standard errors shown are those of the differences between pairs of means using an analysis of variance. (From Callow and Farrant [81], by permission of the authors and publishers)

cooling at different rates, and of using different cryoprotective additives had not been compared. These authors found that optimum survival occurred with a cooling rate of 15.5°C/min in Adler's medium without cryoprotectants, while following the addition of 10 mM glucose or 1.5 M DMSO the optimum cooling rates were 1.84 and 1.9°C/min respectively. Both DMSO and glycerol (1 M) enhanced the survival of the organisms and DMSO appeared to give the best protection (see Figure 8.2).

Trichomonas

An excellent review of the early attempts to cryopreserve *Trichomonas* is contained in the book *Biological Effects of Freezing and Supercooling* by Smith [3]. Despite the large number of studies on this organism, little attention has been given to the cooling rate. Continuous or stepped slow cooling schedules [6, 17, 26, 65, 74, 82–89] resulted in survival, whereas rapid cooling [17, 82] almost always did not. The exception to this was the study of McEntegart [90] where it was found that *T. vaginalis* developed in culture following snap freezing at −79°C in the presence of glycerol (10%), and *T. gallinae* could survive this schedule even without glycerol. Slow cooling was, however, less damaging for both species and also for *T. hominis,* whereas *T. foetus* could not be recovered with any of the schedules or concentrations of cryoprotectant used. In general though, a rate of, or approximating to, 1°C/min appears to have been most favoured [82, 83, 90, 92–94, 194].

Ivey [95], who compared several slow rates, obtained consistently better survival of *T. vaginalis* using 2 or 5°C/min than with 1°C/min. Survival (59%) was also higher when 5% DMSO was present in the samples than when glycerol was used at a concentration of 1 M (0.13% survival).

It appears that the susceptibility of *Trichomonas* to damage, during freezing and thawing, is markedly affected by the phase of growth of the organisms in culture *in vitro.* Levine *et al* [96, 97] found that large numbers of *T. foetus* from the initial or early logarithmic phases of growth were damaged by cooling to −21°C in medium containing glycerol. They did not begin to survive freezing until the cultures were approximately 20.3 hours old, and the optimum recovery was obtained with 32 and 38 hour cultures. Better survival was also obtained in experiments where the population peak was reached relatively early. The failure of McEntegart [90] to obtain survival of cultured *T. foetus* may well have been due to this growth phase effect, since in these studies it was indicated that the organisms had been harvested after 18–20 hours of growth.

Toxoplasma

The sporulated oocysts of *Toxoplasma gondii* deposited in cat faeces can remain viable for a considerable time in soil [98] and their infectivity does not appear to be impaired after many days at −6 or −21°C [99]. In contrast, the vegetative cysts appear to be more vulnerable to refrigeration at these temperatures [100–102].

The review by Dumas [103] catalogues many of the successful and unsuccessful attempts at low temperature preservation. Rapid cooling has been used with only partial success [21, 26, 104–106], and when frozen in this way the organisms appeared to be damaged subsequently during the time spent in storage [26]. Most successful studies have reported the use of slow cooling schedules incorporating cooling rates in the region of 1°C/min [84, 103, 106–111] at least for the initial phase of cooling. Where a comparison of several schedules was made [106] the best survival was obtained with a continuous exponential slow cooling rate giving approximately 0.3°C/min over the range of 0 to −14°C.

Apart from the early studies [21, 26, 104, 105] which made no mention of the stage of parasite used, investigations have been performed largely with the vegetative cysts extracted from mouse brain or peritoneal cells. Bollinger *et al* [107] using culture forms of *T. gondii* maintained in a human fibroblast cell line, observed that successful cryopreservation with a continuous slow (overall approximately 1°C/min) cooling rate, occurred when cultures were between 3 and 6 days old and that this period roughly correlated with the development of a peak in total cell count and in the presence of only small cysts containing 2–8 trophozoids in the infected cells. These observations may indicate that a growth phase effect also operates for this parasite. Alternatively, survival may be low ordinarily and only at peak development is there a sufficient number of cells present to ensure successful passage; this interpretation could also apply to the cryopreservation of *Trichomonas* cultures.

Cryoprotectants were used, in combination with slow rates of cooling, in all successful reports except that of Manwell and Edgett [21]. Success in this study was attributed by the authors to the use of thin walled glass tubes and rapid rotation of the samples in order to distribute the contents of each tube as a thin film; this technique would have produced a faster rate of cooling than in another study [26] where cryoprotectants were also absent.

Eimeria

Sporozoites and sporocysts of *Eimeria spp* both survived either continuous [112–116] or stepped [74, 116, 117] slow cooling, and a rate of 1°C/min to −30°C (followed by 10°C/min to −80°C) was reported to give a higher recovery with each of three species [117] than when the initial cooling step was faster. However, for *E. tenella* the sporocysts were reported to be both considerably easier to prepare and to inoculate than sporozoites and to survive cryopreservation better (4.3 × as measured by the oocyst output from infected chicks). This was confirmed by Doran and Vetterling [115] using the anticoccidial index, although the reverse was true for *E. mealeagrimitis* [115] and for this organism, the survival of excysted sporozoites was considerably greater—35 times—than sporocyts.

Amoebae

Amoebae have been cryopreserved successfully using widely different rates of cooling. Rapid cooling produced by plunging sealed capillary tubes into liquid air has been used to recover viable *Entamoeba invadens* [118] and plunging thin mica slips coated with a suspension of free-living myxamoebae [119, 120] or *Naegleria* [120] into liquid nitrogen yielded viable organisms. Conversely, slow rates of cooling have also been used effectively by several authors for *Entamoeba spp* including *E. invadens* [6, 73, 121–125], for *Dientamoeba fragilis* [126] and for the parasitic *Naegleria, N. fowleri* [127]. These slow rates were usually in the region of 0.5–4°C/min [121–127] and yields of between 2 and 87% of viable organisms, as judged by their motility or their ability to exclude eosin stain [125], were obtained. It does, however, appear that the slow cooling methods currently in use frequently produce inconsistent levels of survival even for replicate samples.

Although the freezing of amoebae has received considerable attention there have been few studies in which a comparison of survival has been made using different cooling rates and the optimum cooling conditions for these parasites have not yet been precisely defined. Diamond [6] found that 1°C/min resulted in good yields (30–40%) of viable *E. histolytica* whereas 8°C/min produced an almost total destruction of the organisms. Fulton and Smith [121] also obtained viable *E. histolytica* with a slow cooling rate of 2.1°C/min from −15 to −79°C and both of the schedules incorporating faster rates which they tested gave no recovery.

Trophozoites of the intestinal flagellate *Giardia* respond somewhat similarly and storage has been reported following rapid cooling to −79°C in a dry ice–alcohol bath [128] and also following stepped slow cooling [129] (30 minutes at −12 to −15°C, −22°C until frozen and then storage at −70°C).

It is interesting that amoebae survive cooling at either slow or very rapid rates, but not apparently at intermediate rates. This effect does not appear to correlate with the use of cryoprotectants, since additives, in most cases glycerol, DMSO or ethanediol, were used in all the successful studies whether the cooling schedule was slow or rapid. Cryoprotectants may well be essential for the preservation of these organisms as studies conducted before the advent of these compounds [26], were unsuccessful.

The effect of freezing on the different stages of amoebae presents an interesting phenomenon since workers who reported success with either slow or rapid rates of cooling indicated that trophozoites were used in their studies. In contrast, an attempt to preserve purified suspensions of mature cysts of *E. invadens* and *E. terrapinae* [124] did not succeed with any of the cooling rates tested. These observations are the opposite of what might be expected as cysts are generally adapted to adverse environmental conditions. The cysts of other unrelated species such as *Eimeria* [112] and *Gregarina* [130] can, however, be frozen.

Helminths

Intestinal Nematodes

It is convenient cryobiologically to consider the intestinal nematodes and tissue dwelling nematodes as two separate groups. The free-living larvae of many of the intestinal nematodes have been shown to be capable of surviving over winter on pasture when the mean daily temperature was below 0°C for many months [131–136] with periods as low as −25.6°C [135], and *Trichostrongylus colubriformis* larvae were shown to survive in the laboratory for considerable periods at temperatures down to −95°C [137] (Figure 8.3). It has also been suggested for *Nematodirus filicollis* [138] that subzero temperatures may actually have a beneficial effect on the ova by stimulating them to develop faster when warmer conditions return.

Different cooling rates have not, however, been investigated extensively with this group of nematodes, but it appears that there may

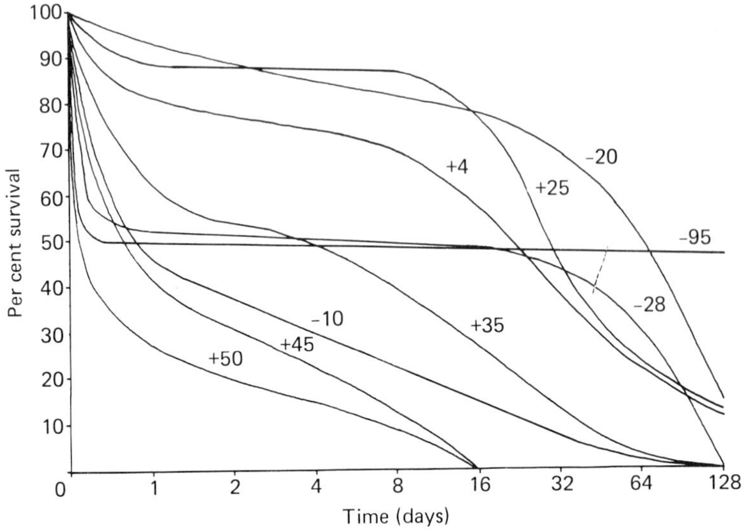

Figure 8.3 Survival of clean desiccated third stage larvae of *Trichostrongylus colubriformis* at constant temperatures. (From Andersen and Levine [137], by permission of the authors and publishers)

be some species which only survive slow cooling, some which prefer rapid cooling, and others which survive using a variety of rates. A cooling rate of 1°C/min gave good survival of *Nematodirus battus* and *N. filicollis* larvae [139] and also *Trichonema spp* [140] and *Trichostrongylus spp* [141], but no infective larvae of *Ostertagia circumcincta* or *Haemonchus contortus* could be recovered using this schedule [139]. Several other authors have used undefined, relatively rapid rates of cooling, produced by placing vials containing larvae, often only in tap water, directly from room temperature into a −70°C refrigerator, or an alcohol–dry ice mixture at −79°C, or into the vapour phase of a liquid nitrogen refrigerator. In this way the infective larvae of *Ancylostoma caninum* [26], *Nippostrongylus brasiliensis* [142] and *Haemonchus contortus* [143] were recovered. Campbell *et al* [143] also successfully used rapid cooling for *T. colubriformis, Ostertagia spp, Cooperia spp* and *Oesophagostomum spp* larvae.

Survival of the infective larvae of many of these nematodes is dramatically enhanced if they are first 'exsheathed' prior to cooling [142–147]. It is paradoxical that to survive adverse weather conditions free in the environment, the infective third stage larva is equipped with a sheath derived from the tegument of the second stage larva, while resistance to freezing increases on removal of this sheath. No satisfactory explanation has so far been advanced to account for this phenomenon, but other changes in the larvae induced at the time of 'exsheathment' may be responsible for this protection. The sheaths of microfilariae when present, do not appear to reduce the resistance of these larvae to low temperatures.

The use of conventional cryoprotective compounds with the third stage larvae of intestinal nematodes has often been shown to confer no advantage [142–144] or even to be detrimental [139. 142], although the addition of DMSO or glycerol has been claimed [139, 140, 147, 149] to increase the survival of some species.

Filarial Nematodes

Survival of the microfilariae of several of these nematodes, following storage at the relatively high subzero temperatures of −25 to −30°C [149–151] as well as at −70 to −79°C [26, 152, 153] has been known for some time. However, no indication of the rate of cooling was given in these studies. More recent studies have also often omitted to mention the cooling rate, but where reports have included this information, the rates used have been slow and in the region of 1°C/min [154–155]. Studies on the cryopreservation of the

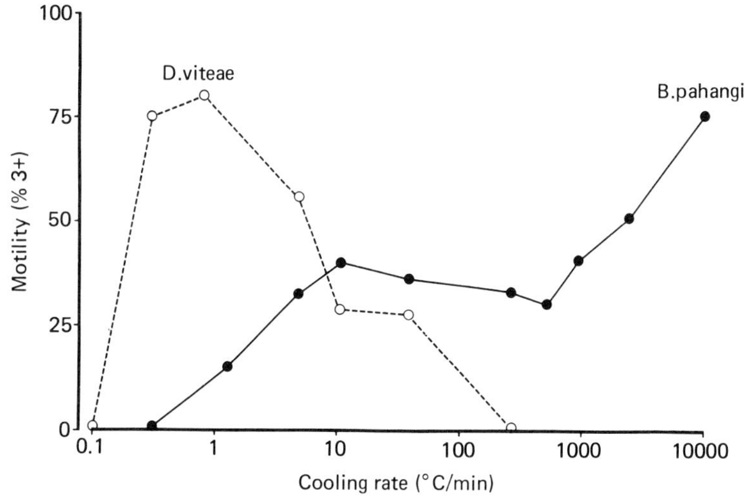

Figure 8.4 Survival of *Dipetalonema viteae* and *Brugia pahangi* third stage larvae after cooling at different rates to −196°C and thawing. Larval motility was assessed on a scale from 0 (dead) to 3+ (highly active) and survival was determined as the percentage of larvae with good (3+) motility. (From Ham [159], by permission of the author)

spiruroid nematode *Dracunculus medinensis* [156–158] indicated that rapid cooling rates in excess of 20°C/min were damaging [157] and a rate of 5°C/min appeared to give the best results for this parasite [158]. Similar results were obtained with microfilariae of the avian parasite *Chandlerella quiscali* [192]; good survival (72–78% motile organisms) occurred when samples were cooled slowly by being placed directly into a −68°C refrigerator (producing a cooling rate of approximately 1.9°C/min between 0 and −30°C whereas survival was reduced to 1.3% using 3–5°C/min and to 0% with 180–300°C/min.

The importance of the cooling rate in the preservation of filarial larvae has been demonstrated by Ham [159] for the third stage larvae of *Dipetalonema viteae* and *Brugia pahangi* which appear to require rates as disparate as 1.3°C/min and 10000°C/min respectively for optimum recovery (see Figure 8.4).

The infectivity of helminths, particularly microfilariae, is often difficult to assay and many authors have chosen to determine survival from motility alone. To assume that motility implies viability and infectivity can be misleading; in the study by Taylor [153], in which up to 50–60% of microfilariae of different species were recovered motile following storage at −79°C, these were, on closer examination, found to be vacuolated. Similar disparity in actual survival and in estimates of survival made by standard assay techniques may also exist for other parasites; it was observed for *E. histolytica* [160] that whereas supravital staining indicated 34%, and motility indicated 24% levels of survival, ultrastructurally almost no trophozoites appeared undamaged.

The infectivity of cryopreserved microfilariae was demonstrated by Bemrick *et al* [161] for *Dirofilaria immitis*. The authors stated that no attempt was made to control the cooling rate to the final temperature of −68°C, although since the sample volumes of infected whole blood were large, 5 ml, this rate must have been fairly slow. With this technique, red cell lysis was reported, and this factor together with the presence of the parasites, was thought to have contributed to the low survival of mosquitoes fed on cryopreserved blood. The infectivity of microfilariae of *Brugia pahangi* [154, 162] and the larvae of *D. medinensis* [157, 158] has similarly been reported following low temperature preservation. Subsequent development of these larvae, and of the cryopreserved third stage larvae of *D. viteae,* into mature infections in their respective mammalian hosts was also demonstrated [155, 158, 161]. Motility is a useful indicator of possible survival, but it is clear from these and other reports [163] that confirmation must come from infectivity studies.

Trematodes and Cestodes

The survival of metacercariae of *Fasciola hepatica* and *F. gigantica* at relatively high subzero temperatures was studied by Boray and Enigk [164] who reported that infectivity was lost by sudden or gradual exposure to −20°C but that −10°C was tolerated for up to 28 days. Taylor [165] had earlier reported survival after 56 days in ice at −4°C. In common with intestinal nematode larvae, *Fasciola* metacercariae

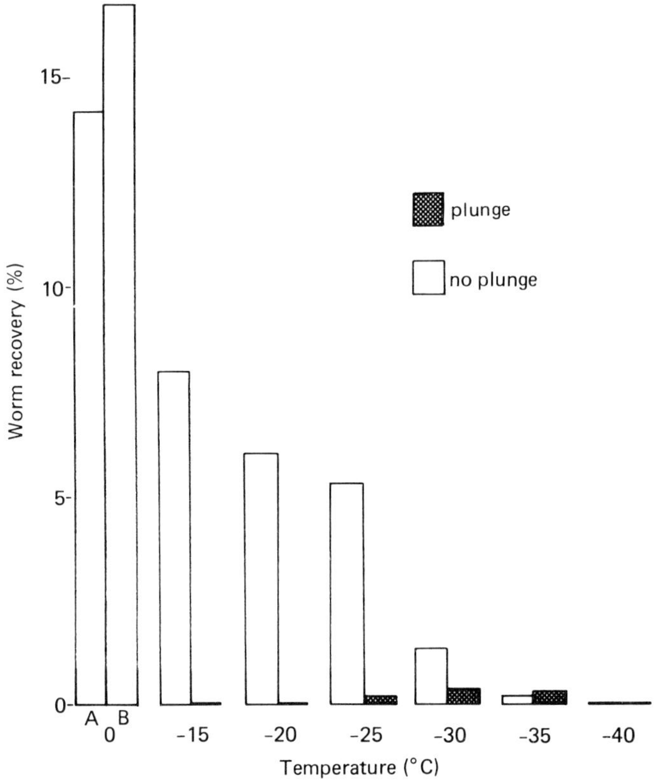

Figure 8.5 Infectivity of schistosomula of *Schistosoma mansoni* following two-step cooling to −196°C and thawing, assessed from recovery of adult worms. Samples from control group A contained schistosomula held at 0°C for 15 min; from control group B schistosomula held at 0°C with 17.5% $^{V}/v$ methanol; samples from other no plunge control groups were held at 0°C with methanol and then cooled slowly (0.65°C/min) to the intermediate temperatures indicated before rewarming. Samples in the experimental groups were plunged into liquid nitrogen (10000°C/min) from the intermediate temperatures indicated before rewarming. Results are expressed as the percentage of those schistosomula injected into mice which were recovered as adult worms by perfusion six weeks later. (From James and Farrant [13])

may often be exposed naturally to cold environmental conditions, and perhaps because refrigeration above zero for many months is possible, there appears to have been little interest so far in preservation at lower temperatures.

Larvae of *Schistosoma spp* are not capable of surviving for long in the environment and they usually become non-infective after a period of only 24–48 hours. Cioli (personal communication) investigated the use of slow and fast cooling schedules with a variety of cryoprotective agents but without success, while an indication that the possible preferred cooling rate for schistosomula lay in the region of 0.3–6.8°C/min was given by James and Farrant [166]. It was found subsequently [13] that a critical temperature zone between −25 and −35°C existed, and that if slow cooling at approximately 0.65°C/min down to this zone, was followed by rapid cooling at approximately 10000°C/min, a small percentage of infective organisms could be recovered (see Figure 8.5). Subsequent results (James, unpublished) indicate that the optimum rate for the initial slow cooling step may be closer to 1°C/min. In the second cooling step survival does not appear to be enhanced by using rates in excess of 10000°C/min, although rates slightly slower result in loss of infectivity.

Cryopreservation of cestodes has apparently not received much attention yet although Ham (personal communication) has found viable onchospheres of *Taenia crassiceps* can be recovered from temperatures down to −35°C.

Dehydration and stepped cooling

As described in Chapter 1, during the process of cooling, water external to the cell or organism freezes first. The developing ice crystals exclude the solutes, which become concentrated in the residual unfrozen water. Thus the medium surrounding the organism becomes increasingly hyperosmotic as the temperature is lowered leading to progressive dehydration of the cells.

Dehydration has been shown to enhance freezing tolerance in those cases where it was induced prior to freezing. *Trichostrongylus colubriformis* larvae, desiccated by exposure to a relative humidity of 65–75% for 16 hours at +30°C were more resistant (approximately 50% survival) to storage at −95°C for 128 days than were undesiccated larvae (survival 2–6%) [137]. Similarly, a considerable increase in recovery of *Trichonema spp* larvae [140] and *Turbatrix spp* [167, 168] following storage at −196°C or −77°C was produced if these organisms were partially dehydrated first. Von Brand and Morris [130] also observed that

Gregarina spp spores in dry cornmeal could withstand direct cooling to −79°C; hydration was not immediately damaging, but after soaking in water for 24 hours no survival was recorded.

With a two-step cooling schedule, dehydration occurs during the period of time spent at the intermediate holding temperature, and optimum dehydration is dependent both on the time and on the particular temperature employed. Similar results can also be obtained with two-stage cooling schedules where dehydration occurs during an initial slow cooling rate. Both techniques terminate the process of dehydration either after a specific time interval or after a specific temperature has been reached. The subsequent cooling step to the final storage temperature may either be rapid enough to convert the intracellular water to a glass phase, or any ice crystals which form may be too small to be damaging. Several examples of the cryopreservation of parasites by stepped cooling schedules help to illustrate these points, and in some cases stepped cooling can produce levels of survival which are greater than can be achieved with simple continuous cooling schedules.

Poole [138] reported that no infective larvae of *Nematodirus filicollis* survived direct cooling to below −40°C, whereas if they were first 'conditioned' at −6.5°C, 2–3% could be recovered in a viable state after rapid cooling, and several hours' exposure to −65°C. The recovery of viable *Turbatrix silusiae* was similarly increased from below 0.8% to more than 6.9% if the larvae were first held at −30°C before plunging into liquid air [167]. Similar observations have been made for several of the protozoa; for malaria, an intermediate temperature of −25°C held for 60 min [37] and for trypanosomes a temperature of −30°C for 3–5 min [63, 64] gave increased survival. With malaria a comparison of two holding temperatures, −25 and −32°C [18], indicated that although at the lower holding temperature red cell lysis was more marked and tended to increase with time, this temperature, held for 30 min, gave a greater yield of undamaged parasites.

For the schistosomes a two-step cooling schedule appears to be the only method so far devised by which infective organisms can be stored successfully at low temperatures [13], and only those schedules incorporating intermediate temperatures in the range of −25 to −35°C were effective. As the rate of cooling is increased there is an apparent shift to progressively lower intermediate temperatures for optimum recovery [193] and the peak survival, judged by observations of motility, for schistosomula cooled at 0.2, 0.4, 0.8 and 1.2°C/min occurred respectively at −26, −32, −35 and −36°C. Temperatures in this region also appear to be especially critical for trypanosomes, and

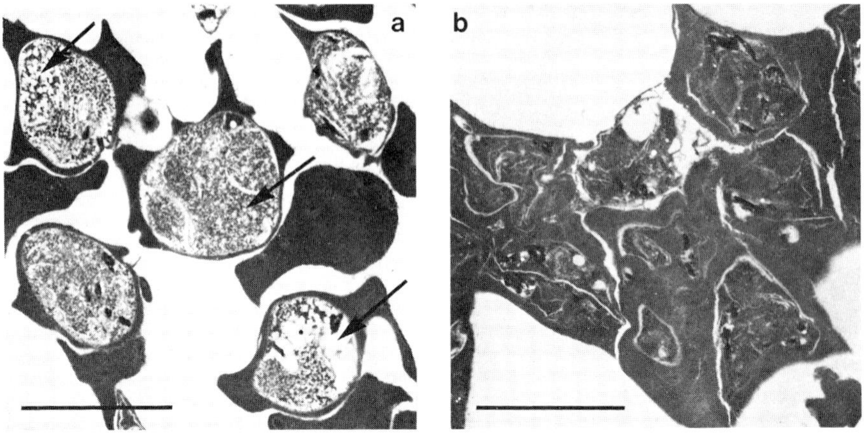

Figure 8.6 Electron micrographs of freeze-substituted trophozoites of *Plasmodium knowlesi* cooled either (a) directly to −196°C or (b) to −196°C from a holding temperature of −25°C. Numerous ice cavities (arrows) were present in parasites plunged to −196°C, but cells prepared by two-step cooling were grossly shrunken and contained relatively little intracellular ice. The scale bar represents 1 μm. (From Wilson *et al* [18], by permission of the authors and publishers)

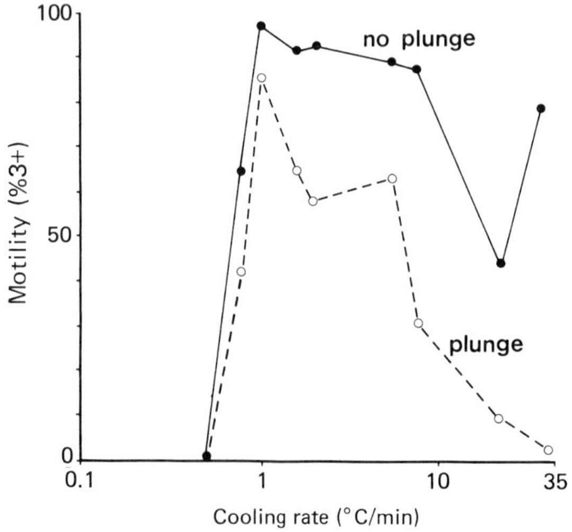

Figure 8.7 Survival (per cent 3+ motile) of *Onchocerca gutturosa* microfilariae following two-step cooling. Samples were cooled at different rates to the intermediate temperature of −17°C and either thawed directly (no plunge) or cooled rapidly to −196°C before thawing. (From Ham *et al* [163], by permission of the author and publishers)

Table 8.4 Infectivity estimates for different cooling schedules (Trypanosoma brucei)

Experiment 1	*Prefreeze control*	*Final intermediate temperature during slow cool before rapid cool to −80°C* −10	−20	−30	−40	−50	−60	−80
Log_{10} no. ID_{63} (organisms per ml)	7.1	≤2.4	5.9	7.0	6.8	7.1	7.3	7.1

Experiment 2	*Prefreeze control* 0 min	40 min	70 min	*Final intermediate temperature during rapid cool before slow cool to −80°C* 0	−10	−20	−30	−40	−50	−60
Log_{10} no. ID_{63} (organisms per ml)	7.6	8.1	7.6	8.1	7.3	7.7	7.1	5.4	3.7	<2.4

Experiment 3	*Prefreeze*	*Time held at −30°C after rapid cool and before rapid cool to -80°C (min)* 0	0.25	0.5	1.0	3.0	5.0	7.0	9.0	14.0	20.0
Log_{10} no. ID_{63} (organisms per ml)	7.6	<2.4	3.7	5.1	6.1	6.8	6.8	6.4	6.4	6.4	6.4

From Cunningham *et al* [63], by permission of the authors and publishers

good survival was obtained only where slow cooling through this zone occurred either continuously, or independently in the first or second cooling steps [63; see Table 8.4].

Ultrastructural studies confirm that dehydration occurs during slow cooling or when samples are held at an intermediate temperature. This has been shown for *Toxoplasma* [169], and electron micrographs of freeze-substituted samples in conjunction with viability assessment have revealed that both for lymphocytes [170] and the intraerythrocytic forms of *Plasmodium* [18; see Figure 8.6], optimum survival is dependent on the extent of cell shrinkage. Similar preliminary studies of schistosomula (Walter and James, unpublished) indicate that cavities in the tegument and musculature, resulting from substitution of intracellular ice, become progressively smaller in samples from lower intermediate temperatures.

A large number of other studies [37, 39, 67, 74, 84, 88, 92, 95, 109, 116, 117, 122, 124, 125, 129, 140, 159, 171] with a variety of parasites, have reported, apparently quite arbitrarily, the use of schedules in which transition from slow to rapid cooling was made in approximately the same temperature zone which has been shown to be critical for schistosomula and trypanosomes. Doran [117] observed also that the degree of survival of *Eimeria* varied with the speed of the first cooling step to −30°C, and optimum recovery was obtained, as appears to be the case with schistosomula, with 1°C/min. Results of all these studies apparently concur with the findings of Asahina [172] who preserved several species of animals and plants by a technique which involved 'prefreezing' at −30°C for a period of 1 hour before plunging into liquid air.

Although temperatures in the region of −30°C appear to be important for the cryopreservation of a number of very different parasites, adaptation of the two-step schedule used for schistosomula to three species of *Onchocerca* [163] indicates that these all require higher intermediate temperatures (between −14 and −17°C) for optimum survival as well as a cooling rate of approximately 1°C/min during the first step (see Figure 8.7). Stepped cooling of parasites has not yet been used extensively, but the studies reported above, and also in particular those of Wilson *et al* [18], suggest that this method is valuable for identifying specific temperature zones which cause damage.

Warming rate and dilution

The rate of warming is probably as important to survival as the rate of cooling, since ice which forms during cooling and which may be present in living cells at low temperatures undergoes phases of reorganisation and recrystallisation between certain critical temperatures during warming. Usually when samples spend an extended period of time in this critical temperature region during warming, the many small ice crystals which may be present grow in size; this may physically damage the cells. Levaditi [58] considered the critical temperature range for warming *T. congolense* and *P. berghei* to be between −12 and 0°C but the recrystallisation zone, and thus the zone of potential damage, actually extends to much lower temperatures.

Ever since *Plasmodium* was successfully recovered from storage at −76°C by Coggeshall [10] using a technique of rapid warming produced by agitating the samples rapidly in a water bath at +37°C, this method has become almost universal for rewarming frozen parasites. Rapid warming in many cases reduces or prevents recrystallisation damage; the rate produced by techniques similar to those of Coggeshall has been estimated [81] to be approximately 600–800°C/min. Recently, even faster warming rates have been used for *Babesia* (4320°C/min) [54] and for schistosomes (approximately 10000°C/min) [13].

Few studies have been performed to test the effect of warming rates on the survival of parasites recovered from low temperatures and many reports make no reference at all to the warming rate or even to the technique used for thawing. It has been reported [71, 173] that trypanosomes can be recovered with partial success using several warming rates (Figure 8.8), although the faster rates do appear to be

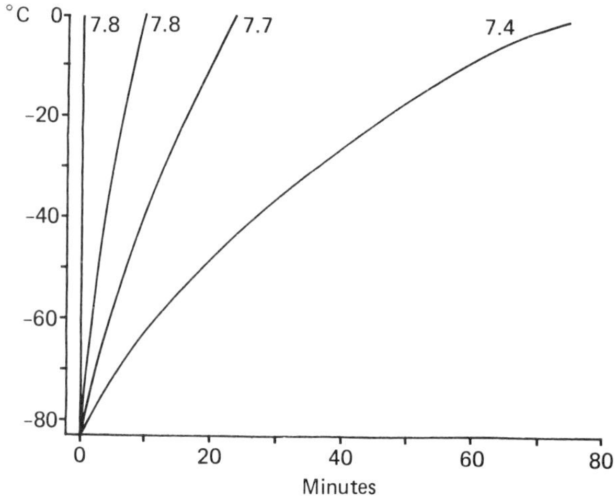

Figure 8.8 Effect of different thawing rates from −80°C to approximately 0°C on the infectivity of *Trypanosoma brucei* subgroup organisms. The figures accompanying the warming curves are the infectivity estimates (Log_{10} no. of ID_{63} per ml) resulting from the different thawing schedules. (From Cunningham *et al* [173], by permission of the authors and publishers)

somewhat better. For malaria [174] the fastest of three warming rates tested led to the shortest latent period of infection following injection, and studies with *Babesia* have indicated, perhaps more conclusively, that whereas survival can be obtained with warming rates between 1.8 and 4320°C/min [54], a significantly reduced latent development time and red cell haemolysis was achieved at the fastest rate.

Intracellular ice recrystallisation is one possible source of damage to organisms during warming. Osmotic stresses due to the presence of salts and cryoprotectants in the suspending medium can also cause damage, not only during the process of warming but as a result of the methods used for the removal of additives. Little attention has so far been paid to these effects, although their importance should not be underestimated, indeed Strome *et al* [36] suggested for malaria, that the loss due to the recovery procedures exceeds any loss that may have occurred during freezing and thawing. Red cells are particularly susceptible to osmotic stress on thawing, and techniques to reduce haemolysis [3] such as stepwise dilution [175, 190] and especially dilution first in 0.5 M glucose saline (Liston, personal communication) also considerably improve the survival of intraerythrocytic stages of malaria.

Final storage temperature

The methods of storage most frequently used are refrigerators giving −70°C or below, solid CO_2 at −79°C, or vapour phase and liquid nitrogen at temperatures between −140 and −196°C. Parasites have been preserved by all of these methods, without any significant loss of infectivity, often for considerable periods, and in the case of *Plasmodium gallinaceum* for more than 10 years at −140°C [36]. However, occasionally deterioration at the higher storage temperatures has been reported. At −95°C the viability of *Trichomonas foetus* was observed to have fallen from 50% to 11% after 5.6 years of storage [176] while at −79°C the time limit for storage of *T. vaginalis* was suggested to be 2 years [177]. Also at −79°C, *Trypanosoma vivax* was found [178] to be non-infective after 4 months, while at −70°C *Toxoplasma gondii* was viable only if thawed immediately after being frozen [26].

In an interesting study of the survival up to 90 days at −75°C of different strains of *T. vaginalis*, Miyata [171] observed that in the presence of different cryoprotectants, glycerol, DMSO or ethanediol, the rate of deterioration was remarkably different. Overall, there was little drop in survival with glycerol during the period of study, while in DMSO or ethanediol, a rapid decline, in some cases from around 80% to almost 0%, occurred over this period.

Cryoprotectants

A large number of this heterogeneous group of compounds have been reported to be useful in the cryopreservation of parasites. The extensive comparative study of O'Connell *et al* [179] revealed that of 83 chemicals, principally alcohols, sugars, amides and amines, the alcohols were most effective for *Trypanosoma gambiense, T. conorhini* and *Crithidia fasciculata* (a non-pathogenic flagellate). Certain sugars also showed some protective activity and glycerol at a concentration of 10% was the most effective cryoprotectant of those studied. Soon after the discovery of glycerol as a cryoprotectant [180] a number of workers had reported its beneficial effect in preserving parasites. Ethanediol [181] and DMSO [61] were similarly found to enhance parasite survival, although ethanediol had been used as early as 1941 [168] to induce partial dehydration of the nematode *Turbatrix aceti* prior to freezing.

Selection of the correct cryoprotectant for each organism is important; for instance glycerol (10%) was found to be marginally better for *Plasmodium berghei* than DMSO (5%) whereas for *P. gallinaceum* the use of DMSO produced significantly higher infectivity titres [33]. For schistosomula it is important that −30°C be attained during the slow cooling phase [13], and it was found that only two compounds out of more than 25 tested [182] were capable of protecting down to this temperature. Cryopreservation of schistosomula was eventually achieved [13] using methanol, and factors which were thought to have contributed to the survival were the relatively low toxicity for schistosomula of this compound, its good freezing-point depression—this being an indirect indication of the reduction in amount of water converting to ice during cooling, and thus on the concentration of the dissolved salts in the residual water—and its low viscocity. Methanol had been described by Lovelock [183] as a compound possessing cryoprotective properties, and it has been used previously for the preservation of trypanosomes [184, 185]. It was also suggested [184] that the low toxicity of methanol to several different cell types warranted further study of its use in cryopreservation, and there is currently a resurgence of interest in this compound.

The cryoprotectants most commonly used for parasites, such as glycerol, DMSO, ethanediol and methanol, are low molecular weight compounds capable of penetrating cells and are thought to provide protection internally. Some non-penetrating high molecular weight compounds have been used successfully in other systems, and are thought to act either by modifying the external physical conditions, or by protecting the outer membranes of cells. Serum, which has been reported to enhance the survival of some parasites, notably nematodes [163, 186] and some protozoa [187, 191] may well act in a manner similar to these non-penetrating cryoprotectants.

The toxicity of cryoprotectants is generally temperature dependent, and thus their addition to a sample is often best carried out in an ice bath. The exception to this is glycerol (and also ethanediol in some systems) which becomes increasingly viscous as the temperature is lowered and therefore takes considerably longer to permeate at 0°C than at room temperature or at +37°C. Cryoprotectants with low viscosities, such as methanol and DMSO, apparently permeate even relatively large organisms within a few seconds at 0°C. These effects of temperature and of the time of equilibration have been illustrated with *T. vaginalis,* for DMSO and glycerol by Miyata [188]. Survival following storage at −75°C increased as the prefreeze incubation

Table 8.5 Recovery of *Dientamoeba fragilis* strain Bipa frozen in liquid nitrogen in the presence of different concentrations of dimethylsulphoxide (DMSO)

	*Mean % recovery**	
Per cent DMSO in freezing mixture	*immediately after freezing*	*after 6 months at −196°C*
2.00	0.00	0.00
2.25	0.50	0.48
2.50	3.00	2.92
2.65	15.50	15.03
2.70	51.10	50.65
2.75	87.30	86.30
2.80	47.60	47.25
2.85	22.81	21.33
2.90	8.60	8.05
3.00	2.72	2.41
3.25	0.31	0.29
3.50–10.00	0.00	0.00

*Mean percentages of recovery with respect to organisms thawed immediately after freezing were based on counts made in 1050 replicate 0.166 mm^2 object fields from 50 individual samples; with regard to the organisms stored for 6 months in liquid nitrogen, the counts involved 525 fields from 25 samples

From Dwyer and Honigberg [126], by permission of the authors and publishers

period in glycerol (10%) was extended, whereas the reverse occurred with DMSO (7.5%). The effect was also markedly temperature dependent and with a constant incubation time of 100 minutes, survival using DMSO increased as the incubation temperature was lowered from +37°C to 0°C, while with glycerol, survival fell from a maximum at +37°C to a minimum between +10°C and 0°C.

Cryoprotectant concentration is often also extremely critical. This has been particularly well demonstrated by the studies of Bemrick [128] using glycerol for *Giardia* trophozoites, and Dwyer and Honigberg [126] for *Dientamoeba fragilis* with DMSO (see Table 8.5). Since the tolerance to cryoprotectants increases rapidly as the temperature is lowered, the toxicity of these compounds at their initial equilibration temperature determines the maximum effective concentration.

Cryoprotectants have played a considerable role in improving the preservation of parasites at low temperatures, although their presence is not always essential to survival; none of these compounds was of course available for use by the earlier workers before 1949, and certain nematodes can be preserved well in their absence [145].

Acknowledgements

The author wishes to acknowledge support from the UNDP/World Bank/WHO Special Programme for Research and Training in Tropical Diseases.

References

1 Power, H. (1663–1664). Quoted by Keilin, D. (1959). *Proceedings of the Royal Society B*, **150**, 149
2 Laveran, A. and Mesnil, F. (1904). *Trypanosomes et Trypanosomiases.* Paris: Masson et Cie
3 Smith, A.U. (1961). *Biological Effects of Freezing and Supercooling.* London: Edward Arnold (Publishers) Ltd
4 Lumsden, W.H.R. (1972). *International Journal for Parasitology* **2**, 327
5 Lumsden, W.H.R. and Hardy, G.J.C. (1965). *Nature* **205**, 1032
6 Diamond, L.S. (1964). *Cryobiology* **1**, 95
7 Muhlpfordt, H. (1960). *Zeitschrift für Tropenmedizin und Parasitologie* **11**, 481
8 Dalgliesh, R.J. (1972). *Australian Veterinary Journal* **48**, 233
9 World Federation for Culture Collections of the International Association of Microbiological Societies (1972). *World Directory of Collections of Cultures of Microorganisms.* New York and Chichester: John Wiley and Sons Inc
10 Coggeshall, L.T. (1939). *Proceedings of the Society for Experimental Biology and Medicine* **42**, 499
11 Barnett, S.F. (1964). *The Veterinary Record* **76**, 4
12 Taylor, M.G., James, E.R., Bickle, Q.D., Doenhoff, M.J., Nelson, G.S., Hussein, M.F. and Bushara, H.O. (1977). *INSERM* **72**, 291
13 James, E.R. and Farrant, J. (1977). *Transactions of the Royal Society of Tropical Medicine and Hygiene* **71**, 498
14 Robson, J., Pedersen, V., Odeke, G.M., Kamya, E.P. and Brown, C.G.D. (1977). *Tropical Animal Health and Production* **9**, 219
15 Bickle, Q.D. and James, E.R. (1978). *Transactions of the Royal Society of Tropical Medicine and Hygiene* **72**, 677
16 James, E.R., Taylor, M.G., Dobinson, A.R., Andrews, B.J. and Bickle, Q.D. In preparation
17 Joyner, L.P. (1954). *The Veterinary Record* **66**, 727
18 Wilson, R.J.M., Farrant, J. and Walter, C.A. (1977). *Bulletin of the World Health Organization* **55**, 309
19 Smith, H.J. (1975). *Canadian Journal of Comparative Medicine* **39**, 316
20 Horsfall, F.L. (1940). *Journal of Bacteriology* **40**, 559
21 Manwell, R.E. and Edgett, R. (1943). *American Journal of Tropical Medicine* **23**, 551
22 Archetti, I. (1941). *Revista di Parassitologia* **5**, 247
23 Manwell, R.D. and Jeffery, G. (1942). *Proceedings of the Society of Experimental Biology and Medicine* **50**, 222
24 Wulfson, F. (1945). *American Journal of Hygiene* **42**, 155
25 Saunders, G.M. and Scott, V. (1947). *Science* **106**, 300
26 Weinman, D. and McAllister, J. (1947). *American Journal of Hygiene* **45**, 102
27 Saunders, G.M., Talmage, D.W. and Scott, V. (1948). *Journal of Laboratory and Clinical Medicine* **33**, 1579

28 Molinari, V. (1960). *Transactions of the Royal Society of Tropical Medicine and Hygiene* **54**, 288
29 Molinari, V. and Tabibzadeh, I. (1961). *Transactions of the Royal Society of Tropical Medicine and Hygiene* **55**, 10
30 Jeffery, G.M. (1962). *Journal of Parasitology* **48**, 601
31 Allen, E.G. (1970). *Applied Microbiology* **20**, 224
32 Pavanand, K., Permpanich, B., Chuanak, N. and Sookto, P. (1974). *Journal of Parasitology* **60**, 537
33 Collins, W.E., Jeffery, G.M., Chester, L.E. and Westbrook, J.R. (1963). *Journal of Parasitology* **49**, 524
34 Dalgliesh, R.J. (1971). *Research in Veterinary Science* **12**, 469
35 Pipano, E. and Senft, Z. (1966). *Journal of Protozoology* **13**, (suppl.), 34
36 Strome, C.P.A., Tubergen, T.A., Leef, J.L. and Beaudoin, R.L. (1977). *Bulletin of the World Health Organization* **55**, 305
37 Booden, T. and Geiman, Q.M. (1973). *Experimental Parasitology* **33**, 495
38 Mieth, H. (1966). *Zeitschrift für Tropenmedizin und Parasitologie* **17**, 103
39 Molinari, V. (1961). *Journal of Tropical Medicine and Hygiene* **64**, 225
40 Warhurst, D.C. (1966). *Transactions of the Royal Society of Tropical Medicine and Hygiene* **60**, 6
41 Schneider, M.D., Johnson, D.L. and Shefner, A.M. (1968). *Applied Microbiology* **16**, 1422
42 Schneider, M.D. and Seal, N. (1973). *Cryobiology* **10**, 67
43 Frerichs, W.M., Johnson, A.J. and Holbrook, A.A. (1968). *Journal of Parasitology* **54**, 451
44 Overdulve, J.P. and Antonisse, H.W. (1970). *Experimental Parasitology* **27**, 323
45 Jeffery, G.M. and Rendtorff, R.C. (1955). *Experimental Parasitology* **4**, 445
46 Kocan, R.M., Kelker, N.E. and Clark, D.T. (1967). *Journal of Parasitology* **14**, 724
47 Ward, R.A. (1962). *Mosquito News* **22**, 306
48 Weathersby, A.B. and McCall, J.W. (1967). *Journal of Parasitology* **53**, 638
49 Bafort, J. (1968). *Annals of Tropical Medicine and Parasitology* **62**, 301
50 Potgeiter, F.T. and VanVuuren, A.S. (1974). *Ondersterpoort Journal of Veterinary Research* **41**, 79
51 Morzaria, S.P., Brocklesby, D.W., Harradine, D.L. and Luther, P.D. (1977). *Research in Veterinary Science* **22**, 190
52 Dalgliesh, R.J. (1972). *Research in Veterinary Science* **13**, 540
53 Dalgliesh, R.J. and Mellors, L.T. (1974). *International Journal for Parasitology* **4**, 169
54 Dalgliesh, R.J., Swain, A.J. and Mellors, L.T. (1976). *Cryobiology* **13**, 631
55 Molinari, V. and Montezin, G. (1956). *Bulletin de la Société de Pathologie Exotique* **49**, 445
56 Molinari, V. and Montezin, G. (1956). *Bulletin de la Société de Pathologie Exotique* **49**, 651
57 Deschiens, R. and Molinari, V. (1963). *Annales de la Société Belge de Medicine Tropicale* **5**, 811
58 Levaditi, J.C. (1952). *Comptes Rendus de la Société de Biologie* **146**, 179
59 Berson, J-P. (1962). *Bulletin de la Société de Pathologie Exotique* **55**, 804
60 Polge, C. and Soltys, M.A. (1957). *Transactions of the Royal Society of Tropical Medicine and Hygiene* **51**, 519
61 Walker, P.J. and Ashwood-Smith, M.J. (1961). *Annals of Tropical Medicine and Parasitology* **55**, 93
62 Webber, W.A.F., Cunningham, M.P. and Lumsden, W.H.R. (1961). *East African Trypanosomiasis Research Organization Annual Report, Jan-Dec,* p. 10
63 Cunningham, M.P., VanHoeve, K. and Grainge, E.B. (1964). *East African Trypanosomiasis Research Organization Annual Report, July 1963-Dec. 1964,* p. 26
64 Cunningham, M.P., Lumsden, W.H.R., VanHoeve, K. and Webber, W.A.F. (1963). *Biochemical Journal* 89, 73
65 Cunningham, M.P. and Harley, J.M.B. (1966). *International Scientific Council for Trypanosomiasis Research 11th Meeting, Nairobi,* publication **100**, 187
66 Resseler, R., LeRay, D. and Goedvriend, J. (1965). *Tropical and Geographical Medicine* **17**, 359
67 Allain, D.S. (1964). *Journal of Parasitology* **50**, 604

68 Minter, D.M. and Goedbloed, E. (1970). *Transactions of the Royal Society of Tropical Medicine and Hygiene* **64**, 789
69 Martinez-Silva, R., Lopez, V.A. and Chiriboga, J. (1970). *Cryobiology* **6**, 364
70 Overdulve, J.P., Antonisse, H.W., Leeflang, P. and Zwart, D. (1970). *Tijdschrift voor Diergeneeskunde* **95**, 678
71 Dar, F.K., Ligthart, G.S. and Wilson, A.J. (1972). *Journal of Protozoology* **19**, 494
72 Taylor, B.J. (1972). *Cryobiology* **9**, 212
73 Diffley, P., Honigberg, B.M. and Mohn, F.A. (1976). *Journal of Parasitology* **62**, 136
74 Raether, W. and Seidenath, H. (1972). *Zeitschrift für Tropenmedizin und Parasitologie* **23**, 428
75 Minter, D.M. and Goedbloed, E. (1971). *Transactions of the Royal Soceity of Tropical Medicine and Hygiene* **65**, 175
76 Cunningham, M.P. (1960). *East African Trypanosomiasis Research Organization Annual Report, Jan–Dec,* p. 6
77 Cunningham, M.P., Lumsden, W.H.R. and Webber, W.A.F. (1963). *Experimental Parasitology* **14**, 280
78 Lumsden, W.H.R., Herbert, W.J. and McNeillage, G.J.C. (1973). *Techniques with Trypanosomes.* Edinburgh and London: Churchill Livingstone
79 Foner, A. (1963). *Israel Journal of Experimental Medicine* **11**, 69
80 Most, H., Alger, N. and Yoeli, M. (1964). *Nature* **201**, 735
81 Callow, L.L. and Farrant, J. (1973). *International Journal for Parasitology* **3**, 77
82 Levine, N.D. and Marquardt, W.C. (1955). *Journal of Protozoology* **2**, 100
83 Lumsden, W.H.R., Robertson, D.H.H. and McNeillage, G.J.C. (1966). *British Journal of Venereal Diseases* **42**, 145
84 Reusse, U. (1956). *Zeitschrift für Tropenmedizin und Parasitologie* **7**, 99
85 Levine, N.D., Mizell, M. and Houlahan, D.A. (1958). *Experimental Parasitology* **7**, 236
86 Levine, N.D., Andersen, F.L., Losch, M.B., Notzold, R.A. and Mehra, K.N. (1962). *Journal of Protozoology* **9**, 347
87 Lindgren, R.D. and Ivey, M.H. (1964). *Journal of Parasitology* **50**, 226
88 Muller, W. (1966). *Zeitschrift für Tropenmedizin und Parasitologie* **17**, 100
89 Bosch, I. and Frank, W. (1972). *Zeitschrift für Parasitenkunde* **38**, 303
90 McEntegart, M.G. (1954). *American Journal of Hygiene* **52**, 545
91 Joyner, L.P. and Bennett, G.H. (1956). *American Journal of Hygiene* **54**, 335
92 Diamond, L.S., Bartgis, I.L. and Reardon, L.V. (1965). *Cryobiology* **1**, 295
93 Jeffries, L. and Harris, M. (1967). *Parasitology* **57**, 321
94 Honigberg, B.M., Stabler, R.M., Livingstone, M.C. and Kulda, J. (1970). *Journal of Parasitology* **54**, 701
95 Ivey, M.H. (1975). *Journal of Parasitology* **61**, 1101
96 Levine, N.D., McCaul, W.E. and Mizell, M. (1957). *Journal of Protozoology* **4**, (Suppl.), 5
97 Levine, N.D., McCaul, W.E. and Mizell, M. (1959). *Journal of Protozoology* **6**, 116
98 Yilmaz, S.M. and Hopkins, S.H. (1972). *Journal of Parasitology* **58**, 938
99 Frenkel, J.K. and Dubey, J.P. (1973). *Journal of Parasitology* **59**, 587
100 Hellesnes, I. and Mohn, S.F. (1977). *Zentralblat für Bakteriologie und Hygiene* **238**, 143
101 Work, K. (1968). *Acta Pathologica et Microbiologica Scandinavia* **73**, 85
102 Dubey, J.P. (1974). *Journal of the American Veterinary Medical Association* **165**, 534
103 Dumas, N. (1974). *Annales de Parasitologie* **49**, 1
104 Weinman, D. and Chandler, A.H. (1954). *Proceedings of the Society of Experimental Biology and Medicine* **87**, 211
105 Manwell, R.D., Coulston, F., Binckley, E.C. and Jones, V.P. (1945). *Journal of Infectious Diseases* **76**, 1
106 Eyles, D.E., Coleman, N. and Cavanaugh, D.J. (1956). *Journal of Parasitology* **42**, 408
107 Bollinger, R.O., Mussalam, N. and Stulberg, C.S. (1974). *Journal of Parasitology* **60**, 368
108 Paine, G.D. and Meyer, R.C. (1969). *Cryobiology* **5**, 270
109 Kasprzak, W. and Rydzewski, A. (1970). *Zeitschrift für Tropenmedizin und Parasitologie* **21**, 198
110 Mackie, M.J. (1972). *Journal of Parasitology* **58**, 846

111 Janitschke, K. and Jorren, H.R. (1975). *Tropenmedizin und Parasitologie* **26**, 307
112 Norton, C.C. and Joyner, L.P. (1968). *Research in Veterinary Science* 9, 598
113 Norton, C.C., Pout, D.D. and Joyner, L.P. (1968). *Folia Parasitologica* **15**, 203
114 Doran, D.J. and Vetterling, J.M. (1968). *Nature* **217**, 1262
115 Doran, D.J. and Vetterling, J.M. (1969). *Proceedings of the Helminthological Society of Washington* **36**, 30
116 Kouwenhoven, B. (1967). *Tijdschrift voor Diergeneeskunde* 92, 1639
117 Doran, D.J. (1969). *Journal of Parasitology* **55**, 1229
118 Pautrizel, R. and Carloz, L. (1952). *Comptes Rendues de la Société de Biologie* **146**, 89
119 Gehenio, P.M. and Luyet, B.J. (1953). *Biodynamica* 7, 175
120 Luyet, B.J. and Gehenio, P.M. (1954). *Journal of Protozoology* **1**, (Suppl.), 7
121 Fulton, J.D. and Smith, A.U. (1953). *Annals of Tropical Medicine and Parasitology* **47**, 240
122 Diamond, L.S., Meryman, H.T. and Kafig, E. (1961). *Journal of Parasitology* 47, (Suppl.), 28
123 Gordon, R.M., Graedel, S.K. and Stucki, W.P. (1969). *Journal of Parasitology* **55**, 1087
124 Neal, R.A., Latter, V.S. and Richards, W.H.G. (1974). *International Journal for Parasitology* **4**, 353
125 Raether, W. and Uphoff, M. (1976). *International Journal for Parasitology* **6**, 121
126 Dwyer, D.M. and Honigberg, B.M. (1971). *Journal of Parasitology* **57**, 190
127 Simione, F.P. and Daggett, P-M. (1976). *Journal of Parasitology* **62**, 49
128 Bemrick, W.J. (1961). *Journal of Parasitology* **47**, 573
129 Meyer, E.A. and Chadd, J.A. (1967). *Journal of Parasitology* **53**, 1108
130 VonBrand, T. and Morris, J.A. (1943). *Biodynamica* **4**, 75
131 Baker, D.W. (1939). *Cornell Veterinarian* **29**, 45
132 Kates, K.C. (1943). *Proceedings of the Helminthological Society of Washington* **10**, 23
133 Seghetti, L. (1948). *American Journal of Veterinary Research* 9, 52
134 Turner, J.H. (1953). *Journal of Parasitology* **39**, 589
135 Rose, J.H. (1956). *Journal of Comparative Pathology* **66**, 228
136 Kates, K.C. (1950). *Proceedings of the Helminthological Society of Washington* **17**, 39
137 Andersen, F.L. and Levine, N.D. (1968). *Journal of Parasitology* **54**, 117
138 Poole, J.B. (1956). *Canadian Journal of Comparative Pathology* **20**, 169
139 Parfitt, J.W. (1971). *Research in Veterinary Science* **12**, 488
140 Bemrick, W.J. (1978). *Cryobiology* **15**, 214
141 Isenstein, R.S. and Herlich, H. (1972). *Proceedings of the Helminthological Society of Washington* **39**, 140
142 Kelly, J.D. and Campbell, W.C. (1974). *International Journal for Parasitology* **4**, 173
143 Campbell, W.C., Blair, L.S. and Egerton, J.R. (1972). *The Veterinary Record* **91**, 13
144 Campbell, W.C., Blair, L.S. and Egerton, J.R. (1973). *Journal of Parasitology* **59**, 425
145 Campbell, W.C. and Thompson, B.M. (1973). *Australian Veterinary Journal* **49**, 111
146 Kelly, J.D., Campbell, W.C. and Whitlock, H.V. (1976). *Australian Veterinary Journal* **52**, 141
147 Vetter, J.C.M. and Klaver-Wesseling, J.C.M. (1977). *Journal of Parasitology* **63**, 700
148 Miller, T.A. and Cunningham, M.P. (1965). *East African Trypanosomiasis Research Organization Annual Report*, p. 19
149 Mazzotti, L. (1953). *Revista del Instituto de Salubridad y Enfermedades Tropicales* **13**, 289
150 Beye, H.K. and Lawless, D.K. (1961). *Experimental Parasitology* **11**, 319
151 Mantovani, A. and Sulzer, A.J. (1967). *American Journal of Veterinary Research* **28**, 351
152 Restani, R. (1968). *Parassitologia* **10**, 75
153 Taylor, A.E.R. (1960). *Journal of Helminthology* **34**, 13
154 Obiamiwe, B.A. and Macdonald, W.W. (1971). *Annals of Tropical Medicine and Parasitology* **65**, 547
155 McCall, J.W., Jun, J. and Thompson, P.E. (1975). *Journal of Parasitology* **61**, 340
156 Bandyopadhyay, A.K. and Chowdhury, A.B. (1965). *Bulletin of the Calcutta Society of Tropical Medicine* **13**, 49
157 Muller, R. (1967). *Transactions of the Royal Society of Tropical Medicine and Hygiene* **61**, 451
158 Muller, R. (1970). *Nature* **226**, 662

159 Ham, P.J. (1977). *MSc Medical Parasitology dissertation,* London School of Hygiene and Tropical Medicine, London University
160 Raether, W., Schupp, E., Michel, R., Niemeitz, H. and Uphoff, M. (1977). *Zeitschrift für Parasitenkunde* **54**, 149
161 Bemrick, W.J., Buchli, B.L. and Griffiths, H.J. (1965). *Journal of Parasitology* **51**, 954
162 Ogunba, E.O. (1969). *Journal of Parasitology* **55**, 1101
163 Ham, P.J., James, E.R. and Bianco, A.E. (1979). *Experimental Parasitology* **47**, 384
164 Boray, J.C. and Enigk, K. (1964). *Zeitschrift für Tropenmedizin und Parasitologie* **15**, 324
165 Taylor, E.L. (1949). *Report of the 14th International Veterinary Congress, London,* p. 81
166 James, E.R. and Farrant, J. (1976). *Cryobiology* **13**, 625
167 DeConinck, L.A.P. (1951). *Biodynamica* **7**, 77
168 Luyet, B.J. and Hartung, M.C. (1941). *Biodynamica* **3**, 353
169 Gartner, L. and Theile, M. (1970). *Zeitschrift für Parasitenkunde* **35**, 55
170 Walter, C.A., Knight, S.C. and Farrant, J. (1975). *Cryobiology* **12**, 103
171 Miyata, A. (1976). *Tropical Medicine* **18**, 143
172 Asahina, E. (1959). *Nature* **184**, 1003
173 Cunningham, M.P., VanHoeve, K. and Grainge, E.B. (1964). *East African Trypanosomiasis Research Organization Annual Report, July 1963-Dec 1964,* p. 29
174 Vargues, R. (1952). *Bulletin de la Société de Pathologie Exotique* **45**, 456
175 Diggs, C., Joseph, K., Flemmings, B., Snodgrass, R. and Hines, F. (1975). *American Journal of Tropical Medicine and Hygiene* **24**, 760
176 Levine, N.D. and Andersen, F.L. (1966). *Journal of Protozoology* **13**, 199
177 McEntegert, M.G. (1959). *Nature* **183**, 270
178 Gitatha, S.K. (1968). *East African Trypanosomiasis Research Organization Annual Report, 1968,* p. 14
179 O'Connell, K.M., Hunter, S.H., Fromentin, H., Frank, O. and Baker, H. (1968). *Journal of Protozoology* **15**, 719
180 Polge, C., Smith, A.U. and Parkes, A.S. (1949). *Nature* **164**, 666
181 Scharf, J. (1954). *Biodynamica* **7**, 225
182 James, E.R. (1977). *Transactions of the Royal Society of Tropical Medicine and Hygiene* **71**, 288
183 Lovelock, J.E. (1954). *Biochemical Journal* **56**, 265
184 Ashwood-Smith, M.J. and Lough, P. (1975). *Cryobiology* **12**, 517
185 Polge, C. and Soltys, M.A. (1960). In *Recent Research in Freezing and Drying,* p. 87. A.S. Parkes and A.U. Smith (eds.). Oxford: Blackwell Scientific Publications
186 Chinery, W.A. and Atiemo, N.A. (1973). *Ghana Medical Journal* **12**, 203
187 Miyata, A. (1973). *Tropical Medicine* **15**, 204
188 Miyata, A. (1975). *Tropical Medicine* **17**, 55
189 Murrell, K.D., Stirewalt, M.A. and Lewis, F.A. (1979). *Experimental Parasitology,* in press
190 Phillips, R.S. and Wilson, R.J.M. (1978). *Transactions of the Royal Society of Tropical Medicine and Hygiene* 72, 643
191 Leef, J.L., Strome, C.P.A. and Beaudoin, R.L. (1978). *Cryobiology* **15**, 681
192 Granath, W.O. and Huizinga, H.W. (1978). *Experimental Parasitology* **46**, 239
193 Stirewalt, M.A., Lewis, F.A. and Murrell, K.O. (1979). *Experimental Parasitology,* in press
194 Miyata, A. (1975). *Japanese Journal of Tropical Medicine and Hygiene* **3**, 161

CHAPTER NINE

Insects and their Cells

R A Ring

Introduction

In this chapter an attempt will be made to appraise the current status of low temperature preservation in the field of entomology and make some comparisons with similar situations in other biological systems. An attempt will be made to emphasise the practical aspects of the subject and to standardise some of the terminology that has been used in the past, often in a confusing way.

The development of insect cell and tissue culture techniques has occurred only during the last two decades, the slow beginning compared to vertebrate techniques being attributed mainly to the lack of a well defined, artificial culture medium which would permit long term growth and survival of the cells [1]. Despite advances made in this area in recent years, relatively little is known about the survival of isolated insect cells and tissues at low temperatures. Most of our knowledge in insect cryobiology lies within the context of the whole organism and some excellent reviews of various aspects of this topic have been published, notably by Salt [2]; Asahina [3, 4]; Baust [5]; Lozina-Lozinskii [6]; and Danks [7]. This chapter will deal with both subjects, first insect cells and some of their subcellular components and second with the insect as a whole organism. The literature in these subjects is now quite voluminous, so only references to the most recent and pertinent findings will be cited.

Effects of low temperature on insect cells and cellular components

Cell and tissue cultures

The main objective of research in insect cell culture has been primarily the culture of cells for the propagation or investigation of viruses associated with arthropods. For instance, lepidopteran cell culture has been designed for the production of viruses as potential agents in biological control programmes against defoliating caterpillars; leaf-hopper and aphid cells for the study of viruses that are plant pathogens; and dipteran cells, particularly mosquitoes, for the cultivation of arboviruses [1], many of which are important human and livestock pathogens. There have been a few attempts, however, to preserve cells and tissues in the living condition for long periods at low temperatures. *Aedes aegypti* cells grown as suspension cultures and *Aedes albopictus* cells grown as monolayer cultures have been frozen and stored in liquid nitrogen with very little loss of viability [8]. Cultures frozen in a medium containing 8% DMSO yielded 93% viable *A. aegypti* cells after storage in liquid nitrogen (−196°C) for one month, and *A. albopictus* cells preserved in a similar manner produced a full cell sheet within 4–6 days of thawing and were successfully carried through three transfers with no apparent loss of viability. Eide *et al* [9] have examined the effects of various rates of cooling and several cryoprotective agents on Grace's *Antheraea* cell line and on dispersed embryonic cells from *Musca domestica.* A regimen including 2–3 hours' acclimation of the cells to the growth medium followed by slow cooling gave the best survival. The cells were cooled slowly to −30°C and either stored at that temperature or placed in liquid nitrogen. After storage for 24 h no differences were noted between these temperatures. After long term storage, however, neither *Antheraea* cells nor *M. domestica* embryonic cells showed good survival at −30°C. Embryonic cells also exhibited very poor survival in liquid nitrogen, survival decreasing with time in storage, whereas in the *Antheraea* cell line a 60 day storage in liquid nitrogen did not affect cell multiplication after thawing. The best cryoprotective agent for the cell line was a poor agent for the embryonic cells. As is the case for whole insects, best cryoprotection is afforded by a combination of several agents.

Lozina-Lozinskii [6] describes some of the phenomena associated with freezing and thawing in isolated salivary gland cells of the larvae of the corn borer, *Pyrausta nubilalis.* The degree of supercooling of cell suspensions and pieces of salivary gland tissue is almost identical to that of the intact larvae (−15 to −24°C). At the supercooling point ice formation occurs in the nuclei of the cells giving them a reticulate or honeycomb appearance. In salivary gland cells from cold tolerant larvae these changes in nuclear structure are reversible on thawing, whereas in cells from non-tolerant larvae they are irreversible. During the period when cold tolerance of the caterpillars is at its maximum (January/February), excised salivary gland cells can survive freezing for considerable periods at temperatures of −78 to −196°C. Irreversible changes in the cell nuclei, however, occur more often at −196°C than at −78°C.

Neurosecretory cell activity

There is evidence to suggest that the effect of low temperature on insect growth and development is not solely a direct one on the rate of metabolism but also an indirect one mediated by the endocrine system.

In the larvae of *Morimus funereus* (Cerambycidae) which feed on oak, winter larvae maintained at −1°C for 15 days have more active neurosecretory cells than those maintained at 23°C [10]. The neurosecretory cells involved are of Type A in the medial group of the protocerebrum and the ventromedial cells of the suboesophageal ganglion, and the morphological parameters used for measurement were the quantity of neurosecretory material, the degree of agglomeration of the granules, the nucleus: cytoplasm ratio, visibility of the nucleolus and size of the cell. Changes in the level of activity of the neurosecretory cells could be correlated with midgut proteolytic activity. The regulation of midgut proteolytic enzyme activity by the endocrine system was first demonstrated in the blowfly *Calliphora erythrocephala* by Thomsen and Moller [11], but this is the first observation on the effect of low temperature on this physiological process. Ivanović *et al* [10] conclude that the midgut protease in *M. funereus* larvae may be considered as a possible temperature effector, the activity of which is regulated by hormonal signals from the neurosecretory cells. On the basis of these changes the insect is enabled to maintain 'vectoric homoeostasis' of metabolic functions [12], temporarily compensating a negative influence of thermal energy. Data of other authors tend to confirm that the level of protein synthesis is elevated in certain animals and tissues following low temperature acclimation (e.g. Applebaum *et al* [13]; Das and Prosser [14]; Hoschemeyer [15]).

In insects, one of the most spectacular physiological effects of exposure to low winter temperatures is in the hormonal control of diapause termination. Lees [16] defines the state of diapause as a spontaneous arrest of growth, development or reproduction which can be regarded as an endocrine deficiency syndrome. The mode of action at the cellular level of low temperature in accelerating diapause termination is controversial, but Van der Kloot [17] and Williams [18] have suggested that it acts by promoting the synthesis of a precursor for acetylcholine production whose absence during diapause results in the electrical silence of the brain and ensures that the neurosecretory cells which produce the 'brain hormone' are inactive. Although this provides an elegant explanation for the effect of low temperature on diapause termination, other workers have been unable to repeat the experiment with the silkmoth, *Antheraea pernyi* (Tyshtchenko and Mandelstam [19]) or with other insects (Smallman and Mansingh [20]; Grzelak *et al* [21]). Way [22] has demonstrated that a cold shock at −20°C to −24°C will terminate diapause abruptly if administered to the late diapause stage of the eggs of the wheat bulb fly (*Leptohylemya coarctata*). At −24°C the rate of diapause termination is about 180 times faster than at −6°C and about 80 times faster than at 3°C. The supercooling points of diapause eggs ranged from −25.3°C to −30.0°C, therefore freezing did not occur. He suggests that a neurohormone may accumulate at higher temperatures during the early stages of hormone diapause but that its actual release is greatly facilitated by short exposures to low temperatures near the supercooling point.

Ring [23] has shown that many hormonally related events such as DNA, RNA and protein synthesis are reduced during diapause at low temperatures (3°C) in the larva of the blowfly *Lucilia sericata,* reflecting the fact that diapause is a stage during which synthetic and mitotic activities are minimal. Recent evidence [24] with the diapause eggs of the silkmoth, *Bombyx mori*, indicates that chilling at 5°C which terminates diapause progressively, accelerates the reconversion of sorbitol to glycogen (see discussion of enzymes below). Some insects will remain in diapause indefinitely and ultimately die unless they experience a period of low temperature [25] and Schneiderman and Horwitz [26] have demonstrated that in the parasitic wasp *Nasonia (Mormoniella) vitripennis* the accelerating effect of low temperature on diapause termination is reversed on exposure to higher temperatures.

Insect enzymes

From the foregoing discussion it can be concluded that low temperatures have profound effects on various hormone-producing cells in insects. Since hormones have a direct or indirect effect on gene expression [27, 28] and hence on protein synthesis, it could also be inferred that low temperatures also play an important role in the activation or inhibition of enzyme production in insects. In *Morimus funereus,* the highest amylolytic as well as proteolytic activity was found in larvae at −1°C and 8°C and the lowest in those at 23°C. The increased amylolytic activity in winter larvae at all experimental temperatures suggests that there is a predominance of carbohydrate metabolism during this season. In recent years there has been a growing body of evidence to indicate that such is the case in other species of insects (see references in Danks [7]). Close correlation exists between the increase in cold hardiness of diapausing and non-diapausing insects under the influence of low temperatures and the increase in production of glycerol and other carbohydrates. Ziegler and Wyatt [29] have reported the activation of glycogen phosphorylase, an enzyme affecting glycerol production, at 0°–4°C in the diapausing pupae of the silkmoth, *Hyalophora cecropia.* Glycerol production is stimulated by low temperatures in other insects including adult carpenter ants, *Camponotus pennsylvanicus* [30], larvae of the mountain pine beetle, *Dendroctonus monticolae* [31], larvae of the leafcutter bee, *Megachile relativa* [32], caterpillars of *Isia isabella* [33], the adult carabid beetle, *Pterostichus brevicornis* [34], the larvae of the viceroy butterfly, *Limenitis archippus* [35], adult cerambycid beetles, *Rhagium inquisitor* [36], and adult blowflies, *Protophormia terranovae* [37]. Accumulation of glycerol can continue even when the insect is frozen, as in the case of the arctic carabid *Pterostichus brevicornis* [38]. Furthermore, Morrissey and Baust [39] have reported that trehalose production continues at −30°C in the gall fly *Eurosta solidagensis,* and Ring [40] has demonstrated that the polyhydric alcohols, glycerol and sorbitol, and the sugars trehalose and glucose can be synthesised or metabolised at subzero temperatures in the larvae of the bark beetle, *Scolytus ratzeburgi.* It has been known for some time that the insect enzyme *trehalase* is apparently tolerant to cold since it can be stored at −12°C for 2 months without loss of activity [41]. *In vitro* studies in other biological systems have shown that freezing may accelerate certain enzymatic reactions, and that many enzymes in the purified state, such as zymase, peroxidase, tyrosinase and catalase, exhibit an increase in activity on freezing in comparison to their activity in a supercooled medium at the same temperature (see Grant and Alburn [42]; Grant [43]; Doscherholmen and Silvis [44]; and Lozina-Lozinskii [6]).

Phosphorylase and aldolase activity have been found in the overwintering prepupae of the slug caterpillar, *Monema flavescens* [45], and Lozina-Lozinskii [6] has shown that in the overwintering caterpillars of the corn borer, *Pyrausta nubilalis,* a sharp lowering of the ambient temperature provokes a fall in the activity of the enzyme system of oxidative phosphorylation, but subsequently, as a result of the adaptation of the caterpillars to low temperature, the coupling of respiration and phosphorylation is restored.

Insect mitochondrial preparations

It is generally accepted that the rapid ageing of mitochondria places severe restraints on the work of researchers, so that mitochondria have to be extracted daily from fresh tissues and used immediately [46]. Even when stored for a few hours at temperatures around 0°C they become functionally unstable, and it is known that mitochondria prepared from insect tissues are even less stable than those of other animals or plants. It is assumed that inactivation is brought about by physicochemical changes in the mitochondrial enzymes, but that some protective function is afforded by agents such as 10% glycerol or an isotonic sugar solution when isolated mitochondria are slowly frozen to −76°C [6]. Abo-Khatwa [46] has described some practical freezing methods for preserving mitochondria isolated from the fat body of the viviparous cockroach *Nauphoeta cinerea.* The insect fat body can be considered as functionally similar to vertebrate liver and cytologically similar to brown fat tissue. DMSO (10% v/v) was chosen as the cryoprotectant since it can preserve energy-linked functions of rat liver and plant mitochondria under freeze-storage conditions. With the insect mitochondria DMSO was found to be ineffective as a cryoprotectant at 0–2°C, a significant aid at −20°C and completely redundant at −196°C. This is the first report of the latter phenomenon, the extreme stability of mitochondria to freezing at liquid nitrogen temperature for at least 3 weeks in the absence of DMSO. Freezing in liquid nitrogen was also found to be satisfactory for the preservation of mitochondrial coupling activities of the ovaries of *Nauphoeta cinerea* and queen termites *(Macrotermes subhyalinus)* as well as for the flight muscles of tsetse flies *(Glossina morsitans)* and winged termites *(Odontotermes spp).* However, the presence of DMSO was essential in preserving mitochondria isolated from insect flight muscles.

In the fat body mitochondria stored at −20°C and the flight muscle mitochondria stored at −196°C, the stabilising effect of DMSO has been attributed to its role as an effective agent in preventing freezing damage caused by rupture of mitochondrial membranes to which the enzymes are linked [47, 48], loss of soluble enzymes, and an increase in latent ATPase activity (see [46] and [49]).

EDTA was found to be an essential ingredient of the mitochondrial isolation medium and, therefore, for all storage regimens. Abo-Khatwa [46] believes that EDTA exerts its stabilising effect via its ability to inhibit latent ATPase activity. The difference in the stability of insect fat body mitochondria against freezing when compared with mammalian liver and plant mitochondria can be attributed to structural differences in the composition of the membrane systems, particularly the degree of unsaturation and fluidity of the membrane lipids.

Effects of low temperatures on insects

There seems little doubt that there is a close connection between the cold tolerance (cold hardiness) of an insect and the climatic and ecological conditions, including microhabitat, under which it lives. Corroborative evidence for such an assumption, however, is difficult to produce, since all the relevant parameters of low temperature survival have rarely been surveyed systematically in a single species [7]. Normally, the term *low temperature* in insect cryobiology refers to temperatures ranging from a few degrees above 0°C to about −75°C, close to the lowest naturally occurring temperature recorded on this planet. Insects are among the few permanent inhabitants of the south polar regions [50] where such temperatures prevail. The majority of cold tolerant insects are found in temperate or polar regions or at high altitudes although this is not invariably true. Some insects from tropical and subtropical areas have considerable supercooling potential (one aspect of cold tolerance) [51], and the renowned African chironomid midge, *Polypedilum vanderplanki,* can survive immersion in liquid air (−190°C) and liquid helium (−270°C) [52, 53]. Even in temperate, cold-hardy species, Salt [54] states that because the overwintering stage happens to be cold tolerant there is no reason to assume that its summer form is not.

Two major methods are employed by cold tolerant insects. In the first category the insect exhibits extensive supercooling ability, but is susceptible to freezing and, therefore, cannot withstand the deleterious effects of conversion of water to ice in the tissues or body fluids. This group is called *frost susceptible* (or freezing susceptible), where frost is defined as the *act or process of freezing.* The second category is called *frost tolerant* (or freezing tolerant), and can survive the formation of ice in the tissues, usually in the extracellular tissue fluids, but frequently exhibits poor supercooling ability.

Factors affecting survival of frost susceptible species (see Figure 9.1)

Acclimation: Acclimation (or acclimatisation or hardening) has been described by Salt [2] as all of the preparations that are made by the insect prior to hibernation or during the prediapause period. In cold tolerant insects the adaptations that are linked to increased super-

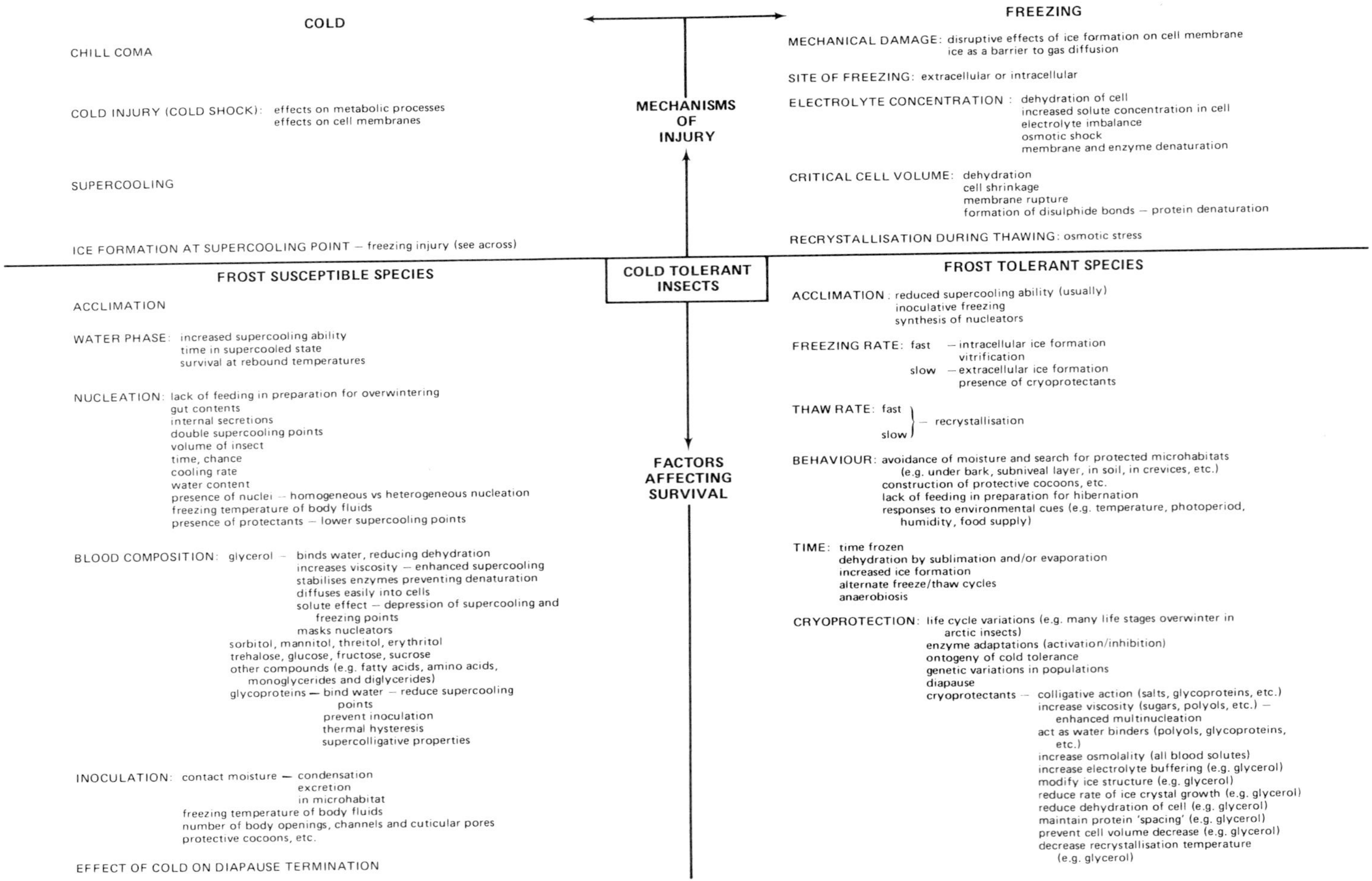

Figure 9.1 Chart showing the main features affecting survival and mortality in

cooling potential and survival at subzero temperatures are mainly metabolic and behavioural reorganisations that are triggered by environmental factors such as temperature, photoperiod, food supply etc. There are many indications that cold tolerance in insects increases under the influence of decreasing temperature, with the process of acclimation being more effective at exposures to temperatures just above zero (3–5°C) rather than at higher or subzero temperatures. The benefits of cold acclimation depend on the future experience of the insect. It is utilised only if the insect is exposed to a lower temperature, and it is measured in terms of advantage over a less acclimated individual or over itself earlier [2].

Supercooling: The majority of insect species studied so far are in the frost susceptible category. Supercooling, which can be defined as the extension of the liquid phase below the equilibrium freezing point of the tissue fluids, is extensive with supercooling points reaching levels of −46°C in the larvae of a beetle, *Pytho sp* from the Canadian Rockies (Ring, unpublished data), −50°C in the eggs of the lepidopteran, *Zeiraphera diniana* [55], −53°C in the larvae of the bark beetle, *Scolytus multistriatus* [6] and −55°C in the larvae of the gall fly *Eurosta solidagensis* [56]. These are considerably lower than the theoretical limit of supercooling of maximally purified water which lies in the region of −40 to −41°C. However, it must be remembered that insects which supercool to temperatures below −40°C have all been found to have low freezing points due to high solute concentrations [2, 57]. For instance, the larva of *Bracon cephi* which supercools to −45°C has a melting point of about −15°C, and thus supercools only about 30°C [58]. In these species the supercooling point represents the lethal limit of exposure to low temperature.

Relationships between supercooling potential and presence of haemolymph solutes such as glycerol, etc. have not yet been clarified. Some insects supercool in the absence of identifiable substances such as glycerol, whereas others do not supercool to any great extent even in the presence of considerable quantities of glycerol [2, 31, 40].

The supercooled state *per se* does not seem to be harmful to the insect [2], and Lozina-Lozinskii [6] has shown that the homogenised tissues of insects supercool to approximately the same degree as the intact tissues and only somewhat less than the whole organism. Similarly, George *et al* [59] found that in plant tissues, living and non-living flower buds of the azalea had the same freezing temperature.

Furthermore, insects may freeze at temperatures above their supercooling points since Salt [2] reports that supercooling is a random

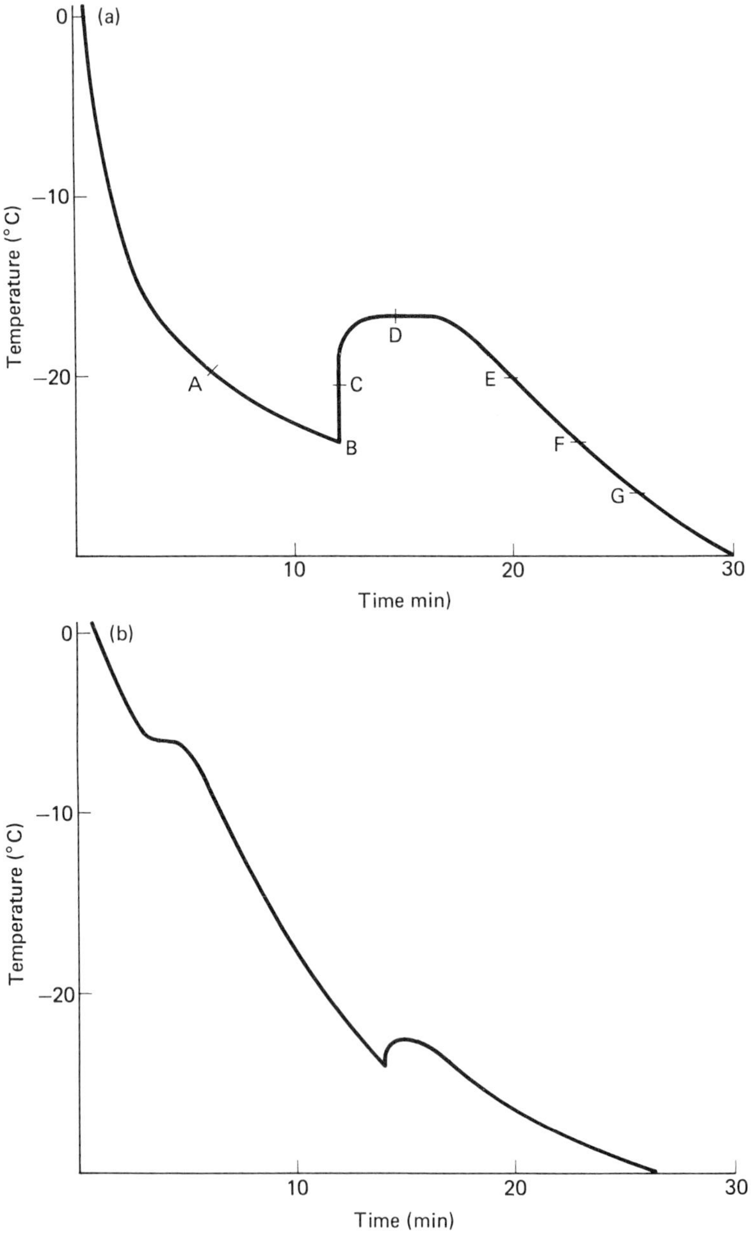

Figure 9.2 Typical cooling curves in frost susceptible insects: (a) single supercooling point e.g. *Scolytus ratzeburgi* [40] ; (b) double supercooling points e.g. arctic carabid beetles (Ring, unpublished)

event and that with prolonged supercooling the probability of spontaneous nucleation increases. Steponkus (personal communication), however, reports that in plant cell chloroplasts there is an overall temperature effect as well as a freezing effect, which results in a loss of ATPase activity in chloroplasts even in the supercooled state.

Nucleation: Salt [2, 60–63] has already given excellent accounts of nucleation theory as it applies to insect survival. At temperatures below zero the molecular arrangement of water becomes progressively more ice-like, and freezing is initiated when molecular aggregations reach a certain critical size. This is termed homogeneous nucleation when only water molecules are involved. However, heterogeneous nucleation, which involves ice crystal formation around a foreign surface, is more likely to occur in insects, particularly when gut contents and/or surface contaminants are present. Gut contents gradually lose their nucleating qualities as digestion proceeds, so that in an overwintering diapause stage this possible site of ice formation is obviated. Ring [64] comments that the phenomenon of diapause contributes to the success of overwintering temperate species because it is normally associated with lack of feeding and, therefore, the absence of nucleators in the gut, dehydration, decrease in secretory and synthetic activity, and increase in synthesis of cryoprotectants. Also, Mansingh [65] states that the phenological and biochemical functions of diapause cannot be achieved unless the diapause individual is also cold tolerant.

Ohyama and Asahina [66], using the technique of differential thermal analysis, have demonstrated that insects may have two separate nucleating events. The first is at a relatively high subzero temperature (approximately −8°C) when the contents of the foregut freeze, and the second is at a much lower temperature (approximately −20°C) when the entire insect freezes solid. The first freezing, due to heterogeneous nucleation in the gut, was shown to be innocuous, whereas the second freezing, perhaps due to homogeneous nucleation in the haemolymph, made the insect hard and brittle, resulting in fatal injury. A similar phenomenon of double supercooling points has been noted during the cooling curves of adult arctic carabid beetles (Ring, unpublished data) (Figure 9.2(b)).

Several other factors are involved in determining the temperature at which nucleation occurs in insects, including probability theory [2, 63]. Under natural conditions great fluctuations in temperature may occur, but Danks [7] points out that there have been few attempts to relate laboratory data to exposures to natural temperatures which would indicate the actual duration of supercooling experienced by insects in the field. Of course, as the temperature is lowered and the exposure to low temperatures is lengthened the probability of nucleation increases. The relationship between probability of nucleation and time is not linear, so that mean values are not reliable when comparing the probabilities of freezing at different temperatures or in different species or populations [2]. Instead, the median time value, which is the time required to freeze half of a given uniform population (FT_{50}), is used. The nucleators within the tissues of an insect or in its gut contents vary widely in efficiency, and the most effective nucleator will determine the supercooling point under the experimental conditions employed. Furthermore, a nucleating agent will initiate ice crystal formation at different temperatures depending on whether it is inside a mass of water or in its surface layer, and on the influence of surface forces on it [2]. Although supercooling points in themselves have no validity as absolute values in determining whether or not insects will survive long exposures in the field to low temperatures above their supercooling points, they do offer a convenient means of comparing cold tolerance between groups or populations or species, provided the methods used in each experiment are uniform. A standard cooling rate of −1°C to −2°C per minute is adopted and care must be taken in selecting only insects of a similar stage, condition, background, etc. to be included in any single experimental group [2, 67].

Apart from cooling rate and time of exposure, other factors which may affect nucleation temperatures are water content, the overall volume of the insect, contact with surface moisture, and concentrations and types of cryoprotectants. Recent evidence [68, 69] indicates that some species of frost tolerant insects actively synthesise a nucleator substance in preparation for overwintering. According to these authors the presence of a nucleating agent in the haemolymph ensures extracellular freezing at high subzero temperatures, thus reducing the possibility of the occurrence of injurious intracellular ice formation. On the other hand, in some frost susceptible species, (the ladybird beetle, *Coleomegilla maculata* [70] and the larvae of the darkling beetle, *Meracantha contracta* [71, 72]), it appears that high molecular weight glycoproteins are produced which act as *antifreezes*. These depress the supercooling point of the frost susceptible insect by acting as maskers of nucleating agents and thus prevent lethal ice crystal formation.

Water content: The water content appears to be considerably reduced in overwintering stages of cold tolerant insects in preparation for hibernation. In species that overwinter in diapause this partial dehydration is often actively controlled and is an integral part of the physiology of diapause induction [65, 73]. This is enhanced by evacuation of the gut contents that normally occurs prior to diapause. Extreme dehydration, when the body water content decreases to 8% or less, occurs in some insects and they are very resistant to temperature extremes. Such desiccation tolerance is taken to its limits in the larvae of the chironomid *Polypedilum vanderplanki* which, in this condition, can be exposed to temperatures down to −270°C and up to 106°C and shows signs of life upon return to normal conditions [52]. A similar situation occurs in some arctic collembolans which, under laboratory conditions, can survive extremely dry conditions for prolonged periods of time (Ring, unpublished data). Also, Lozina-Lozinskii [6] reports that not a drop of liquid can be expressed from the larvae of bark beetles that overwinter in unprotected places; they evidently exist in a supercooled state and can endure temperatures below −40 or −50°C. In frost susceptible species, dehydration plays a twofold role, first by concentrating solutes and thereby reducing supercooling points, and second by reducing the temperature at which inoculative freezing might occur [74].

Inoculation: The waterproofing of the insect cuticle is less than perfect and inoculation by ice from the environment can often occur via contact moisture. This can be in the form of water in the overwintering hibernaculum or it can be produced by condensation of water vapour. It also can take place via wounds or excretory droplets that may remain in contact with the insect, and thus the chance of inoculation increases in proportion to the number of body openings, channels, and cuticular pores [74]. Stressful situations, such as rapid cooling, often lead to the insect ejecting excretory or regurgitated fluids, thus increasing the likelihood of freezing by inoculation. Successful inoculation will also depend on the freezing temperature of the body fluids.

Inoculation by surrounding ice can be avoided or reduced by selection of an appropriate overwintering site and by construction of protective cocoons or similar structures. Such cocoons can provide mechanical protection against surrounding ice or water, and maintain a layer of undisturbed air around the overwintering insect [7].

Blood composition: Extensive supercooling in frost susceptible insects usually occurs in the presence of significant amounts of one or more chemical components of the haemolymph and tissues. This is not invariably so, however, since many insects undergo a considerable degree of supercooling even in the absence of identifiable protective substances. The most common and abundant protective agent is glycerol, first discovered in insect tissues by Wyatt and Kalf [75]. It is a normal metabolite in insects, being an important intermediary in both carbohydrate and lipid metabolism, and is therefore readily made in the cell. Danks [7] notes that glycerol is not a sterile end product of metabolism but can be re-utilised efficiently when its protective role ends, and Ashwood-Smith [76] suggests that the fact that glycerol has antifreeze properties is only coincidental to its main function of being readily available as a source of energy when overwintering is completed.

The exact mode of action of glycerol as a cryoprotectant is as yet unknown, but it evidently plays several different roles and is involved in a multiplicity of interactions (see also discussion of cryoprotection in frost tolerant insects). In frost susceptible species, its solute effect alone will depress the freezing and supercooling points of the haemolymph. Because of its hydrophilic properties and its hydrogen bonding capacity, it may also bond with water leading to increased viscosity of the body fluids. This again would tend to enhance supercooling and prevent freezing. Glycerol could also compete with water for nucleating sites and thus exhibit an ice–nucleator masking effect. Finally, one other possible role which has not been closely studied but which may turn out to be one of glycerol's major functions in insect cold tolerance is that of an enzyme stabiliser [12]. Even at subzero temperatures the insect remains a dynamic system with many catabolic and anabolic reactions taking place [38–40].

Other protective substances, such as the polyhydric alcohols sorbitol, mannitol, threitol and erythritol, and the normal blood sugars trehalose, glucose, fructose and sucrose, may play similar roles and/or act as synergists (for references, see Danks [7]; erythritol, Baust [77], Baust and Edwards [112]; sucrose, Ring and Tesar [57]). Morrissey and Baust [39] and Ring [40] have commented on the advantages of a multifactorial cryoprotectant system which would alleviate the possible accumulation of toxic amounts of any one cryoprotective substance. Perhaps other chemicals could also be implicated in the depression of supercooling points, such as unsaturated fatty acids [78] or amino acids [79].

Duman [71, 72] has recently implicated glycoproteins in enhancing the supercooling ability of a frost susceptible insect. The presence of these macromolecules in the blood of the overwintering larvae of the tenebrionid *Meracantha contracta*, produces the phenomenon of 'thermal hysteresis' whereby the freezing point of the haemolymph is depressed 3–4°C below the melting point. At the same time the supercooling point is depressed to about −11°C, well below the temperature experienced by these beetle larvae in their overwintering microhabitat. Another possible function of the macromolecular antifreeze may be to retard inoculative freezing which might otherwise occur because of moisture in the hibernaculae (in partially decomposed oak logs). Similar glycoproteinaceous antifreezes have been found in some cold-water marine teleosts and in two species of molluscs [72].

Factors affecting survival of frost tolerant insects (*see Figure 9.1*)

Acclimation: As with frost susceptible species, frost tolerant insects must experience a period of acclimation during which the cold intolerant summer form undergoes biochemical, physiological, behavioural and, sometimes, structural changes to avoid injury at temperatures below those at which continued growth and development normally occur.

Freezing rate: The standard cooling rate employed in laboratory studies of frost susceptible and frost tolerant insects is usually 1–2°C cooling per minute [67]. Miller [80] reports, however, that freezing injury to normally frost tolerant arctic carabids, *Pterostichus brevicornis,* can only be avoided at cooling rates of less than 29°C per hour. A more detailed study by Miller [81] on the adult beetle, *Upis ceramboides* (Tenebrionidae) shows that the frost tolerance in this species is *closely* related to the cooling rate. Unusually low cooling rates are required to produce maximum freezing tolerance, and a very slight change in rate can reduce survival from 100% to zero. Cooling to −50°C results in 100% mortality if rates are above 0.35°C/min, but no injury is apparent if the cooling rate is reduced to 0.28°C/min. The lower lethal temperature determined in this beetle with a cooling rate of 0.17°C/min is about −60°C. The maximum cooling rate which allows full survival is almost identical to optimal cooling rates previously found in mouse embryos [82] and in some lymphocytes [83]. However, the very striking sensitivity to such slight changes in cooling rate is unique for *U. ceramboides.*

Reassessment of frost tolerance in some insects is, therefore, necessary, particularly in those reported as *not* frost tolerant but which were inadequately tested at slow rates of cooling. In nature, cooling rates are always much slower than the 1–2°C/min rate employed in the laboratory, particularly when the temperature conditions that prevail in the insect's microenvironment are considered.

Two-step freezing methods have been described whereby a frost tolerant insect is frozen slowly to subzero temperatures in the region of −30°C, where they are held for 1 hour before rapid freezing to liquid gas temperatures (either liquid oxygen at −183°C, liquid nitrogen at −196°C, or liquid helium at −269°C). Under these conditions some frost tolerant insects may survive for long periods of time [3].

Thawing rate: It has been suggested that freezing injury may not be due to the initial freezing or rate of freezing, but related to the rewarming process. During thawing, larger crystals grow at the expense of small ones, and this is thought to increase osmotic stress by rapidly elevating ice and electrolyte content, decreasing water content, and perhaps also causing mechanical damage (see references in Baust [5]). There is very little evidence for this occurring under natural conditions, and no experimental evidence exists for this hypothesis within the field of entomology.

Behaviour: Many behavioural features of frost tolerant insects are also shared by frost suceptible species. For instance, most insects prepare for hibernation by selecting overwintering sites which are devoid of moisture and not liable to be inundated. Such localities would be found in the soil, under litter, under bark, in plant material, rock crevices, etc. In the arctic such microhabitats are afforded by mammal burrows and the subniveal air space. Many exceptions are known, however, and the beetle *Pelophila borealis* overwinters successfully embedded in ice, usually in a state of anaerobiosis [84]. Furthermore, Tanno (personal communication) suggests that some species of sciarids (Diptera, Sciaridae) have to be in contact with ice within their host plants before they can overwinter successfully. This suggests that inoculative freezing occurs and that it does so at fairly high subzero temperatures.

In general, insects overwinter in only one specific stage of their life cycle (egg, larva, pupa or adult), and it is that stage only which is adapted to tolerate cold. Morrissey and Baust [39], however, have recently written an account of the ontogeny of cold tolerance in the goldenrod gall fly, *Eurosta solidagensis,* indicating that this species utilises both cold tolerance strategies. All life stages of the gall fly, except the third instar larvae, demonstrate supercooling points well below the lowest temperature normally experienced by each particular stage. The third instar larvae, however, are frost tolerant, maintain a high supercooling point, but are well protected by a multicryoprotectant system consisting of glycerol, sorbitol and trehalose. In arctic environments well adapted insects are likely to be more opportunistic in their life cycle strategies and may overwinter in more than one life stage.

We can presume that each life stage will exhibit cold tolerance adaptations, although we cannot assume that there will be the same physiological, biochemical and behavioural mechanisms for each stage. Ring and Tesar [57] have recently demonstrated that the arctic beetle, *Pytho americanus,* overwinters either in the larval stage (any instar) or in the adult, and that both stages have about the same degree of frost tolerance. This is the first reported example of an insect species which can tolerate freezing in more than one life stage. Until 1968 it was presumed that only immature stages could tolerate ice formation in their tissues. There are now at least 19 species known where the adult stage is frost tolerant, and the number keeps growing [57, 66, 81, 85, 86].

Time: Prolonged exposures to cold may cause deleterious effects in supercooled, frost susceptible insects only if the temperature is maintained close to the supercooling point. In frost tolerant species, however, freezing injury increases with time spent in the frozen condition. Increased ice formation in the body water is likely to occur with time, as well as increased dehydration by sublimation. In species which overwinter in habitats where they are completely encased in ice, the ability to survive without oxygen for long periods may contribute to overwintering success. Conradi-Larsen and Sømme [87, 88], Sømme [84, 89, 90] and Sømme and Conradi-Larsen [91, 92] have shown that some species of high altitude Collembola and Coleoptera can survive at 0°C in anaerobiosis for part of the winter. During anoxia these insects change to an anaerobic metabolism with lactate as the main end product. It is suggested that an important part of survival depends on the insect's ability to tolerate the concentrations of lactate formed, or on the ability to transform it to other substances such as amino acids.

Alternating freeze-thaw cycles are also likely to be harmful, although Downes [93] reports earlier observations that repeated freezing and thawing in extremely frost tolerant arctic lymantriid caterpillars was without obvious ill effects.

Cryoprotection: Successful protection from freezing injury will depend not only on the combination of the previously mentioned factors working in conjunction but also on the ability of individuals and populations to synthesise and accumulate cryoprotectants. Genetic variations between individuals or populations exist in both the ability to produce appropriate levels and types of cryoprotectants and in the acclimation regimes that have to be experienced in order to trigger the synthesis of cryoprotectants (see Baust and Grandee [94], for example). In some gall wasps (Cynipidae), it has been shown that polyploid races may have a greater temperature tolerance than their diploid types. Also, in the cynipids that are active at low temperatures, it is the diploid female only that is cold tolerant and the haploid male is absent from the winter generation [95].

The cryoprotectant substances that have been recognised in insects have already been listed (see p. 194). They act in a largely undetermined way, although in frost susceptible insects it is believed that an important mode of action is in depressing the supercooling point. In frost tolerant insects, in addition to depressing the freezing point of the body fluids, cryoprotectants assist in combating any deleterious effects of ice formation once freezing does occur, i.e. they act as true cryoprotectants. They may do so in any or all of the following ways: by their colligative action, by increasing viscosity which would enhance multinucleation (many nuclei growing slowly), by increasing osmolality and hence reducing the freezing point, by acting as water binders and thus reducing the rate of ice crystal growth, by reducing dehydration within the cell, by modifying the structure of ice formed, by preventing cell volume reduction below a critical minimal size, by maintaining protein 'spacing', by stablising membranes and enzymes, or by reducing the temperature at which recrystallisation occurs during thawing.

It was formerly believed that frost tolerance was always associated with the presence of high concentrations of glycerol, and that freezing could be tolerated only if ice formed in the extracellular fluids rather than intracellularly. However, there are many exceptions to the former statement and the relationship between glycerol and frost tolerance is not yet clearly understood. Some insects can survive freezing without glycerol, whereas others cannot survive even in the presence of large amounts. There are also exceptions to the statement that intracellular freezing in insects is invariably fatal [6, 96, 97]. Cryoprotection could also ensue from the active synthesis of nucleating agents within the insect [68]. Such a mechanism would allow freezing to occur at relatively high subzero temperatures, usually between -2 and $-10°C$. This, in turn, would tend to encourage the formation of extracellular ice and inhibit the more damaging form of intracellular ice by encouraging cell shrinkage.

It is also known that the frozen state is more economical than the supercooled from an energetic or metabolic point of view. Scholander *et al* [98] found that the larvae of an arctic chironomid begin to freeze at -1 to $-2°C$ and that 85% of their body fluids are frozen at $-10°C$. During this time the metabolism of the frozen larva undergoes a rapid decline as reflected by the Q_{10} value. Between 0 and −5°C, the Q_{10} of the respiration rises from 2–4 to about 20–50. It has been calculated that on this basis nutritive reserves sufficient to last for 10 days at 0°C would enable the insect to survive for 1000 years at −23°C. Similarly, Salt [56] found a sharp drop in oxygen consumption when a larva was frozen as compared to remaining supercooled at the same temperature. His study suggests that fewer food reserves would be utilised by the overwintering insect if it could freeze with a minimum of supercooling [39].

Mechanisms of injury (see Figure 9.1)

Cold: *Chill coma:* An insect usually has a characteristic state for each temperature value within its viable range. Thus, at different temperatures the state of the insect is not just that of an identical system working faster or slower under one set of conditions than under another. Changes occur in the state constants which need input information to trigger and time to complete (i.e. acclimation) [95]. In most temperate species of insects mobility and morphogenesis come to a halt at the chill coma temperature which usually lies somewhere in the region of 5–15°C. Any changes that occur during chill coma are usually completely reversible, although Powell and Parry [99] suggest that chill coma in the green spruce aphid, *Elatobium abietinum,* may lead to death or injury because of starvation. Drescher and Rothenbuhler [100] report that temperatures towards the extremes of the insect's tolerance range may bring about inactivation of cell nuclei. For example, when the egg of the honey bee, *Apis mellifera,* is chilled to 4–6°C for 1–4 hours, adult gynandromorphs appear in the population. The cold apparently inactivates the female pronucleus but allows further development of that of the male.

During low temperature acclimation in the autumn, cold adapted insects usually undergo a temperature compensation of rates of metabolism, activity, etc. in addition to becoming more tolerant to subzero temperatures; they can, therefore, remain more active at low temperatures above zero than their non-acclimated forms. Furthermore, in some arctic, antarctic and nival insects development can continue at temperatures below zero (in Danks [7]). Downes [101] reports a number of low temperature activities in arctic insects, including sightings of bumble bees in flight down to air temperatures of −4°C.

Cold injury: Cold injury can be defined as the damage caused to frost susceptible insects in the absence of freezing. The causes of immobility and death at low temperatures above the freezing point of the tissue fluids are not completely understood, but irreversible disturbances occur in the nervous system of cold-immobilised cockroaches [102], disruption of weak chemical bonds may occur [12], and membrane lipids may be 'frozen'. An insect normally has a characteristic state for each temperature value within its viable range, although the temperature tolerances of each state are not identical [95]. Cold injury is likely to occur, therefore, where desynchronisation of physiological processes occurs due to a lack of or inadequate acclimation. For instance, in non-acclimated insects isolated organs and tissues may supercool more strongly and for a longer time than the whole organism, whereas in fully acclimated insects the reverse is observed and there is more or less complete correlation between the ability of the tissues and cells to supercool with that of the whole organism [6].

Supercooling: Although the supercooled condition *per se* does not seem to have profound adverse effects on the overwintering insect, it is not completely innocuous either. The longer the time spent in the supercooled state the greater the chance of nucleation leading to injurious ice crystal formation and the greater the accumulation of disruptions in cellular functions. The water content of some frost tolerant insects decreases considerably when maintained in the supercooled state for long periods. As a result the probability of homogeneous nucleation decreases, but at the same time there is an increase in concentration of electrolytes and various other solutes which might prove toxic to the insect. Pythid and scolytid adults from the Canadian Rockies lose about 70% and 40% of their water content respectively when maintained for 4 months in the supercooled state. At the end of this period only about half the population of pythids survived, despite their very low supercooling points (*c*—45°C) and none of the scolytids. Most of the observed changes in cellular activities are reversible provided the exposure to low temperatures in the absence of freezing is not very prolonged.

Freezing at the supercooling point: In frost susceptible species of insects the limit of low temperature tolerance is reached at the lowest point of supercooling when spontaneous ice crystal formation occurs in the tissues. Since these species usually have the potential to undergo a considerable degree of supercooling, the onset of freezing is especially disruptive. This is reflected during the 'rebound' of the cooling curve (Figure 9.2a) which indicates very rapid crystallisation within the insect and a rapid loss of heat of crystallisation to the environment. Asahina

[3, 4] notes that an insect can usually be revived if removed from the cooling chamber at times equivalent to points A, B, C, D or E on the cooling curve, but that a frost susceptible insect will not survive exposures to temperatures equivalent to points F or G. Indeed, frost tolerance is usually defined as the ability of the insect to survive recooling after freezing to temperatures equivalent to the supercooling point (i.e. to point F). This is a standard of convenience but may not be applicable as a useful criterion to predict future survival. A more rigorous viability test is required since it is known from the work of Asahina [3], Tanno [103] and Ring and Tesar [57] that although an individual may appear to have survived freezing and appears fully active and coordinated shortly after rewarming, it is possible that the insect may not complete metamorphosis and/or reproduction successfully.

Freezing: Baust [5] has recently reviewed the subject of theoretical mechanisms of freezing injury in frost tolerant animal systems, including insects. It would be redundant to do so here since very little additional information on this topic in insect cryobiology has become available since that time. Suffice it to say that the proposed mechanisms of freezing damage fall into five major categories (see Figure 9.1):

(1) mechanical effects of ice formation;
(2) site of freezing;
(3) electrolyte-concentration;
(4) exceeding of a minimal, critical cell volume;
(5) recrystallisation during thawing.

Conclusions

Insects have the anatomical, physiological, biochemical and behavioural organisation which permits them to survive under a wide range of environmental temperatures. Many of them survive long periods of freezing under natural conditions, and since their cellular organisation is just as complex as that of any other metazoan animal, including mammals and man, the study of their adaptations to low temperatures is not only of great theoretical importance but also could have profound effects on man's economy.

For instance, the study of methods for the successful freezing and storage of insect cells, tissues or organs may assist research in insect bioluminescence (as a source of 'cold light' the photogenic organs of luminous insects are of great theoretical interest, and the photogenic

Table 9.1 Some successful examples and procedures for the freezing and low temperature preservation of insects, insect cells and subcellular components

Subject	*Treatment*	*Cooling rate*	*Cryoprotectants*	*Storage*	*References*
Insect mitochondria (*Nauphoeta cinerea* fat body)	Cells dissected in cold isolation medium of 0.25 M sucrose/ 2 mM EDTA; pH 7.4	Rapid*	DMSO (10% v/v)	1 week @ −20°C	Abo-Khatwa [46]
		Rapid	-	3 wks @ −196°C	Abo-Khatwa [46]
Salivary gland cells (larvae of *Pyrausta nubilalis*)	(1) Cells dissected from frost tolerant diapause cater-pillars (2) Placed in Beadle-Ephrussi physiological solution	Slow* (1-10°C/min)	20% glycerol in Beadle-Ephrussi solution	Brief exposures to −34°C and −46°C	Lozina-Lozinskii [6]
Cell suspensions (*Aedes aegypti*)	Frozen and stored in liquid nitrogen	Rapid	8% DMSO	1 mth @ −196°C	Brown *et al* [8]
Cell monolayer cultures (*Aedes albopictus*)	Frozen and stored in liquid nitrogen	Rapid	8% DMSO	1 mth @ −196°C	Brown *et al* [8]

Embryonic cells (*Musca domestica*)	(1) 2-3 h acclimation of cells to growth medium, followed by slow cooling to −30°C (2) Some cells were also placed in liquid nitrogen	Slow	Combination of cryoprotectants	24 h @ −30°C 24 h @ −196°C	Eide *et al* [9]
Grace's *Antheraea* cell line	(1) 2-3 h acclimation of cells to growth medium, followed by slow cooling to −30°C (2) Some cells were also placed in liquid nitrogen	Slow	Combination of cryoprotectants	24 h @ −30°C 60 days @ −196°C	Eide *et al* [9]
Isolated hearts of the beetle, *Popilius disjunctus*	(1) Hearts dissected in Ringer's soln containing bicarbonate (pH 6.7) (2) Placed in refrigerated chamber (3) Rewarmed rapidly to 31°C	No special attention given to cooling rate	10% glycerol-Ringer soln	11 days @ −20°C	Wilbur and McMahon [104]

Increased cooperation among distantly located scientists in this area of research may be greatly assisted by the recent design of a cheap and convenient apparatus which could be used for the transport of insect cell lines or tissue cultures [105] (see Figure 9.3).

*See Lozina-Lozinskii [6] for classification of cooling rates in biological systems

Subject	*Treatment*	*Cooling rate*	*Cryoprotectants*	*Storage*	*References*
Larvae of *Polypedilum vanderplanki*	(1) Water content of larvae reduced to about 8% (2) Larvae plunged into liquid air or liquid helium	Rapid	None	2-77 h @ −190°C 3-5 min @ −270°C	Hinton [52]
Prepupae of the poplar sawfly (*Trichiocampus populi*)	(1) Cooled slowly to −20°C (2) Rewarmed to −5°C and maintained for several hours (3) Cooled again to −30°C (4) Immersed in liquid nitrogen	Slow (0.4-1°C/min)	Trehalose	24 h immersion in liquid nitrogen (−196°C)	Tanno [106]
Overwintering 3rd instar larvae of *Aporia crataegi* (Lepidoptera, Pieridae)	(1) Cooled to −30°C (2) Maintained for 1 hr @ −30°C (3) Immersed in liquid nitrogen	Slow	-	Brief immersion in liquid nitrogen (−196°C)	Asahina *et al* [107]
Prepupae of the slug caterpillar (*Monema flavescens*)	Cooled to −20°C	Slow	Glycerol	Brief immersion in liquid nitrogen (−196°C)	Asahina *et al* [107]

Frost tolerant, diapause caterpillars of the corn borer (*Pyrausta nubilalis*)	(1) Stepwise cooling first to −30°C and then to −78°C (2) Warmed to 0°C–held for 1 month (3) Returned to room temperature (or any temperature higher than the threshold for morphogenesis)	Slow	Glycerol	Brief exposure to −78°C	Lozina-Lozinskii [6]
Cold acclimated larvae of the European corn borer (*Pyrausta nubilalis*)	Dry-chilling to −20°C	Slow	Glycerol	Several weeks @ −20°C	Hanec and Beck [108]
3rd instar larvae of the goldenrod gall fly (*Eurosta solidagensis*)	Cooled slowly to −55°C	Slow	Glycerol, sorbitol and trehalose	19 days @ −55°C	Salt [109], Morrissey and Baust [39]
Adults of the tenebrionid beetle, *Upis ceramboides*	Linear cooling in glass jars placed in a freezer	Slow (0.17°C/min)	Sorbitol and threitol	10–16 h @ −52°C	Miller [81, 110]

Subject	*Treatment*	*Cooling rate*	*Cryoprotectants*	*Storage*	*References*
Adult carabid beetles (*Pterostichus brevicornis*) collected in winter	Cooled slowly in insulated vials placed in a regulated cooling bath	Slow (4°C/min)	Glycerol	67% survival after 5 h at −87°C	Miller [80, 110]
Adult pythids, *Pytho depressus*	(1) Maintained at 0–4°C for up to 2 weeks (2) Cooled slowly to −27°C in a deep freezer	Slow (20°C/h)	Glycerol	Brief periods at temps down to −27°C	Zachariassen [69]
Adult and larval pythids, *Pytho americanus*	Slowly cooled in a cooling bath to −40°C	Slow (1-2°C/ min)	Mainly glycerol	10 days @ −40°C	Ring and Tesar [57]
Adult ichneumon wasps, *Chasmias spp*	Cooled slowly in petri dishes placed in a freezer	Slow	Glycerol	40 days @ −10°C	Ohyama and Asahina [66]
Adults of the coleopterans: *Damaster blaptoides rugipennis; Pterostichus orientalis; Phosphuga atrata*	Cooled slowly in petri dishes placed in a freezer	Slow	Glycerol	Frozen for 24 h @ −10°C	Ohyama and Asahina [66]

and hymeopterans: *Camponotus obscuripes; Hoplismenus pica japonica; Pterocormus molitorius*					
Diapause larvae of the goldenrod gall fly (*Eurosta solidagensis*)	Cooled slowly within a series of glass tubes placed in a freezer	Slow (1-3°C/ min)	Glycerol and sorbitol	Frozen for 32 days @ −30°C	Sømme [31]
Larvae of the gall fly, *Euura nodus*	Cooled slowly within a series of glass tubes placed in a freezer	Slow (1-3°C/ min)	-	Frozen for 2 h @ −60°C	Sømme [31]
Adults of the carabid beetle, *Pelophila borealis*	(1) Acclimated in a super-cooled state at −5°C for 2-3 weeks (2) Inoculated by placing in contact with a drop of water and freezing in a cooling bath to −5°C	Inoculative freezing	-	12 days @ −5°C	Sømme [84]
Overwintering queens of the bald faced hornet, *Vespula maculata*	Slowly cooled in a cold chamber	Slow (0.4°C/ min)	Glycerol	50% survival after 24 h @ −14°C	Duman and Patterson [111]

organs from the abdominal tips of fireflies are used in the laboratory in a very sensitive bioassay technique for the measurement of ATP activity). Other areas where they may contribute include hormone research, pheromone research (pheromone-producing glands of pest species could be distributed anywhere in the world), biological control of insect pests (specific viruses or bacteria could be raised in tissue culture and accumulated in the frozen state), establishing sperm banks (in apiculture and sericulture) and medical research (e.g. anti-malarial research, or study of the role of arboviruses in human disease).

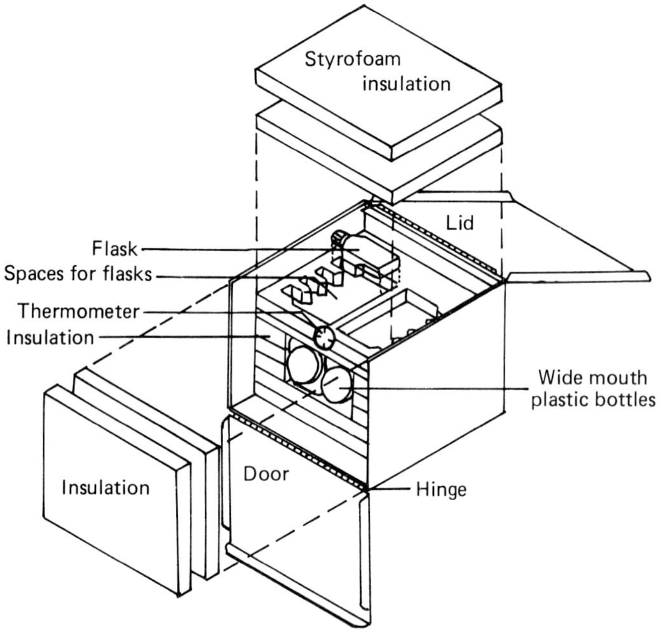

Figure 9.3 A cheap, convenient apparatus for transporting frozen insect cell or tissue cultures (after Anderson and Black [105], by courtesy of the authors)

Freezing methods for whole insects would be of use for the utilisation of insects as food (e.g. locusts, termites, caterpillars, bee brood, etc.), in the production and accumulation of insect hosts and their parasites for biological control programmes, or in the production of frozen predators (e.g. ladybirds, lacewings, etc.) for distribution and release as 'natural enemies' of insect pests. Table 9.1 collates some successful preservation procedures.

References

1 Brooks, M.A. and Kurtti, T.J. (1971). *Annual Review of Entomology* **16**, 27
2 Salt, R.W. (1961). *Annual Review of Entomology* **6**, 55
3 Asahina, E. (1966). In *Cryobiology*, p. 451. H.T. Meryman (ed.). New York: Academic Press
4 Asahina, E. (1969). *Advances in Insect Physiology* **6**, 1
5 Baust, J.G. (1973). *Cryobiology* **10**, 197
6 Lozina-Lozinskii, L.K. (1974). In *Studies in Cryobiology*, p. 259. New York: John Wiley and Sons
7 Danks, H.V. (1978). *Canadian Entomologist* **110**, 1167
8 Brown, B.L., Nagle, S.C., Lehman, J.D. and Rapp, C.D. (1970). *In Vitro* **6**, 242
9 Eide, P.E., Caldwell, J.M. and Dockter, M.M. (1971). *Proceedings of the North Dakota Academy of Science* **25**, 8
10 Ivanović, J.P., Janković-Hladni, M.J. and Milenović, M.P. (1975). *Journal of Thermal Biology* **1**, 53
11 Thomsen, E. and Møller, I. (1959). *Nature* **183**, 1401
12 Hochachka, P.W. and Somero, G.N. (1973). *Strategies of Biochemical Adaptation*, p. 358. Philadelphia: Saunders and Co
13 Applebaum, S., Janković, M., Grozdanović, J. and Marinković, D. (1964). *Physiological Zoology* **34**, 90
14 Das, A.B. and Prosser, C.L. (1967). *Comparative Biochemistry and Physiology* **21**, 449
15 Hoschemeyer, A.E.V. (1969). *Comparative Biochemistry and Physiology* **28**, 535
16 Lees, A.D. (1955). In *The Physiology of Diapause in Arthropods*, p. 151. Cambridge University Press
17 Van der Kloot, W.G. (1955). *Biological Bulletin, Woods Hole* **109**, 276
18 Williams, C.M. (1956). *Biological Bulletin, Woods Hole* **110**, 201
19 Tyshtchenko, V.P. and Mandelstam, J.E.A. (1965). *Journal of Insect Physiology* **11**, 1233
20 Smallman, B.N. and Mansingh, A. (1969). *Annual Review of Entomology* **14**, 387
21 Grzelak, K., Lassota, Z. and Wroniszewska, A. (1971). *Journal of Insect Physiology* **17**, 1916
22 Way, M.J. (1960). *Journal of Insect Physiology* **4**, 92
23 Ring, R.A. (1973). *Journal of Insect Physiology* **19**, 481
24 Yaginuma, T. and Yamashita, O. (1978). *Journal of Insect Physiology* **24**, 347
25 Andrewartha, H.G. (1952). *Biological Reviews* **27**, 50
26 Schneiderman, H.A. and Horwitz, J. (1958). *Journal of Experimental Biology* **35**, 520
27 McEwen, B.G. (1976). *Scientific American* **235**, 48
28 O'Malley, B.W. and Schrader, W.T. (1976). *Scientific American* **234**, 32
29 Ziegler, R. and Wyatt, G.R. (1975). *Nature* **254**, 622
30 Dubach, P., Pratt, D., Smith, F. and Stewart, C.M. (1959). *Nature* **184**, 288
31 Sømme, L. (1964). *Canadian Journal of Zoology* **42**, 87
32 Krunic, M.D. and Salt, R.W. (1971). *Canadian Journal of Zoology* **49**, 663
33 Mansingh, A. and Smallman, B.N. (1972). *Journal of Insect Physiology* **18**, 1565
34 Baust, J.G. and Miller, K.L. (1972). *Journal of Insect Physiology* **18**, 1935
35 Frankos, V.H. and Platt, A.P. (1976). *Journal of Insect Physiology* **22**, 623
36 Zachariassen, K.E. and Pâsche, A. (1976). *Journal of Insect Physiology* **22**, 1365
37 Wood, F.E. and Nordin, J.H. (1976). *Journal of Insect Physiology* **22**, 1665
38 Baust, J.G. (1972). *Nature (New Biology)* **236**, 219
39 Morrissey, R.E. and Baust, J.G. (1976). *Journal of Insect Physiology* **22**, 431
40 Ring, R.A. (1977). *Norwegian Journal of Entomology* **24**, 125
41 Chefurka, W. (1965) In *Physiology of the Insecta*, Vol. II. M. Rockstein (ed.). New York: Academic Press
42 Grant, N.H. and Alburn, H.E. (1966). *Nature* **212**, 194
43 Grant, N.H. (1969). *Cryobiology* **6**, 182
44 Doscherholmen, A. and Silvis, S.E. (1971). *Cryobiology* **8**, 577
45 Takehara, I. (1962). *Low Temperature Science, Series B* **20**, 35
46 Abo-Khatwa, N. (1976). *Life Sciences* **18**, 329

47 Doebbler, G.F., Buchheit, R.G. and Rinfret, A.P. (1962). *Nature* **191**, 1405
48 Doebbler, G.R., Rowe, A.W. and Rinfret, A.P. (1966). In *Cryobiology*, p. 407. H.T. Meryman (ed.). New York: Academic Press
49 Love, R.M. (1966). In *Cryobiology*, p. 317. H.T. Meryman (ed.). New York: Academic Press
50 Gressitt, J.L. (1967). *Entomology of Antarctica, Antarctic Research Series, 10*, p. 395. Washington, DC: American Geophysical Union
51 Cloudsley-Thompson, J.L. (1973). *Journal of Natural History* 7, 471
52 Hinton, H.E. (1960). *Nature* **188**, 336
53 Leader, J.P. (1962). *Journal of Insect Physiology* 8, 155
54 Salt, R.W. (1953). *Canadian Entomologist* **85**, 261
55 Bakke, A. (1969). *Norwegian Journal of Entomology* **16**, 81
56 Salt, R.W. (1956). *Proceedings of the 10th International Congress of Entomology, Montreal* 2, 73
57 Ring, R.A. and Tesar, D. (1980). Submitted to *Journal of Insect Physiology*
58 Salt, R.W. (1959). *Canadian Journal of Zoology* **37**, 59
59 George, M.F., Burke, M.J. and Weiser, C.J. (1974). *Plant Physiology* **54**, 29
60 Salt, R.W. (1958). *Journal of Insect Physiology* **2**, 178
61 Salt, R.W. (1966). *Canadian Journal of Zoology* **44**, 117
62 Salt, R.W. (1968). *Canadian Journal of Zoology* **46**, 329
63 Salt, R.W. (1970). *Canadian Journal of Zoology* **48**, 205
64 Ring, R.A. (1972). *Canadian Journal of Zoology* **50**, 1601
65 Mansingh, A. (1971). *Canadian Entomologist* **103**, 983
66 Ohyama, Y. and Asahina, E. (1972). *Journal of Insect Physiology* **18**, 267
67 Salt, R.W. (1966). *Canadian Journal of Zoology* **44**, 655
68 Zachariassen, K.E. and Hammel, H.T. (1976). *Nautre* **262**, 285
69 Zachariassen, K.E. (1977). *Norwegian Journal of Entomology* **24**, 25
70 Baust, J.G. and Morrissey, R.E. (1975). *Journal of Insect Physiology* **21**, 1751
71 Duman, J.G. (1977). *Journal of Comparative Physiology (B)* **115**, 279
72 Duman, J.G. (1977). *Journal of Experimental Zoology* **201**, 85
73 Tauber, M.J. and Tauber, C.A. (1976). *Annual Review of Entomology* **21**, 81
74 Salt, R.W. (1963). *Canadian Entomologist* **95**, 1190
75 Wyatt, G.R. and Kalf, G.F. (1956). *Proceedings of the 10th International Congress of Entomology, Montreal* 2, 333
76 Ashwood-Smith, M.J. (1970). In *Current Trends in Cryobiology*, pp. 5–42. A.U. Smith (ed.). New York: Plenum Press
77 Baust, J.G. (1978). *Abstracts of the 15th Annual Meeting of the Society for Cryobiology, Tokyo, Japan.* Abstract 73
78 Pantyukhov, G.A. (1964). *Entomological Review* **43**, 47
79 Sømme, L. (1967). *Journal of Insect Physiology* **13**, 805
80 Miller, L.K. (1969). *Science* **166**, 105
81 Miller, L.K. (1978). *Cryobiology* **15**, 345
82 Whittingham, D.G., Leibo, S.P. and Mazur, P. (1972). *Science* **178**, 411
83 Farrant, J. (1972). *Report of International Congress of Cryosurgery, June 1972.* Cited in Whittingham, D.G., Leibo, S.P. and Mazur, P. (1972). *Science* **178**, 411
84 Sømme, L. (1974). *Norwegian Journal of Entomology* **21**, 131
85 Zachariassen, K.E. and Hammel, H.T. (1976). *Norwegian Journal of Zoology* **24**, 349
86 Rains, T.D. and Dimock, R.V. (1978). *Journal of Insect Physiology* **24**, 551
87 Conradi-Larsen, E.M. and Sømme, L. (1973). *Nature* **245**, 388
88 Conradi-Larsen, E.M. and Sømme, L. (1973). *Norwegian Journal of Entomology* **20**, 325
89 Sømme, L. (1974). *Norwegian Journal of Entomology* **21**, 155
90 Sømme, L. (1976). *Norwegian Journal of Entomology* **23**, 149
91 Sømme, L. and Conradi-Larsen, E.M. (1977). *Oikos* **29**, 118
92 Sømme, L. and Conradi-Larsen, E.M. (1977). *Oikos* **29**, 127
93 Downes, J.A. (1962). *Canadian Entomologist* **94**, 143
94 Baust, J.G. and Grandee, R. (1978). *Abstracts of the 15th Annual Meeting of the Society for Cryobiology, Tokyo, Japan.* Abstract 74

95 Clarke, K.V. (1967). In *Thermobiology,* p. 293. A.H. Rose (ed.). London: Academic Press
96 Salt, R.W. (1959). *Nature* **184**, 1426
97 Salt, R.W. (1962). *Nature* **193**, 1207
98 Scholander, P.F., Flagg, W., Hock, R.J. and Irving, L. (1953). *Journal of Cellular and Comparative Physiology* **42**, Suppl. 1, p. 56
99 Powell, W. and Parry, W.H. (1976). *Annals of Applied Biology* **82**, 209
100 Drescher, W. and Rothenbuhler, W.C. (1963). *Journal of Heredity* **54**, 195
101 Downes, J.A. (1965). *Annual Review of Entomology* **10**, 257
102 Calhoun, E.H. (1954). *Nature* **173**, 582
103 Tanno, K. (1968). *Low Temperature Science, Series B* **26**, 71
104 Wilbur, K.M. and McMahan, E.A. (1958). *Annals of the Entomological Society of America* **51**, 27
105 Anderson, J. and Black, L.M. (1975). *In Vitro* **11**, 322
106 Tanno, K. (1968). *Low Temperature Science, Series B* **26**, 79
107 Asahina, E., Ohyama, Y. and Takahashi, T. (1972). *Low Temperature Science, Series B* **30**, 91
108 Hanec, W. and Beck, S.D. (1960). *Journal of Insect Physiology* **5**, 169
109 Salt, R.W. (1957). *Canadian Entomologist* **89**, 491
110 Miller, L.K. (1978). *Journal of Insect Physiology* **24**, 791
111 Duman, J.G. and Patterson, J.L. (1978). *Comparative Biochemistry and Physiology* **59A**, 69
112 Baust, J.G. and Edwards, J.S. (1979). *Physiological Entomology* **4**, 1

CHAPTER TEN

Preservation of Microorganisms by Freezing, Freeze-Drying and Desiccation

M J Ashwood-Smith

Introduction

The first part of this chapter is concerned with some proposed mathematical models that predict the effects of thermal stress on microorganisms. Then follows a brief account of several factors that influence cellular survival after freezing and thawing or desiccation, and of a consideration of the role of oxygen in the development of damage in freeze-dried microorganisms. Some particular attention will be paid to the preservation of *Mycoplasma* before presenting information on the storage of viruses, bacteria and some fungi. These latter data are tabulated for convenience and include brief comments.

No apology is made for not discussing desiccation, freeze-drying and cryopreservation as separate entities. Several papers make reference to all of these methods. In any case the removal of water whether as ice, or by sublimation from ice or from the liquid stage direct to the gas stage is an important prerequisite to all three methods. When freezing and thawing rates are not mentioned in the original papers an estimate, in parenthesis, is included based on probable rates given the stated sample size etc.

Some predictive models of thermal inactivation

The inactivation of microorganisms subjected to various stresses such as freeze-drying, freeze-thawing and thermal inactivation has been studied by Cox [1] using *E. coli* B, *E. coli* commune, *E. coli* Jepp and Semliki forest virus. Loss of viability followed first order denaturation kinetics.

Survival results fitted a curve expressed by the following equation:

$$\ln V_1 = K_1 \ [B(n-x)H_2O]o \ (e^{-k_1 t}-1) + \ln 100$$

where V_1 is the percentage of viability at time t, k is the decay rate constant and K_1 $[B\ (n-x)\ H_2O]o$ is a probability constant multiplied by the concentration of a hydrate of a species B (the concentration of which is related to the probability of death). Cox also reported that both influenza virus and Semliki forest virus underwent viability loss in liquid or frozen suspensions and the decay curves at elevated, ambient and subzero temperatures could be accounted for by the equation. However, the loss of viability of *E. coli* B over 8 years at −70°C was better accounted for by a modification of this equation. Cox suggested that perhaps *E. coli* B contained a more stable sub-population than the major portion of the total population. A review of Strange and Cox [2] 'Survival of Dried and Airborne Bacteria' should be consulted for a recent account of some of Cox's pioneering work in this field.

The predictions of the stability of dried microorganisms was tackled by Greiff and Rightsel [3] who worked with measles virus dried by sublimation *in vacuo*. They suggested that thermal degradation should follow (logarithmically) the Arrhenius plot in relation to the absolute temperature (T) thus:

$$\log k = (\Delta H_a/2.303R)\ (1/T)$$

where k is the specific rate of degradation of the biological material, R is the gas constant and ΔH_a is the heat of activation.

An attenuated Edmonston strain of measles virus was freeze-dried from a mixture of calcium lactobionate and serum albumin in Parker '199' after freezing to −76°C. Residual moisture was less than 1% and the freeze-dried virus was stored at −65°C. Samples were heated at 28°C, 36.2°C and 45°C for a number of days before analysis for viral titre. An excellent 'fit' relating titre to temperature/time indicated a pseudo first order reaction dependent on the concentration of the biological material.

In general, the lower the storage temperature whether for dried, freeze-dried or frozen microorganisms the longer can viability be maintained. In practice, the temperature of liquid N_2 [$-196°C$] is often used for preservation. At temperatures lower than $-120°C$ change is, in all practical terms, not measurable. A theoretical limit to storage does, however, exist as radiation damage occurs (background radiation, depending on geography, geology and altitude is approximately 0.1 rad per year). The LD_{50} for microorganisms is very varied but about 6000 rads would be a low average. Approximately 180000 years would be required to accumulate an amount of radiation sufficient to kill about 50% of the cells upon rehydration. The low temperatures and the presence of either dimethylsulphoxide (DMSO) or glycerol often used to increase survival after freezing and thawing, give a combined protective factor of approximately three times relative to x-ray induced damage at room temperature [4].

The death of starved yeast cells (*Candida utilis*) has been used by Peled *et al* [5] to calculate rate constants. Moats [6] utilised the following equation for this purpose:

$$(d/\sigma)^2 = \frac{N(e^{-2kt} - 2e^{-(kt + kt_{50})} + e^{-2kt_{50}})}{(e^{-kt} - e^{-2kt})}$$

where d/σ is the ratio of deviation from the mean to standard deviation for the fraction of survivors at time t; t_{50} is the time at which the fraction of survivors is 0.5; N is the total number of critical sites; k is the rate constant sought; and e is the Napierian base. Peled *et al* [5] found this equation 'so sensitive to variation in experimental data as to render it impractical for this application'. Their own modification:

$$(d/\sigma)^2 = \frac{X_L\,(e{-2kt}{-2e}{-(kt+kt_{50})} + e^{-2kt_{50}})}{(1-e^{-kt_{50}})\,(e^{-kt} - e^{-2kt})}$$

where X_L obtained for $X_L = N(1{-}e{-}kt_{50})$ gave values which in the author's words 'must be regarded as unacceptable'. Thus it is clear that although some mathematical explanations of inactivation are useful approximations there remain reservations. When one considers, for instance, the several genetic loci controlling x-ray or ultra-violet sensitivity, these reservations are not, perhaps, surprising.

Some postulated mechanisms of freezing and thawing damage in microorganisms

The mechanism by which cells are damaged by freezing and thawing has already been discussed (see Chapter 1) but papers by Mazur [7, 8] are pertinent.

Cooling and warming rates have considerable effect on bacterial survival [9]. The presence of sodium chloride has an adverse effect on survival after freeze-thaw procedures. Bacteria studied, included *Azobacter, Klebsiella, Salmonella, Pseudomonas* and *Streptococcus.* Cooling rates varied from 1°C to 10000°C per minute and thawing rates were either rapid at 1000°C/min or slow at 12°C/min. All five bacterial types survived poorly at cooling rates of about 100°C/min. *Streptococcus faecalis* was not affected by freezing in the presence of saline. Calcott and MacLeod [10] suggested that freeze-thaw damage in *E. coli* was associated with a loss of β-galactosidase. Such a loss was drastically reduced by the inclusion of protective agents such as glycerol, sucrose or Tween 80 to the medium in which cells were frozen and thawed. Thus Calcott and MacLeod concluded that the bacterial membrane was a prime target of freeze-thaw damage. Cell wall damage as well as cell membrane damage was involved in death following freezing and thawing in *E. coli* [11]. The addition of glycerol greatly increased survival of cells frozen in saline and reduced damage both to the cell membrane and the cell wall; Tween 80 prevented membrane damage only. Cell wall damage was not detrimental to cell survival.

Recently, Lee *et al* [12] have shown that the sensitivity of a number of different bacteria to the presence of salt during freezing and thawing is apparently correlated with bacterial cytochrome content. *Lactobacillus casei* and *Streptococcus faecalis* contain no cytochrome and showed no salt sensitivity. Anaerobically grown *E. coli* had a lower cytochrome content than aerobically grown *E. coli* and was less sensitive to freezing and thawing in the presence of salt. It was speculated by Lee *et al* [12] that the cytochromes play an important structural role in the bacterial cell membrane. The loss of 260 and 280 nm absorbing material from the bacterial intracellular pools after the freezing and thawing of aqueous suspensions of *E. coli* has been described by Ray and Speck [13, 14]. 'Repair' or reversibility of some of this damage was evident.

Morichi [15] studied metabolic injury in frozen and thawed *E. coli.*

He concluded that cell density was important in terms of the release of protective or injurious substances and discussed the mechanism by which injured cells proceed to 'irreversible death' perhaps by the activities of 'latent RNases'. Morichi and Irie [16] showed that recovery of frozen and thawed bacteria and freeze-dried bacteria was affected by the conditions pertaining after rehydration or thawing. Some cells of *Streptococcus faecalis* required RNA synthesis after thawing before recovery. Frozen-injured cells of this organism showed no RNA synthesis in the presence of 6% sodium chloride. The recovery of viable cells of freeze-dried and stored *Streptococcus thermophilus* was greatly increased (fivefold) by the addition of cysteine in the plating medium. By contrast recovery of *Vibrio metchnikovii* after freezing and thawing in 10% glycerol was decreased by the presence of peptones in the medium; lactalbumin, for example, decreased recovery by a factor of approximately tenfold.

The lesson from this work of Morichi and Irie is clear. Viability should be measured under different conditions of nutrition, and redox potential if recoveries after freeze-drying or freeze-preservation are very low. Some of the factors, other than cryoprotective additives, that modify damage associated with freezing and thawing or freeze-drying desiccation are referenced in Table 10.1.

The preservation of specific microorganisms

The preservation of mycoplasma

Several papers relating to the cryopreservation of a number of mycobacteria have been published by workers at the Trudeau Mycobacterial Culture Collection at the Trudeau Institute in New York State, USA [17–19]. A number of problems associated with a satisfactory taxonomic classification of the *Mycoplasma* are common to all bacterial genera but the genus *Mycobacterium* has these problems plus some of its own. The prolonged generation time of these organisms and their often 'subtle quantitative differences in response to biochemical tests' requires that good methods for long term preservation be available.

Kubica *et al* [19] have investigated the long term cryopreservation of 154 strains of mycobacteria at −70°C. Optimum results were obtained when bacteria were suspended in either Dubos Tween–albumin broth or in Middlebrook 7 H-9 liquid medium [17]. Rates of cooling were not disclosed (estimated 1–5°C/min) and thawing was rapid (estimated 150°C/min). Storage at −20°C gave very poor results and thus −70°C

Table 10.1 Metabolic factors involved in freeze-drying damage

Factor	*Microorganism*	*Reference*
Oxygen	*E. coli*	Israeli, E. (1975). *Cryobiology* **12**, 15. This reference includes a number of earlier references to oxygen effects on freeze-dried bacteria
Repair	*Salmonella anatum* (Effect of 0.1% milk solids post freeze-drying)	Janssen, D.W. and Busta, F.F. (1973). *Cryobiology* **10**, 386
Repair	*Streptoccocus faecalis, Streptoccocus thermophilus, Vibrio metchnikovii* (Effects of sodium chloride and cysteine, post freeze-drying)	Morichi, T. and Irie, R. (1973). *Cryobiology* **10**, 393
Repair	*E. coli* Repair of freeze-thaw damage. Influence of inorganic phosphate, etc.	Ray, B. and Speck, M.L. (1972). *Applied Microbiology* **24**, 258. See also: *Ibid.* **24**, 585 (1972)

Table 10.1 Metabolic factors involved in freeze-drying damage (continued)

Factor	*Microorganism*	*Reference*
Radiation sensitive mutants	*E. coli* B_s −1 (*uvr B*, $^{-}$*EXR*$^{-}$), *E. coli* (*rec A*$^{-}$). Both these radiation sensitive strains were more sensitive to freeze-drying than the normal *E. coli* B strains	Tanaka, Y. *et al.* (1975). *Biken Journal* **18**, 267. See also: Tanaka, Y. (1978). *Kurume Medical Journal* **25**, 111
Rehydration death	*Lactobacillus bulgaricus* Death of freeze-dried cells during rehydration	Morichi, T. *et al.* (1967). *Agricultural and Biological Chemistry* **31**, 137
Protective effect of racemic threonine (inactivity of optically active forms)	*Lactobacillus arabinosus* Effects on freeze-drying	Morichi, T. *et al.* (1965). *Agriculture and Biological Chemistry* **29**, 66. See also: Morichi, T. (1970). *Proceedings of the Society 1st International Conference on Culture Collections,* Univ. of Tokyo Press, and Morichi, T. (1969). *Freezing and Drying of Microorganisms.* T. Nei (ed.). Tokyo: Tokyo Univ. Press 53 (1969)

was recommended. The tested strains contained mycobacteria from the following groups:

(1) T B complex, including *M. tuberculosis, M bovis, BCG*;
(2) photochromogens;
(3) scotochromogens;
(4) non-photochromogens;
(5) rapid growers, which included *M. fortuitum, M. smegmatis,* etc.

The key taxonomic features were not changed by long term storage and earlier studies [17, 18] suggested 100% viability with no loss in infectivity after cryopreservation.

Boulanger and Portelance [20] cite numerous references to a number of techniques used for the preservation of mycobacteria. Techniques mentioned in the introduction to their paper include suspension desiccation *in vacuo,* rapid desiccation over P_2O_5, deprivation of oxygen by high vacuum, replacement of air by inert gas, storage at 37°C, refrigeration, freezing in liquid air, freeze-drying and lastly freezing at low temperatures. They investigated 14 strains of mycobacteria suspended in Sauton's medium and cooled slowly at about 1.5°C/min (estimated to −55°C). Samples were thawed rapidly (estimated 120°C–150°C/min) or slowly (estimated about 2°C/min) before analysis for viability. Changes in virulence for strain H37Rv (*Mycobacterium tuberculosis* var. *hominis* strain) were not apparent and no detectable change in the immunising activity of *BCG* was seen. The effectiveness of storage varied from strain to strain with these bacilli harvested in midlog growth. Certain diluents were shown to be preferable to others and 5% glycerol was found to be useful; lactose, however, was detrimental.

The stability of *M. paratuberculosis* in bovine faeces to freezing and thawing has been reported by Richards and Thoen [21]. Considerable loss of viable organisms was observed in the first three weeks of −70°C storage. However a significant number of bacteria survived to suggest that the freezing of field samples of faeces would be advantageous in veterinary medicine.

Pattyn [22] investigated the cryopreservation of *M. leprae* at −196° C. This organism is much more sensitive to freeze-thaw damage than other mycobacteria. Samples suspended in Hanks' balanced salt solution were mixed with DMSO and cooled at 1°C/min to −70°C and then at 3°C/min to −100°C before immersion in liquid nitrogen. Thawing was rapid (estimated at about 150°C/min); samples were assayed by injection into mice. Results with 5% DMSO as a cryoprotective agent were irregular and unpredictable. However, when bacteria were cooled slowly in the presence of 7.5% DMSO, results were good and samples displayed the same infectivity as the controls.

The lyophilisation or freeze-drying of mycobacteria has yielded variable results and for *M. leprae* it is not good (Hilson, Shepard, quoted by Pattyn, 1974). Studies on the freeze-drying of mycobacteria by Slosarek *et al* [23] over a long period (8–25 years) have demonstrated excellent survival for some strains such as *M. avium, M. phlei, M. aquae, M. microti, M. fortuitum* and *M. smegmatis*. Poor results were obtained with *M. kansasii, M. tuberculosis* and *M. bovis* (*BCG* strain). Several protective additives were evaluated in these studies and 1% sodium glutamate was considered the best. No changes in properties, including virulence, were observed. However, Dunbar *et al* [24] have reported an increased proportion of domed over-thin colonies after the freeze-drying of *M. intracellulare,* a modification that paralleled some loss of pathogenicity. The recent findings that freeze-drying processes, but not freezing and thawing, produce genetic changes in bacteria [25] and in yeasts [26] suggests that some degree of caution is necessary before assuming that freeze-drying is the universal panacea for all preservation needs.

No organism can be successfully freeze-dried if it cannot first be successfully frozen and thawed. In general, few papers are concerned with fundamentals as the authors, quite naturally, are more interested in simple, practical preservation systems to aid their own fundamental studies on disease, etc. Work by Mazur [7, 8], by Postgate and Hunter [27] and Cox and his colleagues [28] should be consulted for such information. We may return to *Mycoplasma* preservation, a subject in need of fundamental research to illustrate a sound 'cryobiological approach', in a paper by Raccach *et al* [29]. The test organisms, *Acholeplasma laidlawii* (oral strain) and *M. mycoides* var. *capri* (PG3) were suspended in a modified Edward medium. Cells were cooled and thawed at different rates. Results indicated a very pronounced dependence on initial cell concentration; the lower the cell density the lower the survival by a large factor. The effect of cooling rate is shown in Figure 10.1, thawing rate was also important with the best survival being at a rate of 67°C/min. Both glycerol and DMSO at 1.5 M gave excellent cellular survival and the authors concluded that DMSO was preferable to glycerol.

Another interesting aspect of this paper was the modification of membrane lipids by growth conditioning. A marked increase in oleic acid content of *A. laidlawii* gave a significant increase in survival. By contrast, decreased cholesterol increased freezing resistance. For example, slow cooling to −70°C gave a survival of 0.2% for high cholesterol as against a survival of 48% for the cholesterol poor strain. Raccach *et al* [29] end their interesting paper, 'encourage a more intensive modification on the effects of membrane lipid composition on freezing injury, utilizing mycoplasmas as convenient tools for this purpose'.

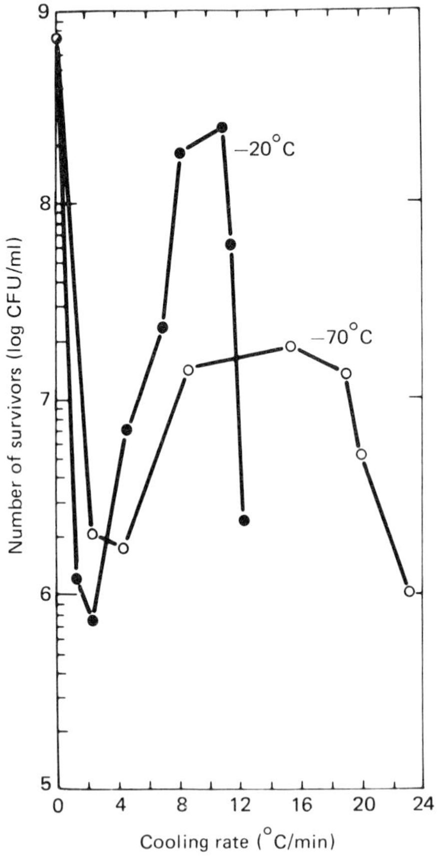

Figure 10.1 Effect of the cooling rate on the survival of *A. laidlawii* cells suspended in β-buffer and cooled to −20°C or to −70°C. (From Raccach *et al.*, 1975 [29], by permission of the authors and publishers)

The preservation of viruses

Bacteriophages: Early studies on the inactivation of bacterial viruses (phages) were pioneered by Leibo and Mazur [30] amongst others.

In recent years the work of Steele deserves comment [31–34].

Standard cryoprotective additives such as glucose, sucrose, glycerol, DMSO, PVP, dextran and ammonium acetate protected $T_4 B_0$ phage from damage associated with freezing to and thawing from temperatures as low as −30°C. Samples were cooled at 1°C/min with seeding at −3°C; slow thawing was used. Cryoprotection was essentially almost

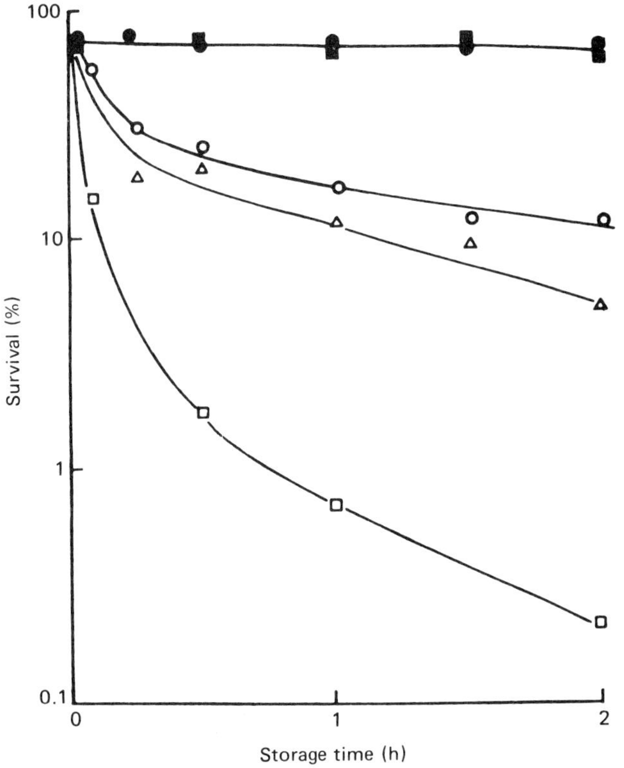

Figure 10.2 Effect of the eutectic phase change on the survival of $T_4 B_0$ phage suspended initially in 0.1 molal NaCl. In the single storage cycle samples were stored at −23°C, with the eutectic unseeded (●) or seeded (○) and thawed rapidly (in a +37°C water bath) at the indicated times. Survivals for all points in this figure were determined by the standard plaque assay procedure. In double storage cycles samples were seeded and stored at −23°C with the eutectic unseeded (■) or seeded (□) and thawed rapidly at the indicated times. One set of samples (△) was incubated at 37°C for 3 h between the first and second storage periods. For these double storage cycles the survivals shown in the figure are expressed as a percentage of plaque forming units remaining after the first storage cycle. (From Steele, 1976 [33], by permission of the author and Cambridge University Press)

complete when any of the above mentioned compounds was used at 0.1 mol/kg. Steele was able to show that phages suspended in sodium bromide (3.5 mol/kg) and cooled *without freezing* to −20°C and then warmed at +4°C were killed but this destruction was prevented, in large measure by 'cryoprotective additives'. Glucose and PVP gave 100% survival. He concluded from these experiments [32] that cryoprotection, at least with phages cannot be accounted for in terms of colligative action.

When $T_4 B_0$ phages were frozen to temperatures below the eutectic temperature of the sodium chloride of the suspending medium about 10% of the viruses survived. Exposure of these phages to either ultrasonic vibration or to repeated freezing and thawing indicated that they

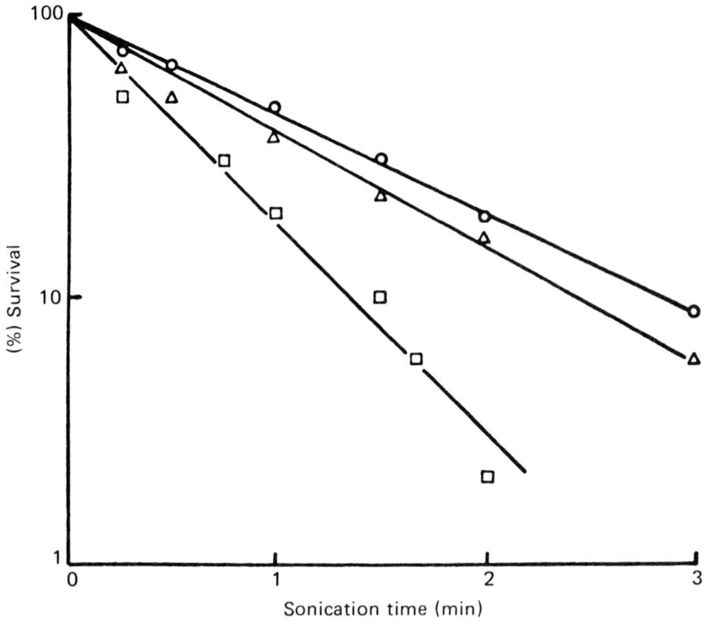

Figure 10.3 Effect of the eutectic phase change on the sensitivity of $T_4 B_0$ phage to ultrasonic vibration (20 kHz, 66 μ). Samples of $T_4 B_0$ phage suspended in 0.1 molal NaCl were ultrasonicated for the indicated times subsequent to the following pretreatment: (1) ○, control, no pretreatment; (2) □, stored at −23°C for 2 h with the eutectic seeded and thawed rapidly; (3) △, stored at −23°C for 2 h with the eutectic seeded, thawed rapidly and then incubated at 37°C for 3 h. In (2) and (3) survivals are expressed as a percentage of the plaque forming units remaining after the pretreatment procedures. (From Steele, 1976 [33], by permission of the author and Cambridge University Press)

were hypersensitive and, therefore, latently injured (Figures 10.2 and 10.3). It is not clear, however, how an incubation for 3 hours at 37°C returns the viruses to an almost normal sensitivity [33].

Morphological changes characteristic of changes seen with low temperature salt denaturation and of those produced by eutectic changes are illustrated in Figure 10.4 [34]. Salt denaturation caused contraction of tail sheaths but eutectic phase changes caused two separate lesions. The percentage of normal virus particles, counted electron microscopically, was essentially the same as the number of viable viruses measured by standard phage assays (Figure 10.5). Two separate mechanisms of injury have been suggested to account for eutectic damage namely, (1) a subunit dissociation of the tail sheath (Figure 10.4(d)) and later damage (Figure 10.4(f)) and (2) a dissociation of subunits of the head membrane to produce a tear allowing DNA to escape (Figure 10.4(l)–(k)).

Cryoprotective properties of peptides to T_4 bacteriophage have been investigated by Steele *et al* [35, 36]. Protection against electrolyte effects was shown to be a function of peptide concentration but protection against eutectic effects was dependent on peptide molecular weight [31].

The drying of bacteria, bacteriophages and viruses has been employed for many years as an inexpensive and reasonable way of long term preservation. The original antirabies vaccine prepared by Pasteur in 1885 was produced by desiccating virus infected rabbit spinal cord over potassium hydroxide; the attenuated virus was active as an antigen [37]. Extensive research by Iiyima and Sakane [38, 39] on the simple desiccation of 18 different bacteria and 24 DNA and RNA bacteriophages based on the methods of Annear [40] attested to the general applicability of the method. Residual moisture and storage temperatures were important factors. Shapira and Kohn [41] have published a comprehensive account on the freeze-drying of bacteriophage T_4 and have shown that freeze-drying damaged the head coat leading to a loss of DNA. Often ghosts would be formed, completely empty of DNA. These studies should be consulted together with those of Steele (see above). Shapira and Kohn found that the rate of reconstitution was very important in terms of the overall final activity, that the addition of protective additives was not important, that after freeze-drying the viral activity was greatly reduced in contrast to simple freezing and that the higher the initial phage concentration the greater was the survival.

The preservation of mycobacteriophages by freeze-drying in the presence of 5% sodium glutamate and 0.5% gelatine has proved success-

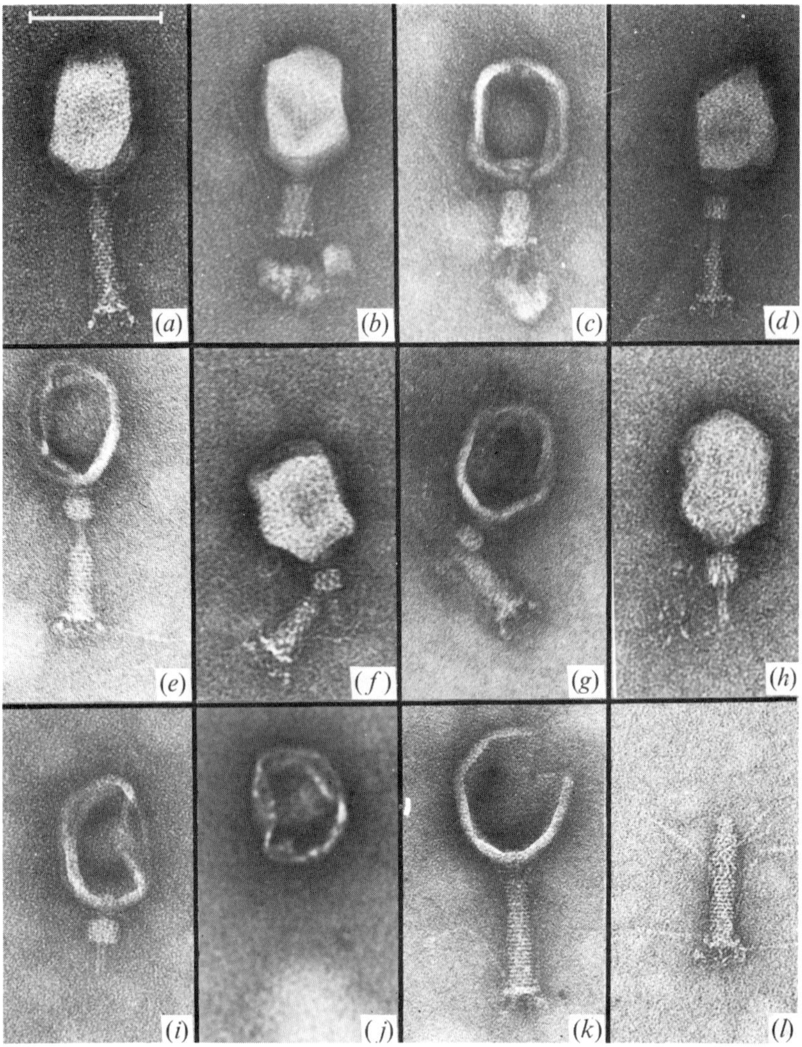

Figure 10.4 Electron microscopic appearances of frozen-thawed T_4B_0 phage particles. The white bar in (a) represents 100 nm. (b)–(1) are at the same magnification. (a) Normal virus particle; (b) contracted tail, full head; (c) contracted tail, empty head; (d) disjoined tail sheath, full head; (e) disjoined tail sheath, empty head; (f) fractured tail core, full head; (g) fractured tail core, empty head; (h) disjoined full head; (i) disjoined empty head; (j) disjoined empty head, without tail remnant; (k) empty torn head, normal tail; (1) disjoined tail. (From Steele, 1976 [34], by permission of the author and Cambridge University Press)

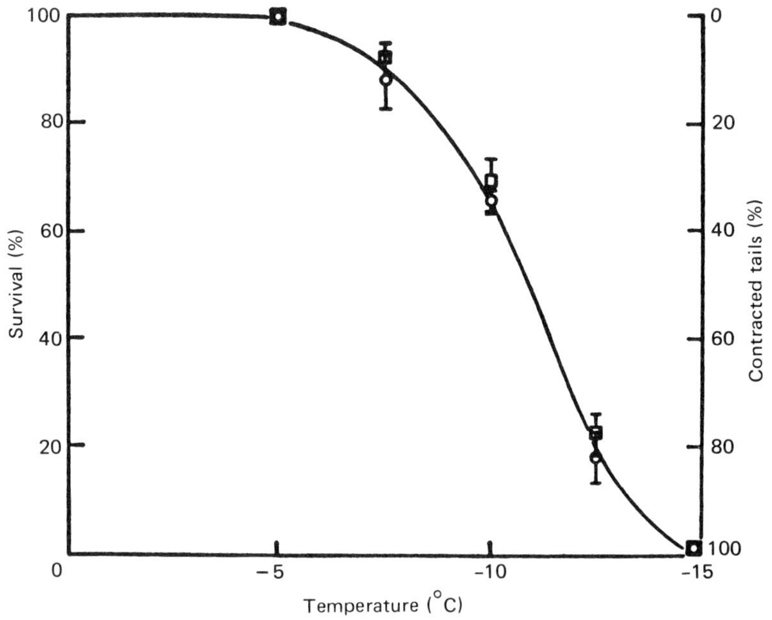

Figure 10.5 T_4B_0 phage suspended in 0.1 molal NaBr were cooled to the indicated temperatures at 1°C/min and then thawed slowly. The results show the percentage viabilities of the frozen-thawed samples as determined by plaque assay (○) and the percentage of particles with contracted tails counted electron microscopically (□). (From Steele, 1976 [34], by permission of the author and Cambridge University Press)

ful in the hands of Engel *et al* [42] and is routinely used, according to the authors, for mycobacterial phage typing.

Before tabulating the preservation by desiccation, freeze-drying or freezing of other viruses and bacteria a word of caution regarding the genetic effects of some of these processes is necessary. Freezing and thawing, *per se*, are not mutagenic [25, 43]. Desiccation, however, has been reported to be mutagenic to yeasts [26], and freeze-drying processes produce mutation in *E. coli* at several loci [25]. Mutation in *E. coli* is illustrated in Figure 10.6 but the mechanism is not understood although there is evidence for DNA strand breakage (Figure 10.7). The occurrence of mutations in *Serratia marcescens* during freeze-drying has been reported [44].

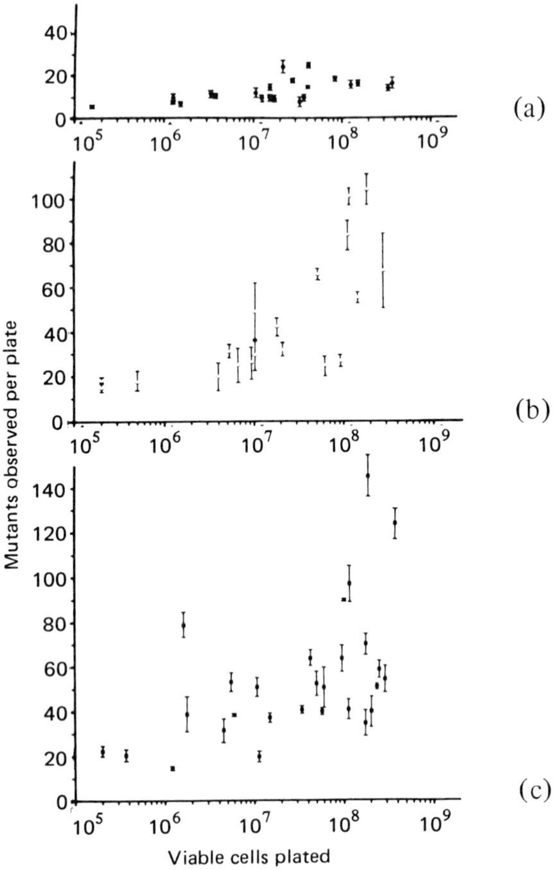

Figure 10.6 The effect of freeze-drying on mutation in *E. coli WP2 try-*. Data are from a number of different experiments. Mutants are revertants from tryptophan auxotrophy to prototrophy. (a) (●), control (not freeze-dried); (b) (△), freeze-dried in brain heart infusion broth (BHIB), with no additives; (c) (■), freeze-dried in BHIB with various additions. Additives included 5 and 10% dextran (10000 daltons), 12.5% sucrose, 5 and 10% methanol, 3 and 6% monosodium glutamate, 3% skimmed milk, and 10% peptone. Bars indicate standard errors. (From Ashwood-Smith and Grant, 1976 [25], by permission of the authors and publishers)

Viruses: The paper by Greiff and Rightsel [3] on the stability of viruses to freezing and freeze-drying is most useful. Recent papers on animal viruses are referenced in Table 10.2.

Before detailing various preservation methods that have proved use-

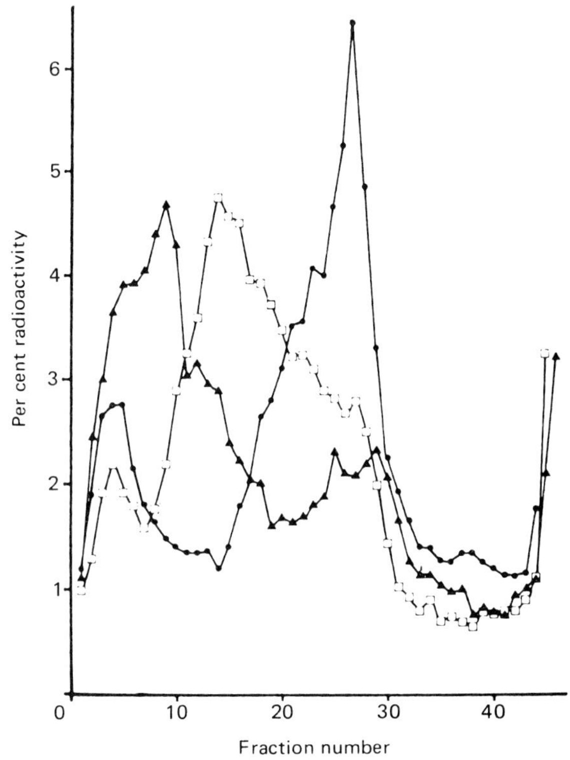

Figure 10.7 DNA profiles from freeze-dried bacteria after alkaline sucrose gradient analysis. (●), Control, non-freeze-dried bacteria (frozen and thawed bacteria have a similar distribution of DNA molecules [2, 3]); (□), freeze-dried bacteria analysed immediately after rehydration; (▲), freeze-dried bacteria analysed 30 min after rehydration and incubation at 37°C. Ordinate, percentage of radioactivity (^{3}H) incorporated into trichloroacetic acid-insoluble fraction. Abscissa, fraction number, 5-20% sucrose from left to right. (From Ashwood-Smith and Grant, 1976 [25], by permission of the authors and publishers)

ful for bacteria, an important point regarding contaminated liquid nitrogen was recently made by Schafer *et al* [45]. In many laboratories a number of different bacteria, viruses and cell lines are held in either glass or plastic ampoules under liquid nitrogen. Schafer and his colleagues noted that several glass ampoules containing vesicular stomatis virus (VSV) had shattered. An examination of the refrigerant liquid nitrogen revealed the presence of infectious virus. The paper states, 'The recovery of an infectious virus of minor clinical importance portends the potential biohazard of storing more virulent viruses under similar conditions'.

Table 10.2 Preservation of viruses by freezing or freeze-drying

Virus	*Comments*	*Reference*
ENTEROVIRUSES Poliovirus type I (PO-1) Coxsackie virus type A9 (CA-9) Poliovirus type 3 (PO-3) Echo virus type 7 (EC-7)	General stability data relative to pH, ions and above freezing temperatures	Salo, A.J. and Oliver, D.O. (1976). *Archives of Virology* **52**, 269
ENTEROVIRUSES *ECBO, VG(5)27*	Freeze-drying from aqueous TRIS gave excellent initial results but there was a rapid decrease in titre with storage time	McCafferty, D.F. and Woodside, W. (1977). *Journal of Pharmacy and Pharmacology* **29**, 10P
ENTEROVIRUSES (non-poliomyelitic strains) ECHO types 1–32 Coxsackie types B1-6 and A9	Viruses were stored at –18 to –20°C without stabilisers for 1-6 years with no apparent loss in activity	Florentyna, Z. and Taytsch-Kapulkin, (1972). *Experimental Medicine and Microbiology* **24**, No. I, 66
Rhinovirus 2 (HGP) Influenza virus B (B/Eng/13/65) Coxsackie virus A21 (strain loe)	Samples suspended in Agar–charcoal were stored at +4°C, –20°C and –70°C and compared with results obtained with conventional Hanks' medium. Hanks' medium was superior to the new medium for rhinoviruses	Chaniot, S.C.M. *et al.* (1974). *Archiv für Virusforschung* **44**, 401

Table 10.2 Preservation of viruses by freezing or freeze-drying (continued)

Virus	*Comments*	*Reference*
Bovine C-type leukaemia virus	The syncytia-inducing action was destroyed by freezing and thawing to $-70°C$ (this, with other evidence suggested the fusion factor was part of the virus particle)	Diglio, C.A. and Ferrer, J.F. (1976). *Cancer Research* **36**, 1056
Herpes simplex (HSV-2) (MS and EK strains)	A study in which several cooling and warming rates together with glycerol or DMSO yielded good cryopreservation. Fast cooling rates were best	Ashland, R.J. and Barnhart, E.R. (1975). *Journal of Clinical Microbiology* 2, 270
Herpesvirus (from wild turkeys) WTV	The virus was completely inactivated by one cycle of freezing and thawing. However, infected cells frozen with 7% DMSO allowed recovery of active virus	Grant, H.G. (1975). *Journal of Wildlife Diseases* **11**, 562
Vaccinia	Desiccation was studied. The effects of 5% sodium glutamate or 5% peptone were investigated. Stability was unaffected when residual moisture content was kept below 4%	Suzuki, M. (1973). *Cryobiology* **10**, 432. See also: *Ibid.* **10**, 435 (1973)

Table 10.2 Preservation of viruses by freezing or freeze-drying (continued)

Virus	*Comments*	*Reference*
Measles	See text reference [3]. Data on thermal inactivation. Equations for predicting 'shelf-life' of freeze-dried viruses	Greiff, D. and Rightsel, M.A. (1965). *Journal of Immunology* **94**, 395
Duck hepatitis virus	Thermostability data at room temperature, –20°C and –60°C. Useful in terms of elucidating strain differences	Hwang, J. (1975). *American Journal of Veterinary Research* **36**, 1683
PLANT VIRUSES		
Southern bean mosaic virus (SBMV)	Irreversible changes in SBMV virus were produced after freezing and thawing in water. *However, isolated RNA was fully active.* Considerable information on protein RNA interactions provided	Sehgal, O.P. and Das, P.D. (1975). *Virology* **64**, 180
Turnip yellow mosaic virus (TYMV) and several other simple plant viruses	Ultracentrifuge data/freezing and thawing study	Kaper, J.M. and Siberg, R.A. (1969). *Cryobiology* **5**, 366 Kaper, J.M. and Siberg, R.A. (1969). *Virology* **3**, 407
Chicory yellow mottle virus (CYMV)	Freezing and thawing studies	Quacquarelli, A. *et al.* (1972). *Journal of General Virology* **17**, 147

Table 10.2 Preservation of viruses by freezing or freeze-drying (continued)

Virus	*Comments*	*Reference*
Red clover mottle virus (RCMV)	Freezing, thawing and storage studies using ultracentrifugational analysis	Marcinkia, K. and Musil, M. (1977). *Acta Virologia* **21**, 71
Tobacco rattle virus (TRV)	Storage studies related to protein degradation	Mayo, M.A. and Cooper, J.I. (1973). *Journal of General Virology* **18**, 281

Table 10.3 Preservation of bacteria and fungi by freezing, freeze-drying or desiccation

Organisms	*Comments*	*Reference*
GENERAL REVIEW	The preservation of culture collections by freezing and drying	*Cryobiology* (1973). **10**, 347
GENERAL	Desiccation/freeze-drying as preservation methods. Bacteria suspended on glass beads prior to desiccation	Segerstrom, N. and Sabiston, C. B. (1977). *Journal of American Medical Technology* **39**, 22
GENERAL BACTERIA Genera and 130 species	Freeze-drying. Best results for most bacteria were obtained with a mixture of calf serum and lactose solution (5–10%). Different strains required careful and individual attention. High temperature testing (75 - 100°C) used as a good indicator for predicting survival. See reference 3 for predicting viral survival	Sourek, J. (1974). *International Journal of Systematic Bacteriology* **24**, 358
GENERAL Bacteria Fungi	Cryopreservation at −79°C over a 5 year period. Discussion of freezing rate, suspending medium and storage time. High survival and no change in selected biochemical characteristics. *Pseudomonas aeruginosa* was difficult to preserve. Glycerol (20%) was used for the preservation of moulds and dermatophytes	Kessler, H.J. and Redman, U. (1974). *Infection* **2**, 29

Table 10.3 Preservation of bacteria and fungi by freezing, freeze-drying or desiccation (continued)

Organisms	*Comments*	*Reference*
GENERAL BACTERIA	Desiccation. Preservation of bacteria on cellulose and alginate films	Annear, D.I. (1957). *Journal Applied Bacteriology* **20**, 17 Annear, D.I. (1962). ref. [40]
GENERAL Bacteria, yeast, fungi	Cryopreservation studies on solid media at −20°C similar to conditions associated with the freezing of foods	Defives, C. and Catteau, M. (1977). *Annals of Microbiology* **128A**, 239
GENERAL Microorganisms	Methodology	Cox, C.S. (1968). *Nature* **220**, 1139
MICROORGANISMS	Basic theory with relation to water content, etc.	Nei, T. (1973). *Cryobiology* **10**, 403

Table 10.3 Preservation of bacteria and fungi by freezing, freeze-drying or desiccation (continued)

Organisms	*Comments*	*Reference*
S. albus *Diplococcus pneumoniae* *Erysipelothrix* *Neisseria gonorrhoeae* *N. meningitidis* *Haemophilus influenzae* *Salmonella typhimurium* *Vibrio* *Leiptospira grippotyphosa* *Cl. tetani* *Cl. tetanomorphum* *Cl. botulinum* *Cl. perfringens* *B. thetaiotaomicron* *Treponema* *Mycoplasma* (various) *Candida albicans*	Vacuum-dried cells for preservation (fetal bovine serum + 10% saccharose). Cotton wool used as a support	Eyer, H. *et al.* (1975). *Zentralblatt fur Bakteriologie, Parasitenkund, Infektionskrankheiten und Hygiene, Abteilungen I Originale* **A231**, 243
BACTERIA 259 (strains belonging to 32 genera and 135 species including Gram+ aerobic; aerobic Gram−; psychrophilic bacteria, lactic acid bacteria, acetic acid bacteria)	Cryopreservation in 10% glycerol at −53°C for 16 months. Details of liquid paraffin storage. Discussion on the more sensitive strains such as *Pseudomonas, Acetobacter, Sarcina.* Several of the psychrophilic bacteria were very sensitive	Yamasato, K. *et al.* (1973) *Cryobiology* **10**, 453

Table 10.3 Preservation of bacteria and fungi by freezing, freeze-drying or desiccation (continued)

Organisms	*Comments*	*Reference*
Borrelia anserina (fowl tick fever)	Cryopreservation in serum, citrated blood +15% glycerol. Storage at −196°C. Storage times up to 150 days	Hart, L. (1971). *Australian Veterinary Journal* **46**, 455
ENTEROBACTERIACEAE *Gonococcus* Diphtheric bacteria Plague bacilli	Storage in solid agar (temperature data in this paper are not fully reported)	Suassuna, I. *et al.* (1977). *Review of Microbiology* 8, 16
Lactobacillus leichmannii	Cryopreservation. Cryoprotective agent was 6% malt extract. Younger cells were more susceptible than older cells	Johannson, E. (1972). *Journal of Applied Bacteriology* **35**, 415 See also: Johannson, E. (1972). *Journal of Applied Bacteriology* **35**, 423
GONOCOCCUS *Neisseria gonorrhoeae*	Desiccation over P_2O_5 on gelatin disks. Storage at −20°C for one year. Viability was good (paper in Japanese, summary in English)	Yamai, S. *et al.* (1975). *Japanese Journal of Bacteriology* **30**, 589
Vibrio cholerae	Cryopreservation at −60°C of infected mice (carcasses). Isolation of bacteria from stored, frozen mouse intestines was a useful method of cryopreservation	Lee, M. *et al.* (1975). *Australian Journal of Experimental Biology and Medical Science* **53**, 361

Table 10.3 Preservation of bacteria and fungi by freezing, freeze-drying or desiccation (continued)

Organisms	*Comments*	*Reference*
SALMONELLAE (4 different serotypes)	Low temperature (+4°C) storage in physiological saline for 6–8 years. No biochemical or serological changes were observed. Large decrease in viable titres. One strain not viable at end of observation period	Wilson, C.R. (1974). *Journal of the Association of Official Analytical Chemists* **57**, 1403
Salmonella typhi (str-dependent strain)	Attenuated oral strain used as a vaccine after freeze-drying was clinically useless. Vaccines were lyophilised at the Walter Reed Army Institute of Research, USA. No details of process included in paper	Levine, M.M., Dupont, H.L., Hornick, R.B., Snyder, M.J., Woodward, W., Gilman, R.H. and Libonati, J.P. (1976). *Journal of Infectious Diseases* **133**, 424
GASTROINTESTINAL BACTERIA *(lactobacilli, bifidobacteria, coliforms, anerobes)*	Cryopreservation at −70°C after freezing to −196°C. Segments of the colon and distal ileum were used. NB: preserved in their natural habitat. Interesting scanning electron micrographs	Davis, C.P. (1977). *Applied and Environmental Microbiology* **31**, 304
Bordetella pertussis (18 strains)	Cryopreservation at −70°C in 15% glycerol supplemented broth	Eckert, H.L. and Flaherty, D.K. (1972). *Applied Microbiology* **23**, 186

Table 10.3 Preservation of bacteria and fungi by freezing, freeze-drying or desiccation (continued)

Organisms	*Comments*	*Reference*
MYCOPLASMA *Mycoplasma gallisepticum* *Mycoplasma synoviae*	Freeze-drying from 12.5% sucrose. Growth stage important for *M. galli septicum*	Yugi *et al.* (1973). *Cryobiology* **10**, 464 See also page 223 for detailed discussions
E. coli (competent cells for transformation assays)	Cryopreservation at −82°C in 15% glycerol. No viable cells lost. Useful for DNA transformation studies	Morrison, D.A. (1977). *Journal of Bacteriology* **132**, 349
Leptospira (133 different strains)	Room temperature preservation of organisms suspended in Korthof's medium in sealed ampoules (paper in Chinese, summary in English)	Weiji, C. *et al.* (1976). *Acta microbologica sinica* **16**, 58
CHLAMYDIAE *Chlamydia trachomatis* (3 separate strains)	Cryopreservation at −196°C. Glycerol, DMSO and sucrose solutions investigated as cryoprotectives. Best results were obtained with 15% sucrose in the presence of 10% serum. Glycerol and DMSO were less good	Prentice, M. and Farrant, J. (1977). *Journal of Clinical Microbiology* **6**, 4
YEASTS *Saccharomyces cerevisiae*	Cryopreservation of yeast cells for antibiotic assays and growth studies. No cryoprotectives used	Beezer, A.E. *et al.* (1976). *Journal of Applied Bacteriology* **41**, 197

Table 10.3 Preservation of bacteria and fungi by freezing, freeze-drying or desiccation (continued)

Organisms	*Comments*	*Reference*
Saccharomyces cerevisiae *Pseudomonas aureofaciens* *Streptomyces tenebrarius*	Cryopreservation at −196°C. Additives used were 10% glycerol, with 5% of either lactose, maltose or raffinose	Daily, W.A. and Higgins, C.E. (1973). *Cryobiology* **10**, 364
RUST FUNGI (25 genera and species)	Spores retained viability for several years at −196°C. Details on heat and cold shock. Information on moisture content is considered important for the freezing of *P. striiformis*	Cunningham, J.L. (1973). *Cryobiology* **10**, 361
BASIDIOSPORES *Laccaria laccatta*	Cryopreservation at −196°C followed by desiccation. 25% of the cells remained viable	Stack, R.W. *et al.* (1975). *Mycologia* **67**, 167
Neurospora crassa (three mutant strains)	Freeze-dried cells. Characteristics not changed	Jong, S.C. and Darris, E.E. (1976). *Canadian Journal of Microbiology* **22**, 1062

The preservation of bacteria

A selected number of more recent papers dealing with storage, low temperature storage, cryopreservation and either desiccation preservation or freeze-drying preservation are detailed in Table 10.3.

The role of oxygen in modifying damage associated with water and the removal of water in microorganisms has been intensively investigated in recent years. It is a subject of some controversy although the bulk of the evidence suggests that in aerosols of microorganisms at certain relative humidities oxygen damage is important. In the development of 'damage' in freeze-dried microorganisms oxygen, probably through the involvement of free radicals, is assumed to play an important role. However, Cox and Heckly [46] working with freeze-dried preparations of *Serratia marcescens* have produced convincing evidence to the contrary (see Figure 10.8). In damage associated with freezing and

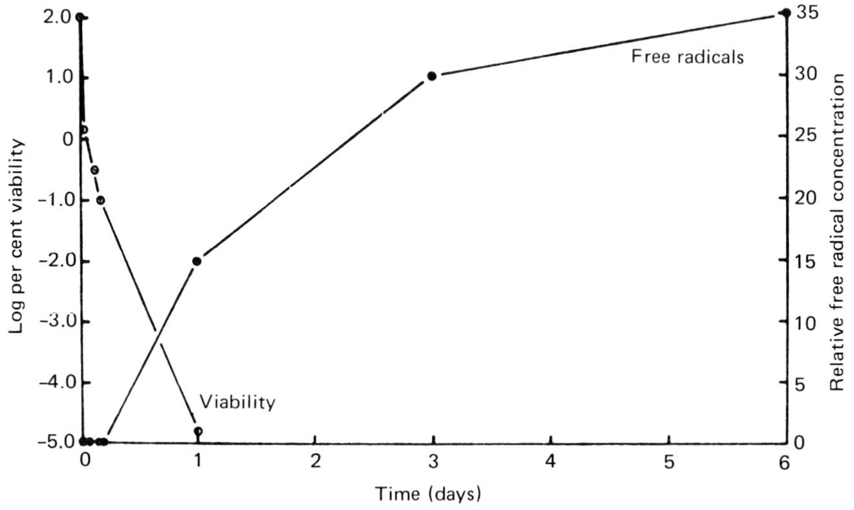

Figure 10.8 Free radical formation and loss of viability as a function of time in freeze-dried *Serratia marcescens* 8UK/2 under 10% oxygen at zero per cent relative humidity. (Reproduced by permission of the National Research Council of Canada from the *Canadian Journal of Microbiology*, Volume 19, pp. 189–194, 1973)

thawing, however, the evidence does not allow positive conclusions. A number of important papers concerned with oxygen are summarised in Table 10.4.

Table 10.4 Oxygen effects in freeze-dried, desiccated or frozen microorganisms

Microorganism	*Comments*	*Reference*
E. coli	Freeze-dried. O_2 effects on freeze-dried bacteria. Extreme sensitivity to oxygen even at low temperatures	Lion, M.B. and Bergmann, E.D. (1961). *Journal of General Microbiology* **24**, 191
E. coli	Freeze-dried. Electron spin resonance signals from freeze-dried cells in presence of O_2	Lion, M.B. *et al.* (1961). *Nature* **192**, 34
E. coli	Freeze-dried. Protection against O_2 mediated damage in freeze-dried cells. Serum albumin and lacto-protone are protective agents	Lion, M.B. (1963). *Journal of General Microbiology* **32**, 321
E. coli	Freeze-dried/aerosol. Data on effect of relative humidity on aerosols produced from freeze-dried bacteria	Cox, C.S. (1970). *Applied Microbiology* **19**, 604
E. coli	Aerosols. Toxic effects of O_2 on *E. coli* in aerosols. Effect of genes controlling x-radiation sensitivity	Cox, C.S. *et al.* (1971). *Journal of Hygiene* **69**, 661

Table 10.4 Oxygen effects in Freeze-dried, desiccated or frozen microorganisms (continued)

Microorganism	*Comments*	*Reference*
E. coli	Freeze-dried. Effect of Colicin E treatment on O_2 damage in freeze-dried cells	Israeli, E. and Kohn, A. (1972). *FEBS Letters* **26**, 323
E. coli/ bacteriophage T_4	Freeze-dried. Formation of phage from freeze-dried bacteria; effect of O_2	Israeli, E. and Shapira, A. (1973). *Journal of General Microbiology* **79**, 159
E. coli	Freeze-dried. Membrane malfunction in freeze-dried bacteria. O_2 associated damage to bacterial cytoplasmic membrane	Israeli, E. *et al.* (1974). *Cryobiology* **11**, 473
E. coli	Freeze-dried. Toxicity of O_2 to freeze-dried bacteria. Damage to DNA synthesis imitation complex?	Israeli, E. *et al.* (1975). *Cryobiology* **12**, 15
Pasteurella tularensis	Aerosol/freeze-dried. Survival in air and N_2 measured disseminated into aerosols from wet and freeze-dried preparations. Effect of relative humidity was measured	Cox, C.S. (1971). *Applied Microbiology* **21**, 482 See also: Cox, C.S. and Goldberg, L.J. (1972). *Applied Microbiology* **23**, 1

Table 10.4 Oxygen effects in freeze-dried, desiccated or frozen microorganisms (continued)

Microorganism	*Comments*	*Reference*
Serratia marcescens	Freeze-dried. O_2 was toxic to freeze-dried cells (1st order kinetics at low concentration; zero order kinetics at high concentration). Detected free radicals *not* involved in O_2 dependent loss of viability	Cox, C.S. and Heckly, R.J. (1972). *Journal of Microbiology* **19**, 189 See also: Cox, C.S. *et al.* (1974). *Canadian Journal of Microbiology* **20**, 1529, for aerosol data. See also: reference [44]
Bacteriophages T_3 and T_7	Freeze-thawing; freeze-drying and aerosolisation. Studies on viability and electron microscopy. Results indicated that dehydration associated with aerosolisation or freeze-drying did not cause death. Death was associated with rehydration	Cox, C.S. *et al.* (1974). *Journal of General Microbiology* **81**, 207

The part played by metabolic factors on damage produced by freeze-drying in microorganisms is complex. Papers by Postgate and Hunter [27], Morichi [15, 16] and Ray and Speck [13, 14] should be consulted. Whether or not cells are in the logarithmic or stationary growth phase when harvested for lyophilisation is important and well recognised. However, a number of other factors which have been reported recently are presented in Table 10.4. The association between sensitivity to freeze-drying and radiation damage is of considerable interest [47] and may not be unrelated to observations that DNA single strand breaks are produced by desiccation processes [25].

Recent reviews

Two excellent and comprehensive reviews have recently been published: by Heckly [48] on the preservation of microorganisms and by Schipper [49] on the preservation of yeasts and filamentous fungi.

Conclusion

It is clear from nearly all the published work that the freeze-drying of bacteria, viruses and some vaccines does not always result in perfect results. Clearly storage of lyophilised material at room temperature or at +4°C is inexpensive and very convenient, especially for field work. However, the decreasing titres with increasing storage time and the possible induction of mutations in the surviving population are far from ideal. Storage of desiccated microorganisms is associated with similar problems.

Cryopreservation in 10% DMSO or glycerol is more expensive over the long term but has the advantage of giving high survival that does not change with duration of storage at −196°C and is possessed of no genetic hazards. Indeed radiation damage is protected against by the low temperatures and the presence of cryoprotective agents which are known radioprotectors.

References

1 Cox, C.S. (1976). *Applied and Environmental Microbiology* **31**, 836
2 Strange, R.E. and Cox, C.S. (1976). In *The Survival of Vegetative Microbes.* Symposium Society for General Microbiology, p. 111
3 Greiff, D. and Rightsel, W.A. (1965). *Journal of Immunology* **94**, 395
4 Ashwood-Smith, M.J. and Grant, E.L. (1977). In *The Freezing of Mammalian Embryos.* Ciba Foundation Symposium, **52**, p. 251. K. Elliott and J. Whelan. (ed.). Amsterdam: Elsevier

5 Peled, O.N., Salvadori, A., Peled, U.N. and Kidby, D.K. (1977). *Journal of Bacteriology* **129**, 1648
6 Moats, W.A. (1971). *Journal of Bacteriology* **105**, 165
7 Mazur, P. (1966). In *Cryobiology*, p. 213. H.T. Meryman (ed.). New York: Academic Press
8 Mazur, P. (1970). *Science* **168**, 939
9 Calcott, P.H., Lee, S.K. and MacLeod, R.A. (1976). *Canadian Journal of Microbiology* **22**, 106
10 Calcott, P.H. and MacLeod, R.A. (1975). *Canadian Journal of Microbiology* **21**, 1724
11 Calcott, P.H. and MacLeod, R.A. (1975). *Canadian Journal of Microbiology* **21**, 1960
12 Lee, S.K., Calcott, P.H. and MacLeod, R.A. (1977). *Canadian Journal of Microbiology* **23**, 413
13 Ray, B. and Speck, M.L. (1972). *Applied Microbiology* **24**, 585
14 Ray, B. and Speck, M.L. (1972). *Applied Microbiology* **24**, 258
15 Morichi, T. (1969). In *Freezing and Drying of Microorganisms*, p. 53. T. Nei (ed.). University of Tokyo Press
16 Morichi, T. and Irie, R. (1973). *Cryobiology* **10**, 393
17 Kim, T.H. and Kubica, G.P. (1972). *Applied Microbiology* **24**, 311
18 Kim, T.H. and Kubica, G.P. (1973). *Applied Microbiology* **25**, 956
19 Kubica, G.P., Gontijo-Filho, P.P. and Kim, T. (1977). *Journal of Clinical Microbiology* **6**, 149
20 Boulanger, R.P. and Portelance, V. (1975). *Canadian Journal of Microbiology* **21**, 694
21 Richards, W.D. and Thoen, C.O. (1977). *Journal of Clinical Microbiology* **6**, 392
22 Pattyn, S.R. (1973). *Annales de la Société Belge de Médecine Tropicale* **53**, 645
23 Slosarek, M., Sourek, J. and Mikova, Z. (1976). *Cryobiology* **13**, 218
24 Dunbar, F.P., Pejovic, I., Cacciature, R., Peric-Golia, L. and Runyon, E.H. (1968). *Scandinavian Journal of Respiratory Diseases* **49**, 153
25 Ashwood-Smith, M.J. and Grant, E.L. (1976). *Cryobiology* **13**, 206
26 Hieda, K. and Ito, T. (1974). In *1st Intersessional Congress of International Associations of Microbiology. Tokyo, Japan. Section on Freezing and Drying*
27 Postgate, J.R. and Hunter, J.R. (1961). *Journal of General Microbiology* **26**, 367
28 Anderson, J.D. and Cox, C.S. (1967). Microbial survival. In *Symposium of the Society for General Microbiology XVII, Airborne Microbes*
29 Raccach, M., Rottem, S. and Razin, S. (1975). *Applied Microbiology* **30**, 167
30 Leibo, S.P. and Mazur, P. (1969). *Virology* **38**, 558
31 Steele, P.R.M. (1972). *Journal of Hygiene* **70**, 465
32 Steele, P.R.M. (1976). *Journal of Hygiene* **76**, 453
33 Steele, P.R.M. (1976). *Journal of Hygiene* **77**, 113
34 Steele, P.R.M. (1976). *Journal of Hygiene* **77**, 119
35 Steele, P.R.M., Davies, J.D. and Greaves, R.I.N. (1969). *Journal of Hygiene* **67**, 107
36 Steele, P.R.M., Davies, J.D. and Greaves, R.I.N. (1969). *Journal of Hygiene* **67**, 679
37 Pasteur, L. (1885). *Comptes Rendus des Séances de l'academie des Sciences 2'Semestre* (T. Cl. no. 17. p. 765)
38 Iiyima, T. and Sakane, T. (1973). *IFO Research Communications* **6**, 4
39 Iiyima, T. and Sakane, T. (1977). *IFO Research Communications* **8**, 60
40 Annear, D.I. (1962). *Australian Journal of Experimental Biology and Medical Science* **40**, 1
41 Shapira, A. and Kohn, A. (1974). *Cryobiology* **11**, 452
42 Engel, H.W.B., Smith, L. and Berwald, L.G. (1974). *American Review of Respiratory Disease* **109**(5), 561
43 Ashwood-Smith, M.J. (1965). *Cryobiology* **2**, 39
44 Servin-Massieu, M. and Cruz-Camirillo, R. (1969). *Applied Microbiology* **18**, 689
45 Schafer, T.W., Everett, J., Silver, G.H. and Came, P.E. (1976). *Science* **191**, 24
46 Cox, C.S. and Heckly, R.J. (1973). *Canadian Journal of Microbiology* **19**, 189
47 Tanaka, Y., Ohnishi, T., Takeda, Y. and Miwatani, T. (1975). *Biken Journal* **18**, 267
48 Heckly, R.J. (1978). In *Advances in Applied Microbiology* **24**, p. 1. D. Perlman, (ed.). New York: Academic Press
49 Schipper, M.A.A. (1978). In *Advances in Applied Microbiology* **24**, p. 215. D. Perlman (ed.). New York: Academic Press

CHAPTER ELEVEN

Plant Cells

G J Morris

Introduction

A serious loss of plant genetic resources is occurring because of the replacement of diverse primitive strains by introduced cultivars. At the same time there has been increasing interest in traditional crop varieties and their wild relatives for the immense range of genetic variation that they contain [1]. It is necessary, therefore, to conserve these resources so that as broad a genetic base as possible can be maintained for future breeding and crop development. Storage of seeds in the dried state (water content 2–5%) is applicable to many species. However, seed storage is not suitable for maintenance of discrete clones or for species with strong heterozygosity or a high probability of sterility. Also, there is an important group of plants with 'recalcitrant' seed [2], where a reduction in water content below a critical level (12–30% depending on species) decreases the period of viability; these seeds are thus difficult to preserve by conventional methods. Examples of 'recalcitrant' seeds include those of many of the large seeded trees; especially important are tropical crops such as rubber, oil palm, coffee and cocoa.

Shoot-tip and tissue cultures are now routinely used for the propagation of many species [3]; cryopreservation of these cultures may be an alternative method of gene banking. Biological systems stored in liquid nitrogen are genetically stable [4, 5], thus, the selection pressures inherent in any subculturing programme would be avoided. Many mutants of plant suspension cultures have been selected and characterised, a large number of which are genetically unstable. These novel genotypes are of potential importance in producing new cultivars of high yielding or disease resistant crops. The economic importance of such plants may take several years to evaluate, storage in liquid nitrogen would thus be a simple method of stabilising these cultures. Beside the need to conserve the gene pool of higher plants it is also essential to develop methods for the preservation of bryophytes, pteridophytes and algae.

The cryobiology of plant cells is an extensive subject encompassing the topics of chilling injury, frost hardiness, cold acclimatisation and cryopreservation. An understanding of the biochemical mechanisms of injury induced by environmental chilling and freezing may also allow practical methods of developing an increasing hardiness to chilling and freezing. This would be of major economic importance, as for example a 2°C increase in frost hardiness of wheat could extend production to new areas with a potential increase in world wheat production of between 25 and 40% [6]. However, this review will concentrate on the long term cryopreservation of plant cells and will only refer to the other related topics when they are of direct relevance. Some aspects of the cryobiology of plant cells have been covered in recent reviews [7–13]. Articles on the practical details of the cryopreservation of seeds [14, 15], plant tissue cultures [16–19] and algae [20] have been published.

Classical cryopreservation studies were carried out with animal cells and tissues. Direct application of the methods evolved was generally unsatisfactory for recovery of plant material. It has since become apparent that novel methods are required for the optimal recovery of plant cells from liquid nitrogen. For the long term storage of material in gene banks it is essential that the loss of viability induced by freezing and thawing is minimal, otherwise practical difficulties may arise in establishing new cultures when the viable cell concentration falls below a minimal inoculum level. Although freezing and thawing *per se* have not been demonstrated to be mutagenic the possibility remains that at low rates of recovery the selection of pre-existing freezing resistant mutants may occur. Finally, cryopreservation may select subpopulations of a cell type [21, 22], different stages of the cell cycle [23] and levels of cell ploidy [24]. This selectivity may have practical applications, i.e. in eliminating organised structures from cell suspensions, but must be avoided for purposes of genetic conservation.

The many variables which determine the response of cells to the stresses of freezing and thawing can be separated into two classes. First there are the intrinsic or cellular factors including the stage in the cell cycle, age of culture, growth temperature and post-thaw culture conditions. Second, there are the extrinsic or physical determinants such as the type and concentration of cryoprotective additive, rate of 'linear' cooling, two-step methods, warming rate and the final temperature attained. Although many of these factors are interrelated and all contribute to the viability of cells following freezing and thawing they will be covered as separate topics. Finally, theories of the mechanism of injury to plant cells induced by freezing and thawing will be discussed.

Viability assays

Many parameters have been used to determine the survival of plant cells following freezing and thawing [25]. Regrowth is a definitive indicator of viability and with some suspension cultures of algae and plant tissues the cell number can be quantitatively determined by colony formation in agar [26, 27]. In this assay system a colony can arise from either a single cell or from a cell clump. As many plant cells in liquid culture form aggregates, which are broken into smaller units by freezing and thawing [28, 29], it is essential to estimate cellular multiplicity before and after freezing (see Chapter 12). With suspension cultures that do not form colonies in agar, cell growth can be quantified by the increase in dry weight [30] or by most probable number techniques [31]. Alternatively the growth rate upon thawing may be examined spectrophotometrically [32], by direct cell counts [31, 33] or by the synthesis of chlorophyll [34].

In multicellular systems growth need not directly reflect cell survival. In embryogenic carrot suspension cultures the regrowth following freezing and thawing occurs by the formation of secondary embryos from a few viable meristematic cells; the majority of the primary embryo cells are non-dividing [19].

A number of indirect post-thawing tests are used as indicators of viability. These rapid tests allow the screening of a large number of thawed specimens, and, by monitoring their progress, aid the development of post-thawing techniques. Two assay systems are in common use; the practical details are included in Chapter 12.

Reduction of tetrazolium salts

Viable cells have the ability to reduce tetrazolium salts by mitochondrial action to formazan, a water-insoluble compound [35]. This reduction is quantitative and the formazan can be extracted and assayed spectrophotometrically.

Fluorescein diacetate staining

This assay applied to tissue culture cells by Widholm [36] involves the uptake and intracellular breakdown of fluorescein diacetate by esterase activity to fluorescein. Application of the stain is followed by comparative examination in tungsten and ultra-violet light. Only viable cells are visible under ultra-violet illumination and can therefore be estimated as a percentage of the total cells present.

Following a freeze-thaw cycle there is a quantitative agreement between the results obtained by colony formation and reduction of tetrazolium salts for *Haplopappus* cells [27, 37]. However, in other cell systems no such correlation occurs, the indirect assays usually overestimate recovery potential [17, 19, 38]. The results obtained by the reduction of tetrazolium salts and by fluorescein diacetate staining often do not provide correlating data [23, 32, 39]. Therefore, indirect assays must be examined critically and wherever practical, cellular viability should be assayed by some parameter of cell division.

Cell material

Selection of the cellular material to be cryopreserved is important, whenever possible a naturally resistant stage such as seeds [14, 15, 40, 41], zygotes [42], microspores [43] or pollen [44–48] should be used. It has been reported that ampoules containing dense cell suspensions show a higher percentage recovery compared to those with a lower cell density [38, 49]. However, these results are difficult to interpret due to the problem of minimum inoculum density.

Protoplasts of cells can be successfully cryopreserved [19, 50–52], often with a higher recovery rate than untreated cells [19, 51]. Results of this type suggest that the shrinkage of the plasmalemma away from the cell wall is a damaging factor of freezing injury.

Isolated meristems have several practical advantages over callus or suspension cultures. They can be cultured to produce shoots directly [3], without the problem of inducing organogenesis in hitherto undifferentiated tissues. Isolation and culture techniques for meristem tips can also be used to eliminate viral infections. Some of the control mechanisms of the whole plant are also believed to be retained by the isolated meristem tip, which may lead to greater genetic stability than is present in either callus or suspension cultures. However, the greater cellular organisation of these systems causes practical problems in developing cryopreservation protocols [53–58].

Growth before freezing

The cellular responses to the stresses of freezing and thawing are partly determined by the growth conditions prior to freezing. To obtain cells in a physiological state most resistant to the mechanisms of freezing injury the effects of age of culture, composition of the growth medium and temperature of incubation have been examined.

Age of culture

When plant tissue cultures are grown by a batch method, cells taken from the end of the lag phase or during the exponential phase of growth are more resistant to freezing injury than those in the early lag or stationary phase [16, 23, 32, 38, 49, 59–62]. It has been suggested that the susceptibility of stationary and early lag phase cells to freezing injury is associated with their large cell volume and the concomitant increase in the degree of cellular vacuolation and water content. By reducing the interval between subcultures the lag and stationary phases of growth can be virtually eliminated, the cells then obtained have a lower cell volume and a higher cytoplasm to vacuole ratio and are more uniform in their response to freezing injury [59, 62]. Using synchronous cultures, Withers [23] has shown that cells newly entered into the G1 phase of the cell cycle are particularly resistant to freezing injury. It has been suggested that for plant tissue cultures, exponential growth, with a long cell doubling time to permit the existence of an extended G1 phase of mitosis, is likely to produce a culture most suitable for cryopreservation [23].

In contrast, many strains of unicellular algae cells from the exponential phase of growth are more sensitive to the stresses of freezing and thawing than those from older cultures [20, 28, 63, 64]. Similar results have been obtained for batch cultures of bacteria (see Chapter 10). In algae the extent of vacuolation decreases in the stationary phase of culture as a result of the accumulation of lipid and other storage material [65].

Thus for plant suspension cultures it appears to be a general rule that a growth phase with minimal vacuolation has the maximum resistance to injury induced by freezing and thawing.

Growth medium

Growth media supplemented with penetrating cryoprotective additives [66, 67] or sugars [32, 59, 62, 68] have been used prior to freezing. The beneficial effects of growth in cryoprotectants may be due to the slow permeation of these compounds, especially of glycerol, into plant cells (see section on cryoprotective additives). Also, biochemical and ultrastructural alterations may occur following long periods of exposure to these additives [69, 70].

Growth of *Acer* and *Capsicum* cultures in medium with added mannitol (1.4–5.2%) combined with rapid subculturing, reduced the cell volume and increased cellular freezing tolerance [62]. Upon subculturing into mannitol-free medium the cells returned to normal volume, but retained a transient increased level of freezing resistance; therefore, the reduction in cell volume cannot fully explain the observed cryoprotective effects. Growth of plant cells in hypertonic solutions modifies the composition of the membrane fatty acids [71] and induces the intracellular accumulation of osmotically active compounds such as glycerol [72, 73], proline [74] and sugars [75, 76]; these may account for the protection against freeze-thaw injury. In contrast, the addition of mannitol to suspension cultures of *Daucus carota* did not result in either a reduction in mean cell volume or an increase in freezing tolerance. Thus species differences occur in the metabolic adaptations of cell cultures to hypertonic solutions.

With photosynthetic cultures the incorporation of glucose into the culture medium induces a shift from autotrophic to heterotrophic nutrition. Following such a metabolic shift cells of *Chlorella protothecoides* [77] and *Spirulina platensis* [64] become more sensitive to the stresses of freezing and thawing. In *C. protothecoides* the reduction in freezing tolerance is associated with the formation of a large vacuole and a decrease in the degree of unsaturation of the membrane fatty acids. With *C. emersonii* growth at suboptimal nutrient concentrations increased in the recovery upon thawing from −196°C [78]. In parallel with the increase in freezing resistance there was an accumulation of glycolipid, a reduction in the vacuole size and an increase in fatty acid unsaturation.

Growth temperature

The natural frost resistance of some species of freshwater algae [79], benthic seaweeds [80, 81] and higher plants [9] increases with decreasing environmental temperature. Poplar callus when cultured by a stepwise hardening process survived freezing to −196°C [82]. The resistance to freezing injury of callus cultures derived from fruit twigs [83, 84] *Chrysanthemum* [85] and *Helianthus* [86] was also increased following cold hardening. Cold acclimatisation has been found useful in increasing recovery from −196°C for woody twigs [87], meristems [57, 58] and algae [88–91]. However, there are no reports of successful hardening of suspension cultures of higher plant cells prior to liquid nitrogen storage.

Typical patterns of freeze-thaw survival developing during a period of cold acclimatisation are illustrated using a strain of *C. emersonii.*

At all growth temperatures examined there was a peak in the resistance to freezing injury. If these peak values are plotted against growth temperature (Figure 11.1), it can be seen that cells grown at 20°C are the most sensitive to the stresses of freezing and thawing. At higher and lower culture temperatures resistance to freezing injury developed.

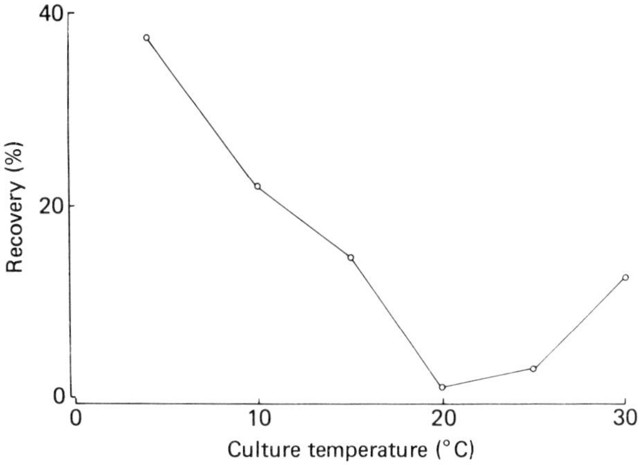

Figure 11.1 The effect of growth temperature on the subsequent recovery (%) of *Chlorella emersonii* from −196°C

The optimal growth temperature for this strain of *Chlorella* is in the range 20–25°C, there thus appears to be an inverse correlation between growth rate and resistance to freezing injury. This hypothesis has been confirmed for *C. emersonii* by reducing the concentration of nutrients or by adding metabolic inhibitors at 20°C. Treatments which reduced growth rate increased freezing resistance [78]. Therefore during the cold hardening of *C. emersonii* it is not the effects of low temperatures *per se* which are important but simply those of reduced metabolic rate. A similar relationship has been made for higher plants [9].

Extensive attempts to determine the nature of the cold hardening process have been made from comparisons of the biochemical composition of the frost-hardy and non-frost-hardy states. Many biochemical [9] and ultrastructural [86, 92, 93] alterations have been described. However, in all these studies it is impossible to differentiate between the adaptations necessary for cells to metabolise at the reduced temperature, those which are a consequence of the reduced growth rate and the specific modifications conferring freezing tolerance. Many biochemical models of the cold-hardening process utilise photosynthetic intermediates [94]. However, the apoplastidic alga *Prototheca*

[91] and the ciliate *Paramecium* [95] can be cold hardened. Therefore it is possible that several independent mechanisms are responsible for the cold-hardening phenomenon.

Chilling injury

Some plant cells are damaged following a reduction in temperature *per se*, this chilling injury is of two distinct types:

(1) Direct chilling injury or cold shock, which is expressed immediately upon a reduction in temperature. This direct injury is cooling rate dependent, with more damage occurring at rapid rates of cooling than at slow rates [28, 96]. This has been described for rapidly dividing cells of Cyanophyta [97, 98] and *Chlorella* [28, 99]. Some species of higher plants [9] and suspension cultures derived from them are also sensitive to cold shock injury.

(2) Indirect chilling injury, which requires a long period, often days, at the reduced temperature for cellular damage to be expressed. This cellular injury is independent of the rate of temperature reduction. Indirect chilling stress may occur during a period of cold acclimatisation (see section on growth temperature) or following a long time of exposure to a cryoprotective additive at reduced temperatures. Selection of atypical chilling resistant cells has been demonstrated to occur [100]. Therefore, cultures with a significant loss of viability due to the effects of indirect chilling injury should not be used for the conservation of genetic material.

Brandts [101] suggested that protein denaturation is the cause of chilling injury. The hydrophobic interactions which are important in maintaining the stability of many proteins become weaker at reduced temperatures resulting in irreversible denaturation. However, it is now generally accepted that both types of chilling injury are the result of cell membrane phase transitions [102–104]. The membrane phase change from a liquid–crystalline to a solid–gel state, that has been demonstrated in many cell types [105–108], is apparently damaging to plant cells which are sensitive to chilling. Cold shock injury is due to the direct loss of selective permeability of the membranes with a leakage of intracellular constituents occurring at temperatures below that of the phase change [97, 98]. Indirect chilling injury is a consequence of membrane phase changes upon cellular metabolism. The effects of phase changes on the membrane-bound enzymes of the mitochondria have been considered to be especially important [109]. However, ultrastructural studies indicate that damage to the tonoplast and rough endoplasmic reticulum occur before visible injury to the mitochondria or cellular membranes [110].

With cell types that are sensitive to the stress of chilling the extent of the damage can be reduced by altering the growth conditions before cooling. Growth in hypertonic solutions of mannitol reduces chilling sensitivity of some plant tissue culture cells [23]. Injury is a function of the growth temperature; cells of the blue-green algae *Anacystis nidulans* are extremely susceptible to cold shock when cultured at 35°C but are resistant when grown at 25°C [98]. This is possibly associated with the increase in the degree of unsaturation of the phospholipid fatty acids at the reduced growth temperature, which would result in a more fluid membrane with a lower temperature of phase transition [111]. Compounds can be added to alter membrane fluidity either directly by being incorporated into the membranes [112] or indirectly by modifying the degree of fatty acid unsaturation. Ethanolamine increases the degree of unsaturation of phospholipid fatty acids of tomato cells [113]. Plants with the modified membrane composition are more resistant to the stress of chilling [114].

Chilling injury *per se* may occur at temperatures below 0°C, even in plants not damaged by cooling above 0°C. Exposure to hypertonic solutions, formed by the removal of liquid water as ice during freezing, increase the susceptibility to a subsequent reduction in temperature [115, 116].

Cryoprotective additives

Even when the maximum intrinsic freezing tolerance has been induced by modifications in the growth conditions very few plant cell types survive freezing to and thawing from −196°C. Therefore it is necessary to include compounds to reduce cellular injury during a freeze-thaw cycle. The effects of a large number of potential cryoprotectants on the recovery of mammalian cells, especially erythrocytes and spermatozoa, have been examined. The results of such studies have provided insights into the mechanisms of freezing injury (see Chapter 1). Unfortunately, few basic studies have been reported for plant cells and it has been generally assumed that those additives used for the cryoprotection of mammalian cells, e.g. glycerol and DMSO will also be optimally effective for plant cells.

Practical methods for the cryopreservation of plant cells have been arrived at empirically and the details of the type and concentration of additive employed together with the rate of cooling have been tabulated [16, 17, 19]. However, in many of these studies the effects of different additives have been examined at a single rate of cooling. This can lead to false conclusions due to the complex interactions between type and concentration of additive and the rates of cooling and warming on cell survival.

Cryoprotective additives can be divided into three classes on the basis of their molecular weight and permeability to cells.

Penetrating additives

This class of additive includes the most commonly employed cryoprotectants, glycerol and DMSO. At the relatively high concentrations that they are used cytotoxic effects may occur. These are of two

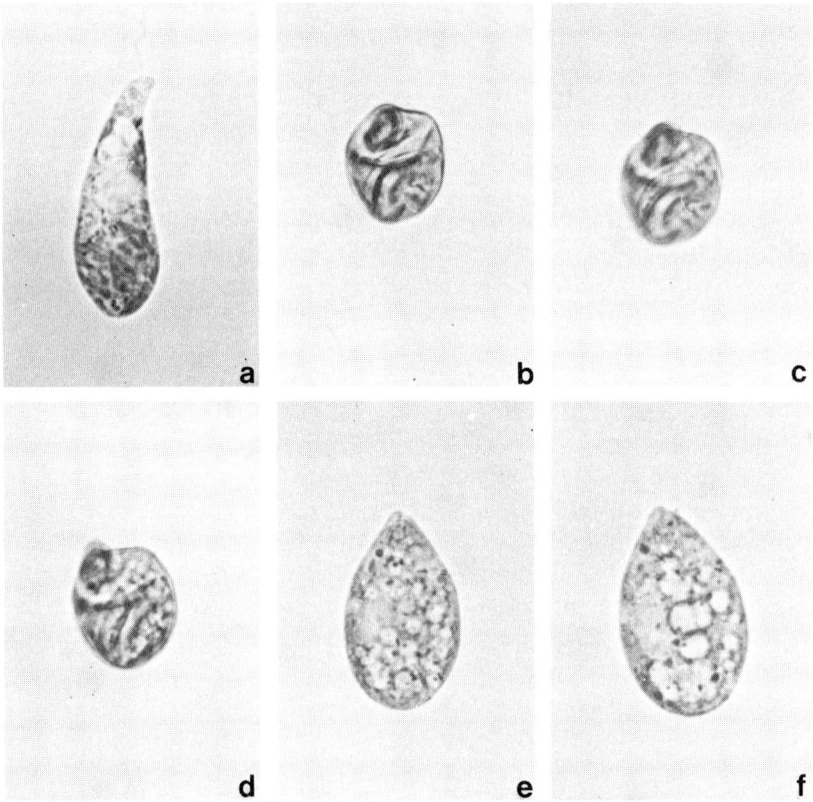

Figure 11.2 The morphology of *Euglena gracilis* after exposure to glycerol (1.5 M) for 5(b), 10(c), 15(d), 30(e) and 60(f) minutes at 20°C. Untreated control cell (a)

possible types, osmotic and biochemical. The osmotic effects include dehydration following the initial exposure to the additive and rehydration upon subsequent dilution. The extent of these water movements will be determined by the cellular permeability to the additive, one with a high permeability will not be as damaging on an osmotic basis as an additive with lower permeability. The biochemical consequences of high intracellular levels of cryoprotectants are poorly understood and will depend on the type of additive, temperature and duration of exposure. The observed toxicity is then a resultant of the osmotic and intracellular effects of these additives. The effects of three different penetrating additives at the same final concentration (1.5 M) on the morphology of *Euglena gracilis* at 20°C are illustrated in Figures 11.2 and 11.3. The recovery of cells following rapid dilution from equivalent treatments are shown in Figures 11.4 and 11.5.

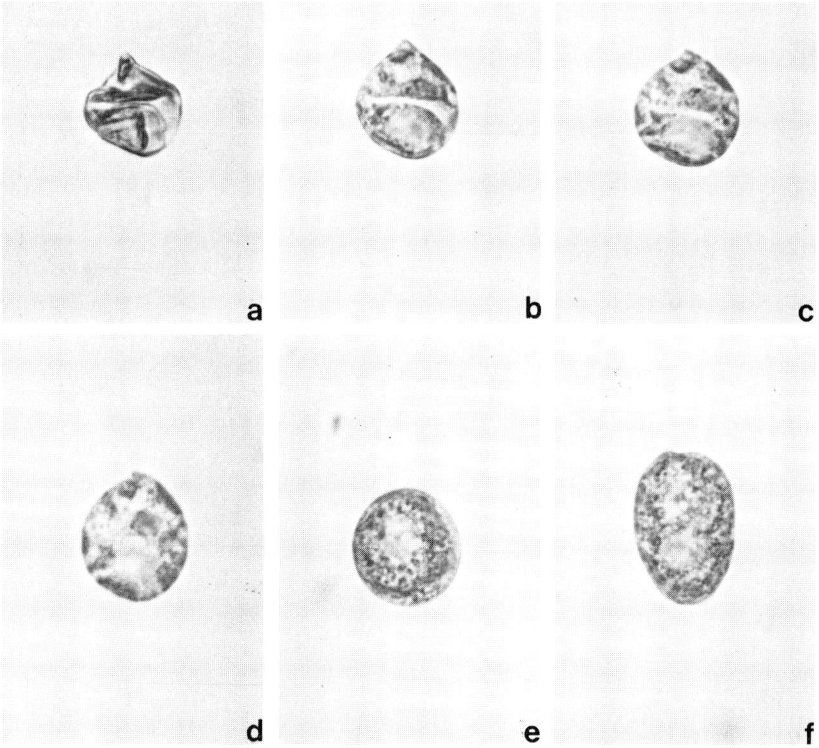

Figure 11.3 The morphology of *Euglena gracilis* after exposure to either DMSO (1.5 M) for 3(a), 15(b), 30(c) and 60(d) minutes, or methanol (1.5 M) for 3(e) and 5(f) minutes at 20°C

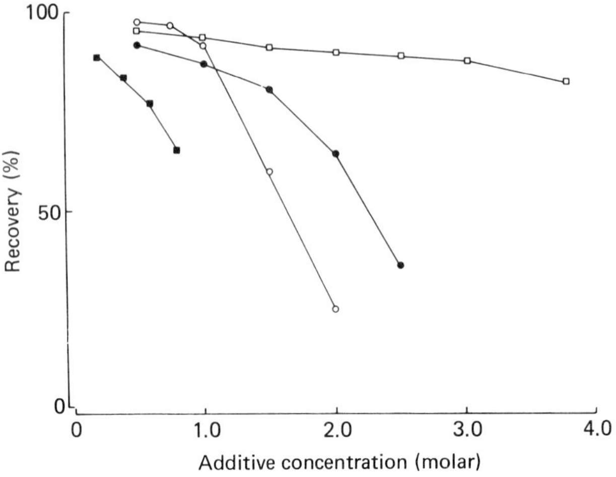

Figure 11.4 Recovery (%) of *Euglena gracilis* following a 15 minute exposure at 20°C to various concentrations of methanol (□), DMSO (●), glycerol (○) or glucose (■)

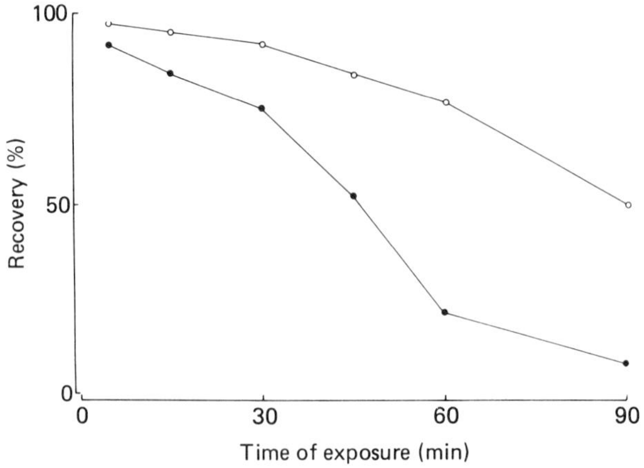

Figure 11.5 Recovery (%) of *Euglena gracilis* after exposure to DMSO (1.5 M) for different times at 20°C (●) or 0°C (○)

Glycerol: With *Euglena* this additive is apparently non-penetrating, cells remain shrunken until some cytotoxic event occurs, which results in uptake of glycerol and return to normal volume (Figure 11.2). With *Chlorella,* membrane integrity is destroyed following exposure to 1.5 M glycerol [117]. Towill and Mazur [27] have reported that glycerol is non-permeating for tissue culture cells of *Acer and Haplopappus.* This is confirmed by freeze-fracture studies in which a reduction in the size of intracellular ice crystals is only observed following incubation of plant tissues in glycerol for several days before freezing [118, 119]. In contrast, glycerol apparently penetrates suspension cultures of carrot cells [66]. Thus, species differences in cellular permeability may explain why this additive is protective for some plant cell types, but not others.

Dimethylsulphoxide: *Euglena* has a higher permeability coefficient for DMSO than for glycerol. Following an initial shrinkage there is an uptake of additive; after 15 minutes the cell resumes normal volume (Figure 11.2). The cells are non-motile and morphologically altered; these effects are reversible upon the removal of DMSO when motility, typical euglenoid morphology and cell viability are retained (Figure 11.4). Prolonged exposure to the intracellular effects of DMSO is cytotoxic (Figure 11.5).

Methanol: This additive penetrates cells very rapidly; *Euglena* returns to normal volume with 1 minute of exposure to the additive (Figure 11.3) and as with DMSO there are reversible morphological changes. Permeability studies on the characean alga *Nitella* indicate that methanol has a permeability coefficient similar to that of tritiated water and some 1.65×10^5 times greater than that of glycerol [120]. Unfortunately no values were given for DMSO. When compared with other additives on a molar basis methanol is surprisingly non-toxic to plant [63] (Figure 11.4) and other cell types [121, 122].

The cytotoxicity of penetrating additives is a function of the temperature and period of exposure. At low temperatures these compounds are less damaging (Figure 11.5), but this may only be a factor of the temperature coefficient of cellular permeability. With erythrocytes, glycerol does not have to penetrate before freezing to be cryoprotective [123]. However, the possibility arises that this additive may penetrate during freezing and thawing, as has been demonstrated for the normally impermeant compound sucrose [124]. To protect intracellular organelles it has been suggested that glycerol must penetrate cells before freezing [125]. Many plant cell types are damaged by exposure to low temperatures *per se,* thus it may be beneficial to add penetrating cryoprotectants to plant cells for relatively short periods of time at 20°C to ensure uptake of the additive and then to continue with the cooling procedures.

Non-penetrating additives of low molecular weight

Although sugars and amino acids have been used extensively in studies on the mechanisms of freezing injury to isolated chloroplasts [126–128], they have not been used to protect whole cells. However, as many of the growth media used for culturing plant cells contain both sugars and amino acids in relatively high concentrations they may be protective during freezing and thawing.

Non-penetrating additives of high molecular weight

Polyvinylpyrollidone (PVP) has been demonstrated to have some cryoprotective effects for plant suspension cultures [66] and algae [20, 28]. As yet it is not widely employed for the practical storage of plant cells. The effects of other potentially cryoprotective polymers such as hydroxyethyl starch and dextran on the survival of plant cells following freezing and thawing have not been reported.

Mixed additives

Many combinations of additives have been used, especially glucose + DMSO [32, 59, 129, 130] and glycerol + DMSO [19, 61, 66] and it has been reported that combined cryoprotective effects are greater than the sum of the component protective effects. The permeability characteristics of membranes are modified by DMSO [69] and the possibility then occurs that the normally impermeant or slowly penetrating additives will be taken up more rapidly in the presence of DMSO.

Cooling rates

The rate of cooling is an important factor in determining cell recovery upon subsequent thawing. With many cell types an optimum rate of cooling occurs, with cellular injury increasing at both faster and slower cooling rates (see Chapter 1). Results of this type have led to the formulation of the 'two-factor' hypothesis of freezing injury. However, an optimal rate of cooling has been described for a cell-free system, that of a suspension of the enzyme catalase [131]. Therefore in this system factors other than shrinkage and intracellular ice formation must account for the observed cooling rate dependence. Similar mechanisms of injury may be operational at the cellular level of freezing injury. The rate of cooling also affects membrane lipid lateral phase separations and the subsequent alterations to membrane protein topography [132]. This may also be a determinant of freezing injury.

The freezing process in plant suspension cultures has been studied by light microscopy [133] and freeze-fracture electron microscopy [134]. At low rates of cooling two types of cellular response may occur. The most commonly observed is that of total cell shrinkage due to the osmotic gradient between the cell contents and the hypertonic extracellular medium. Alternatively, ice may form between the cell wall and the protoplast. Due to the differences in chemical potential between supercooled water within the protoplast and the ice outside, shrinkage of the protoplast away from the rigid cell wall then occurs. At faster rates of cooling the probability of intracellular ice formation increases. Many workers have suggested that formation of intracellular ice *per se* is damaging for plant cells. However, following rapid thawing, cellular recovery has been demonstrated in some cultures with a high incidence of intracellular ice [77, 134].

There are two common methods by which cells are cooled, either by linear cooling rates or stepwise methods. These two protocols offer different practical approaches to cryopreservation and also give different types of information on freezing injury.

Linear cooling rates

Very few plant cell types survive the stresses of freezing to and thawing from −196°C in the absence of pretreatments or the addition of cryoprotectants. Some exceptions occur in the Cyanophyta [26], green algae [20, 26, 28], pollen [44–48], mosses [135, 136], dessicated embryos [137] and seeds [14, 15, 18, 40, 41]. Examples of the different types of response of cells to a range of cooling rates are illustrated in Figure 11.6. With *Chlorella protothecoides,* providing that thawing is rapid, cell recovery is independent of the rate of cooling. This alga is resistant to stresses of shrinkage and rehydration [138], a possible damaging factor at low rates of cooling [27]. At high rates of cooling, cell survival has been demonstrated despite the presence of intracellular ice [77]. This response to different rates of cooling is unusual but not unique, similar results have been reported for rotifers [139] and isolated lysosomes [140]. With a strain of *Scenedesmus* an optimal rate of cooling occurs, with survival decreasing at both faster and

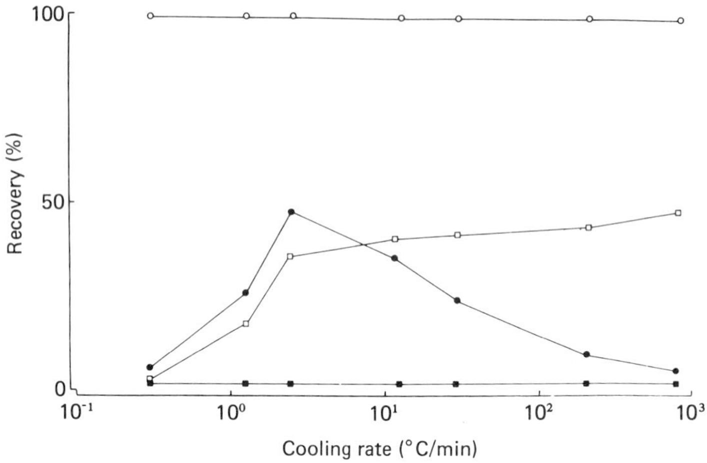

Figure 11.6 Recovery (%) of algal cells after cooling at different rates to −196°C. *Chlorella protothecoides* (○), *Scenedesmus quadricauda* (●) and *Chlorella emersonii* grown for either 7 days at 20°C (■) or 21 days at 4°C (□)

slower rates of cooling. The rate of cooling at which the optimal recovery occurs and the maximum survival obtained will be related to the cell volume and the water permeability. The diffusional water permeability coefficient has only been accurately determined for a few plant cell types. Stout *et al* [141] have developed a nuclear magnetic resonance technique for measuring the water permeability of plant cell membranes which avoids the problem of 'unstirred layers'. This technique, when applied at different temperatures, will enable the temperature coefficient of water permeability to be determined. This value, when integrated into thermodynamic equations [142, 143] will allow predictions to be made on the extent of dehydration at any temperature during cooling and the probability of intracellular ice formation.

The response to cooling of the majority of plant cells is typified by that of *Chlorella emersonii*. Survival is less than 0.1% at all rates of cooling and the recovery is only significantly increased by the addition of cryoprotectants or cold hardening. With *C. emersonii* cold hardening increases recovery at the faster rates of cooling examined. In contrast the recovery of the apoplastidic alga *Prototheca* after a similar period of cold acclimatisation is maximal at the lowest rates of cooling [91]. Therefore in any comparison of growth treatments upon cellular freezing tolerance a range of cooling rates must be examined as protective effects may not be observed at a single rate of cooling. The cellular reponse to a spectrum of cooling rates may also provide insights into the mechanism of acquired protection.

Effects of cryoprotective additives

The effects of cryoprotective additives on the recovery of cells over a range of cooling rates are illustrated with *Chlorella emersonii* using rapid thawing (for untreated controls see Figure 11.6). The cryoprotective effects of the penetrating additive DMSO are illustrated in Figure 11.7. The optimal recovery was observed with 0.75 and 1.5 M

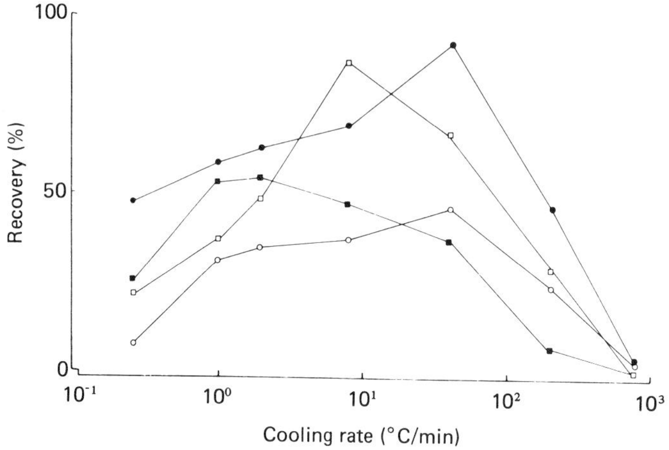

Figure 11.7 Recovery (%) of *Chlorella emersonii* after cooling at different rates to −196°C in the following concentrations of DMSO, 0.375 M (○), 0.75 M (●), 1.5 M (□) and 2.5 M (■)

DMSO, at higher (2.5 M) and lower (0.375 M) concentrations the recovery was lower. As the concentration of DMSO increased the cooling rate giving optimal recovery moved to lower values. Similar results were obtained with the other rapidly penetrating additive examined, methanol (Figure 11.8). On a molar basis methanol was a more effective cryoprotectant, especially at low rates of cooling.

Penetrating additives may protect cells by reducing the extent of the increase in extracellular salt concentration at any temperature during freezing [144, 145]. However, the extracellular osmolality will not be affected. As these additives can move into and out of the cells during freezing and thawing the degree of cellular dehydration will be reduced. The actual extent of cell shrinkage at any cooling rate

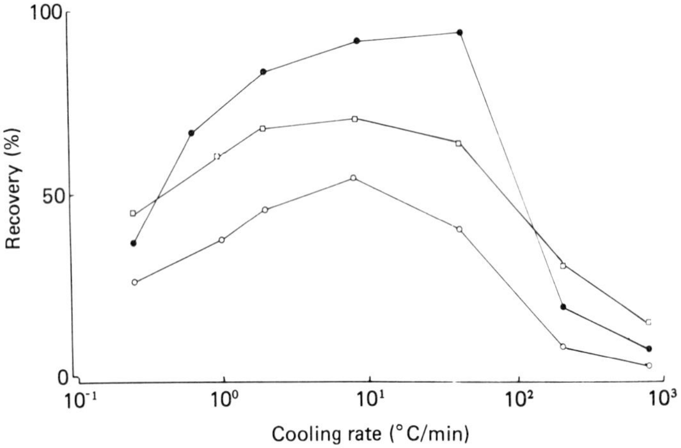

Figure 11.8 Recovery (%) of *Chlorella emersonii* after cooling at different rates to −196°C in the following concentrations of methanol, 0.5 M (○), 1.25 M (●) and 2.5 M (□)

will be determined by the salt concentration and the permeability of the cells to the additive. The temperature coefficient of permeability and the effect of the additive upon cellular water permeability will also be important. By reducing the loss of cellular water that occurs at any cooling rate there will be an increase in the probability of intracellular ice formation. Thus increasing the concentration of penetrating additive results in the optimum cooling rate moving to lower values.

With glycerol the optimal recovery following freezing and thawing was obtained with a concentration of 0.75 M (Figure 11.9). Glycerol will reduce the salt concentration at any temperature during freezing [144] as it does not freely permeate *Chlorella* [70, 117] and the extent of cellular dehydration will not be affected. As glycerol is a cryoprotective agent, even at low rates of cooling, it must be assumed that cellular shrinkage *per se* is not the sole damaging mechanism of freezing injury at low rates of cooling. Other stresses such as alterations in extracellular pH [146, 147] or thermal and dilution shock [115] may be important. Exposure to glycerol causes an osmotic loss of cellular water before freezing, therefore the probability of intracellular ice formation at rapid rates of cooling is reduced.

The high molecular weight additive PVP (average molecular weight 40000) does not induce a significant loss of water from cells before freezing. At rapid rates of cooling no protective effects were observed indicating that the probability of intracellular ice formation is not

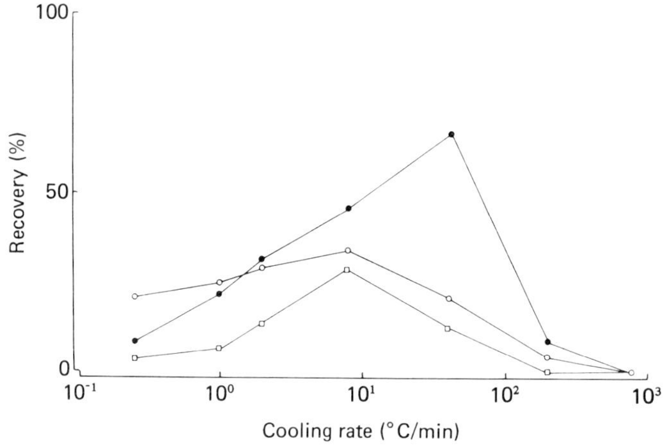

Figure 11.9 Recovery (%) of *Chlorella emersonii* after cooling at different rates to −196°C in the following concentrations of glycerol: 0.375 M (○), 0.75 M (●) and 1.5 M (□)

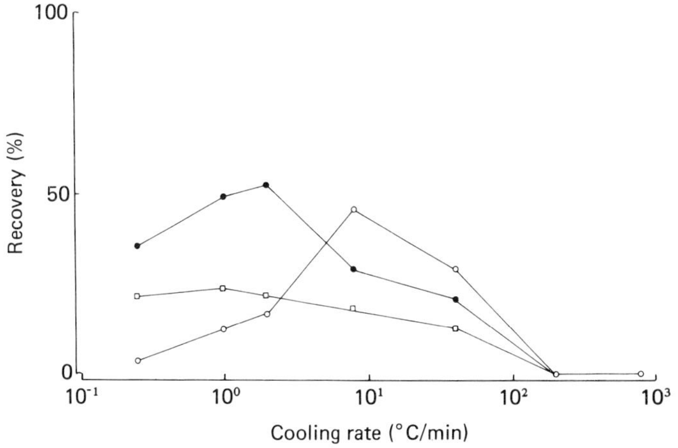

Figure 11.10 Recovery (%) of *Chlorella emersonii* after cooling at different rates to −196°C in the following concentrations of PVP (% w/v): 5 (○), 10 (●) and 15 (□)

affected (Figure 11.10). As the concentration of PVP is increased the cooling rate giving the optimal survival moves to lower rates. A similar pattern is observed with the penetrating additives, but the mechanism of cryoprotection afforded by PVP is not clear [148].

Combinations of cryoprotective additives have been reported to have a synergistic effect. This comparison has usually been made at one rate

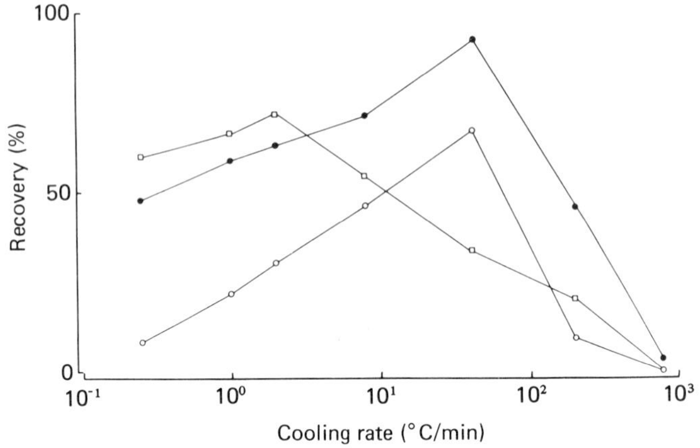

Figure 11.11 Recovery (%) of *Chlorella emersonii* after cooling at different rates to −196°C in glucose, 0.75 M (○), DMSO, 0.75 M (●) and glucose 0.75 M + DMSO 0.75 M (□)

of cooling, and in Figure 11.11 it can be seen that this comparison is not valid. With a mixture of glucose and DMSO the protective effects are only greater at low rates of cooling, whilst at faster rates an increase in injury, compared to that with a single additive, is observed. The net effect of combining additives is to move the optimal cooling rate to a lower value.

A comparison of three different higher plant cell types frozen at different rates to −196°C in DMSO (5%) shows differences in the optimal rates of cooling and the maximum survival obtained (Figure 11.12). The cooling rate giving optimal survival will be determined by the cell volume and permeability to water and DMSO. The actual recovery obtained will be a function of the cellular resistance to the mechanisms of injury with slow and fast cooling and to what extent these two factors overlap.

The previous methods have all been applicable to suspension cultures of cells. For more organised systems there are two methods that have been found to be effective.

'Dry' freezing

Somatic embryos and clonal plantlets of *Daucus carota* [58] and immature (hydrated) zygotic embryos of *Zea mays* [18] can be recovered from −196°C following a 'dry' freezing method. After uptake

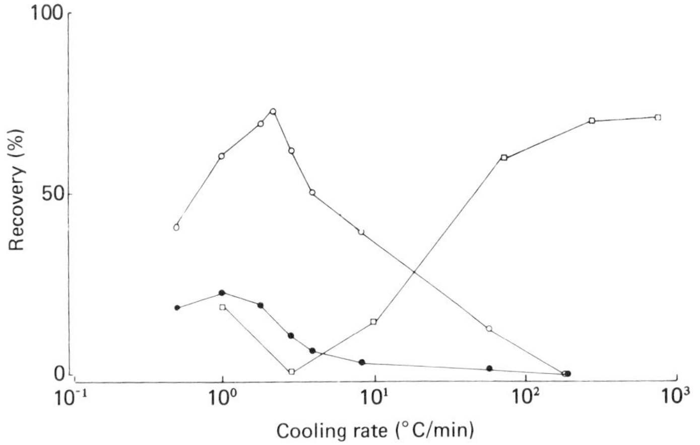

Figure 11.12 Recovery (%) after cooling at different rates to −196°C in DMSO (5% v/v) of suspension cultures of *Daucus carota* (○) and *Atropa belladona* (●) and shoot apices of *Dianthus caryophyllus* (□). (Redrawn from Nag, K.K. and Street, H.E. (1975). *Physiologica Plantarum* **34**, 260 and Seibert, M. and Wetherbee, P.J. (1977). *Plant Physiology* **59**, 1043)

of the cryoprotectant (e.g. DMSO 5%) the specimen is drained dry and enclosed in an envelope of aluminium foil. This is cooled at a rate of 1°C/min to −196°C. Upon thawing specimens are recovered simply by placing them on semi-solid medium. Viability is completely lost by immediate post-thaw washing, indicating that deplasmolysis is a critical factor in determining survival. Dried and partially hydrated seeds can also be successfully preserved by this method [14, 15, 18, 40, 41].

Rapid freezing

The optimum recovery of most plant suspension cultures occurs at low rates of cooling. However, successful recovery of meristems has only been reported following rapid cooling. For *Dianthus* the optimum rate of cooling is in the range 850–1100°C/min [57] and for *Lycopersicum* 20–55°C/min [54]. A 'dry' method involving direct plunging of cryoprotected meristems mounted on hypodermic needles into liquid nitrogen (cooling rate > 1000°C/min) is the only protocol by which meristems of *Solanum goniocalyx* have been recovered from −196°C [53]. Pollen [44–46] and woody twigs [87, 149–152] can also be recovered following rapid cooling. At these rates of cooling, intracellular ice formation is inevitable. However, the ice crystal size would be small and therefore, provided that thawing is rapid, recrystallisation injury will be minimised. Thus, rapid freezing may be the most appropriate method for preservation of small samples of low water content.

Intracellular ice crystal size may be further reduced by using a slurry of solid nitrogen or primary cooling fluids, e.g. isopentane, Freon or propane as coolants. Vitrification of *Chlorella* has been reported following spray freezing [70]. Intracellular ice could not be detected by freeze-fracture electron microscopy and following rapid thawing the cell recovery was high. The advantage of ultra-rapid cooling is that cryoprotectants are not essential. However, many practical problems exist in the spray freezing of large quantities of axenic cultures. An alternative approach is the application of high pressure during freezing, which reduces the amount of ice present at any temperatures [153, 154].

Two-step cooling

This method was initially described by Luyet and Keane [155] and subsequently applied to a number of cell types [156]. In this technique freezing occurs during the initial period of rapid cooling to a constant holding temperature. After maintaining the sample at this temperature it is then cooled to the storage temperature. This method has several practical advantages: it is simple to carry out, requires no controlled cooling rate equipment and low concentrations of cryoprotective additive are effective [157]. Two-step methods have been applied successfully to a number of plant suspension cultures [16, 18, 20, 24, 32, 158].

The recovery following two-step cooling of *Chlamydomonas reinhardii* is presented in Figure 11.13. The cells were suspended in methanol (2.5 M) and frozen rapidly to −30°C before either thawing, or plunging to −196°C and then thawing. The survival of cells thawed directly from −30°C decreased with increasing time of exposure. The recovery of cells warmed from −196°C increased with time at −30°C to a maximum value then slowly decreased in a manner parallel to the loss of viability at the holding temperature. At both higher (−20, −25°C) and lower (−35°C) temperatures a similar effect was observed, but the recovery was lower than that obtained at −30°C. At other holding temperatures (−10, −15, −40°C) there was no significant recovery of cells from −196°C. The optimal recovery obtained after two-step cooling is often greater than that following linear cooling rate methods. This increase in survival is not due to a convenient rate of cooling produced by immersing the sample in a constant temperature

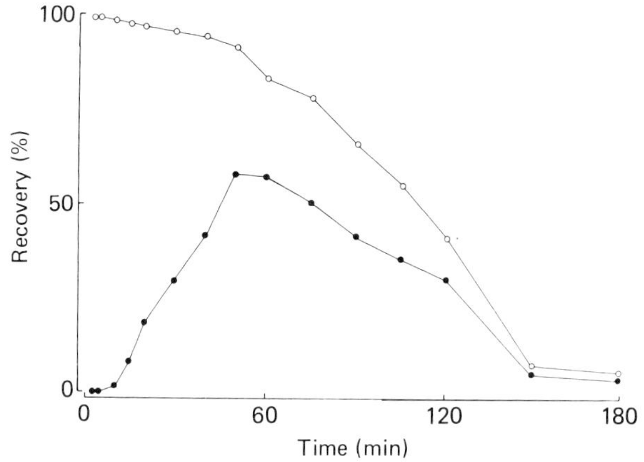

Figure 11.13 Recovery (%) of *Chlamydomonas reinhardii* frozen in methanol (2.5 M) at −30°C for different times before either thawing (○) or plunging to −196°C before thawing (●)

bath since protection is acquired with time at the holding temperature, once that temperature has been reached. Because of the rapid cooling to the holding temperature it is assumed that the cells are little affected by the potentially damaging hypertonic solutions. Shrinkage then occurs at the holding temperature [149, 150, 159] and it is this shrinkage that reduces the probability of intracellular ice formation on further rapid cooling to −196°C. At too high a prefreezing temperature there is insufficient shrinkage to reduce this probability. Cooling at too low a holding temperature results in intracellular ice formation at the holding temperature before sufficient cellular dehydration occurs. For a more detailed discussion of this cooling method see Chapters 1 and 12.

Warming rates and post-thaw manipulations

In all studies where the effect of warming rate on the survival of plant suspension cultures frozen to −196°C have been examined rapid warming gave maximal recovery. Slower warming has been found in isolated cases to be as effective as, but never superior to, rapid thawing [30, 32]. In contrast, following slow freezing of woody twigs [149, 150], callus [82] and meristems [58, 88], low rates of warming gave maximal survival. Slow warming would be predicted to reduce deplasmolysis injury. However, cells may be damaged by the recrystallisation of intracellular ice (if present) or the effects of long times of exposure to hypertonic solutions during the warming phase.

Upon thawing, cryoprotective additives are potentially cytotoxic. Polymers, sugars, glycerol and alcohols can be simply diluted out as plant cell growth is not inhibited by low levels of these additives. However, DMSO inhibits cellular metabolism, even at low concentrations and thus must be effectively removed by washing and centrifugation. A stepwise dilution of the cryoprotectant may avoid deplasmolysis injury [27].

With some cell types, division is resumed directly upon thawing [39], in others there is a prolonged lag phase between thawing and regrowth [38, 61, 62]. During the immediate post-thaw period, a transient [39] or permanent loss of viability [62] may occur. Latent damage incurred during freezing and thawing may contribute to this decline. It has also been suggested that post-thawing conditions are failing to 'rescue' some potential survivors [160]. It appears that the osmotic conditions are particularly critical in the early recovery phase. Protoplasts isolated from newly thawed cells of *Oryza sativa* are less stable than those from unfrozen cells, or from thawed cells tested after a recovery period [161].

The growth requirements of frozen and thawed cells are often different from those of unfrozen cells, indicating that some form of reversible metabolic injury occurs during freezing and thawing. This has been examined in detail for bacteria (see Chapter 10). With unicellular algae the recovery upon thawing is a function of the composition of the plating medium used to determine cellular viability. Survival is always higher when a complex medium is employed than when a minimal medium is used. For tomato meristems, hormonal control of some developmental processes is altered upon thawing [54], the presence of gibberellic acid being essential for the outgrowth of meristems only in frozen and thawed material. A period of reduced lighting aids the recovery of cryopreserved meristems of *Solanum* [53].

After thawing in cells which have the potential to divide, certain patterns of ultrastructural changes are consistently observed [18, 62, 160]. There is an increase in the average number of spherosomes per cell; these structures decline in number at the same time as there is an increase in smooth endoplasmic reticulum membrane. Rough endoplasmic reticulum becomes more prominent in cells as they approach the dividing state. It has been suggested that the spherosomes are a store for lipoprotein released by sublethal freezing injury. Minor modifications are also observed in the structure of the golgi bodies, mitochondria and plastids, but their appearance returns to normal morphology during the recovery lag phase. These ultrastructural observations suggest that different levels of freezing injury occur in cells, some of which may be reversible, depending upon the culture conditions upon thawing [18, 160, 162].

Sites of freezing injury

Few of the physicochemical theories of the mechanism of freezing injury describe a specific biochemical site for injury. Levitt [8, 9, 163], suggested that as water is converted to ice during freezing, structural proteins are forced into closer proximity. Because of this compaction, exposed sulphydryl groups in adjoining proteins or in adjacent strands of the same protein may become linked by disulphide bonds. When water returns to the cell during thawing, competition between hydration forces and the newly formed disulphide bonds cause the protein to become denatured. It has been argued that the SH ⇌ SS hypothesis accounts for all biochemical alterations that occur in cells killed by freezing and thawing and that the changes that accompany frost hardening would reduce the likelihood of disulphide bond formation. However, with the possible exception of higher plants which have been cooled slowly and thawed rapidly this theory is not now generally accepted [11, 12].

It is generally assumed that the primary site of freezing injury is the cellular membranes [92]. Upon thawing from a damaging temperature, alterations to the selective permeability of the cellular membranes

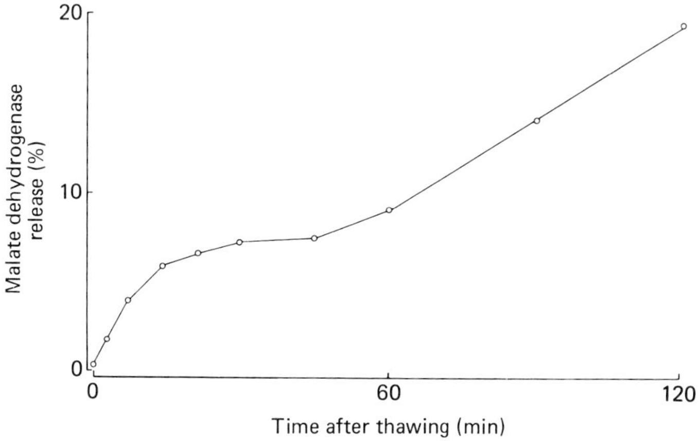

Figure 11.14 Release (%) of malate dehydrogenase from *Euglena gracilis* at different times after freezing to −6°C and thawing. Enzyme loss to the supernatant is expressed as a percentage of total intracellular activity

are observed [164–166]. The kinetics of the loss of membrane integrity, as determined by the leakage of a cytoplasmic enzyme (malate dehydrogenase), are biphasic (Figure 11.14). There is an initial rapid loss, which may be a direct reflection of freezing damage, then after an extended plateau level a secondary leakage of enzyme is observed. This secondary loss may be due to the activation of intracellular lytic enzymes. The reduction in cellular viability is directly proportional to the extent of the primary loss of membrane selective permeability (Figure 11.15). Palta and Li [167] have suggested that the first sign of injury in plant cells following freezing and thawing is inactivation of ion pumps.

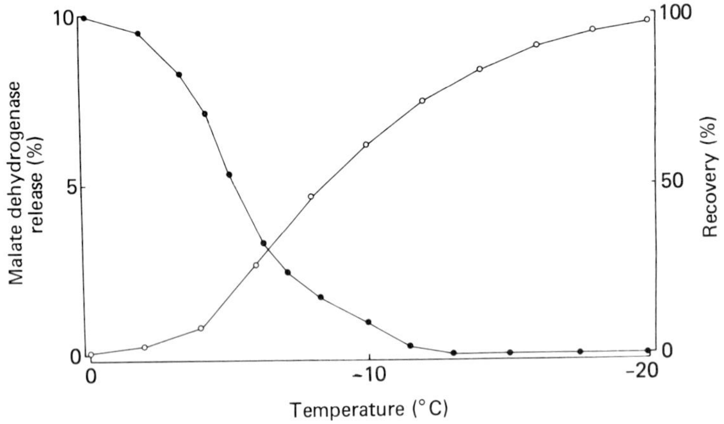

Figure 11.15 Cellular recovery (●) and release (○) of malate dehydrogenase from *Euglena gracilis* following cooling at 0.25°C/min to different temperatures before thawing

Modifications to the composition of the membranes which alter its biochemical characteristics would be predicted to modify the response of cells to freezing and thawing. An increase in membrane fluidity would reduce the temperature at which lipid lateral phase separations occur during cooling [105, 168]. Resistance to hypertonic stress is also correlated with a high value for membrane fluidity [71, 169, 170]. Cooling *per se* and the effects of hypertonic solutions, singly or in combination are possible damaging mechanisms during freezing and thawing [27, 115, 144]. However, there is no simple relationship between whole cell fatty acid composition and stress tolerance for a number of unicellular algae [158] and wheat cultivars [171]. It is possible that it is the composition of the lipids at a critical microenvironment within the membrane or of specific 'target' organelles which are important in determining the cellular response to freezing and thawing. From biochemical studies of the process of frost hardening it has been suggested that it is the total membrane content of a cell which is important rather than the actual membrane composition [172, 173].

In *Chlorella* there is a direct correlation between the presence of a large vacuole and the sensitivity to the stresses of rapid freezing and thawing [78]. With many other cell types, pretreatments which reduce the size of the vacuole, such as growth in hypertonic solutions, selection of exponential phase cells and cold acclimatisation, result in an increase in freezing tolerance. The response of cells to shrinkage and dehydration is correlated with vacuolar injury [174, 175]. Phospholipases which are compartmentalised within plant vacuoles [176] are activated upon thawing with the subsequent liberation of free fatty acids [78, 158, 177–179]. Other plant vacuolar enzymes are released following freezing and thawing [180–182]. Therefore, damage to the limiting membrane of the vacuole must occur at an early stage of freezing injury. This is confirmed in ultrastructural studies of potato leaf parenchyma, following freezing and thawing the tonoplast breaks down before the plasmalemma does [167].

The use of protoplasts offers a model system for elucidating the mechanisms of chilling and freezing injury. Removal of the cell wall allows direct biochemical access to the outside of the plasmalemma and overcomes the problem of mechanical interactions between the plasmalemma and cell wall. Studies on the response of protoplasts to hypertonic solutions [92] indicate that membrane alterations occur during dehydration. These are due to the disruption of intermolecular forces within the membrane and are either irreversible or slowly reversible. The maximum surface area that can be attained during osmotic dilution, before lysis occurs, is proportional to the extent of contraction previously experienced by the protoplast. Further studies of this type or of simpler model systems such as liposomes [183] may allow the primary site of injury induced by freezing and thawing to be determined. Only when this specific biochemical lesion is understood can efficient methods of cryopreservation be developed. Until then cryobiological experimentation must be of an empirical nature.

References

1 Frankel, O.H. and Hawkes, J.G. (eds.) (1975). *Crop Genetic Resources for Today and Tomorrow*. Cambridge: Cambridge University Press
2 Roberts, E.M. (1973). *Seed Science Technology* **1**, 499

3 Murashige, T. (1974). *Annual Review of Plant Physiology* **25**, 135
4 Ashwood-Smith, M.J. and Grant, E. (1977). In *The Freezing of Mammalian Embryos*, p. 251. K. Elliot and J. Whelan (eds.). Amsterdam: Elsevier
5 Lyon, M. (1976). In *Basic Aspects of Freeze Preservation of Mouse Strains*, p. 57. O. Muhlbock (ed.). Stuttgart: Gustav-Fisher Verlag
6 Weiser, C.J. (1978). In *Plant Cold Hardiness and Freezing Stress*, p. 391. P.H. Li and A. Sakai (eds.). New York: Academic Press
7 Burke, M.J., Gusta, L.V., Quamme, M.A., Weiser, C.J. and Li, P.H. (1976). *Annual Review of Plant Physiology* **27**, 507
8 Levitt, J. (1966). In *Cryobiology*, p. 495. H.T. Meryman (ed.). New York: Academic Press
9 Levitt, J. (1972). *Responses of Plants to Environmental Stresses.* New York: Academic Press
10 Li, P.H. and Sakai, A. (eds.). (1978). *Plant Cold Hardiness and Freezing Stress.* New York Academic Press
11 Mazur, P. (1969). *Annual Review of Plant Physiology* **20**, 419
12 Olien, C.R. (1967). *Annual Review of Plant Physiology* **18**, 387
13 Weiser, C.J. (1970). *Science* **169**, 1269
14 Sakai, A. and Noshiro, M. (1975). In *Crop Genetic Resources for Today and Tomorrow*, p. 317. O.H. Frankel and J.G. Hawkes (eds.). Cambridge: University Press
15 Stanwood, P.C. and Bass, L.N. (1978). In *Plant Cold Hardiness and Freezing Stress*, p. 361. P.H. Li and A. Sakai (eds.). New York: Academic Press
16 Sakai, A. and Sugawara, Y. (1978). In *Plant Cold Hardiness and Freezing Stress*, p. 345. P.H. Li and A. Sakai (eds.). New York: Academic Press
17 Bajaj, Y.P.S. and Reinert, J. (1977). In *Applied and Fundamental Aspects of Plant Cell, Tissue and Organ Culture*, p. 757. J. Reinert and Y.P.S. Bajaj (eds.). Berlin: Springer Verlag
18 Withers, L.A. (1978). In *Proceedings of the International Association for Plant Tissue Culture* (in press)
19 Withers, L.A. and Street, H.E. (1977). In *Plant Tissue Culture and its Biotechnical Applications*, p. 226. W. Barz, E. Reinhard and M.H. Zenk (eds.). Berlin: Springer Verlag
20 Morris, G.J. (1978). *British Phycological Journal* **13**, 15
21 Farrant, J., Knight, S.C. and Morris, G.J. (1972). *Cryobiology* **9**, 516
22 Knight, S.C., Farrant, J. and Morris, G.J. (1972). *Nature (New Biology)* **239**, 88
23 Withers, L.A. (1978). *Cryobiology* **15**, 87
24 Shillito, R.D. (1978). PhD Thesis, University of Leicester
25 Palta, J.P., Levitt, J. and Stadelmann, E.J. (1978). *Cryobiology* **15**, 249
26 Holm-Hansen, O. (1963). *Physiologia Plantarum* **16**, 530
27 Towill, L.E. and Mazur, P. (1976). *Plant Physiology* **57**, 290
28 Morris, G.J. (1976). *Archives of Microbiology* **107**, 57
29 McGrath, M.S. and Daggett, P.M. (1977). *Canadian Journal of Botany* **55**, 1794
30 Dougall, K. and Wetherell, D.F. (1974). *Cryobiology* **11**, 410
31 Saks, N.M. (1978). *Cryobiology* **15**, 563
32 Sugawara, Y. and Sakai, A. (1974). *Plant Physiology* **54**, 722
33 Hwang, S. and Horneland, W. (1965). *Cryobiology* **1**, 305
34 Quatrano, R.S. (1968). *Plant Physiology* **43**, 2057
35 Steponkus, P.L. and Lanphear, F.O. (1967). *Plant Physiology* **42**, 1423
36 Widholm, J.M. (1972). *Stain Technology* **42**, 189
37 Towill, L.E. and Mazur, P. (1975). *Canadian Journal of Botany* **53**, 1097
38 Bajaj, Y.P.S. (1976). *Physiologia Plantarum* **37**, 263
39 Sala, F., Cella, R. and Rollo, F. (1979). *Physiologia Plantarum* **45**, 170
40 Gresshoff, P.M. and Gartner, E. (1977). *Arabidopsis Information Service* **14**, 12
41 Mumford, P.M. and Grout, B.W.W. (1978). *Annals of Botany* **42**, 255
42 Bennoun, P. (1972). *Compte rendu hébdomadaire des séances de l'Académie des sciences, Paris, Series D* **275**, 1777
43 Coulibaly, Y. and Demerly, Y. (1978). *Compte rendu hébdomadaire des séances de l'Académie des sciences, Paris, Series D* **286**, 1065
44 Anderson, J.O., Nath, J. and Harner, E.J. (1978). *Cryobiology* **15**, 469
45 Barnabas, B. and Rajki, E. (1976). *Euphytica* **25**, 747

46 Nath, J. and Anderson, J.O. (1975). *Cryobiology* **12**, 81
47 Visser, T. (1955). *Mededelingen van de Landbouwhoogeschool te Wageningen* **55**, 1
48 Weatherhead, M.A., Grout, B.W.W. and Henshaw, G.G. (1979). *European Potato Journal* (in press)
49 Nag, K.K. and Street, H.E. (1975). *Physiologia Plantarum* **34**, 260
50 Siminovitch, D. (1977). *Cryobiology* **14**, 690
51 Siminovitch, D., Singh, J. and de la Roche, I.A. (1978). *Cryobiology* **15**, 205
52 Wiest, S.C. and Steponkus, P.L. (1977). *Plant Physiology* **58**, 4
53 Grout, B.W.W. and Henshaw, G.G. (1978). *Annals of Botany* **42**, 1227
54 Grout, B.W.W., Westcott, R.J. and Henshaw, G.G. (1978). *Cryobiology* **15**, 478
55 Seibert, M. (1976). *Science* **191**, 1178
56 Seibert, M. and Wetherbee, P.J. (1974). *In Vitro* **10**, 350
57 Seibert, M. and Wetherbee, P.J. (1977). *Plant Physiology* **59**, 1043
58 Withers, L.A. (1979). *Plant Physiology* **63**, 460
59 Finkle, B.J., Sugawara, Y. and Sakai, A. (1975). *Plant Physiology* **56**, 80
60 Hollen, L.B. and Blakely, L.M. (1975). *Plant Physiology* **56**, 39
61 Nag, K.K. and Street, H.E. (1973). *Nature* **245**, 270
62 Withers, L.A. and Street, H.E. (1977). *Physiologia Plantarum* **39**, 171
63 Morris, G.J. and Canning, C.E. (1978). *Journal of General Microbiology* **108**, 27
64 Takano, M., Suedo, J.I., Takahira, O. and Gyozo, T. (1973). *Cryobiology* **10**, 440
65 Pyliotis, N.A., Goodchild, D.J. and Grimme, L.H. (1975). *Archives of Microbiology* **103**, 259
66 Nag, K.K. and Street, H.E. (1975). *Physiologia Plantarum* **34**, 254
67 Gresshoff, P.M. (1977). *Plant Science Letters* 9, 23
68 Latta, R. (1971). *Canadian Journal of Botany* **49**, 1253
69 Lyman, G.H., Preisler, H.D. and Papahadjopoulus, D. (1976). *Nature* **262**, 360
70 Plattner, H., Fischer, W.M., Schmitt, W.W. and Bachman, L. (1972). *Journal of Cell Biology* **53**, 116
71 Stuiver, C.E.E., Kuiper, P.J.C. and Marschner, H. (1978). *Physiologia Plantarum* **42**, 124
72 Ben-Amotz, A. and Avron, M. (1973). *Plant Physiology* **51**, 875
73 Borowitzka, L.J. and Brown, A.D. (1974). *Archives of Microbiology* **96**, 37
74 Stewart, G.R. and Lee, A.R. (1974). *Planta* **120**, 279
75 Hellebust, J.A. (1976). *Canadian Journal of Botany* **54**, 1735
76 Kirst, G.O. (1975). *Zeitschrift für Pflanzenphysiol* **76**, 316
77 Morris, G.J., Clarke, K.J. and Clarke, A. (1977). *Archives of Microbiology* **114**, 249
78 Morris, G.J. and Clarke, A. (1978). *Archives of Microbiology* **119**, 153
79 Scholm, H.E. (1968). *Protoplasma* **65**, 97
80 Bird, C.J. and McLachan, J. (1974). *Phycologica* **13**, 215
81 Lyutova, H.I., Zavadskeya, I.G., Luknitskaya, A.F. and Feldman, N.L. (1967). In *The Cell and Environmental Temperature,* p. 116. A.S. Troshin (ed.). Oxford: Pergamon Press
82 Sakai, A. and Sugawara, Y. (1973). *Plant Cell Physiology* **14**, 1201
83 Ogolevets, I.V. (1976). *Fiziologiya Rastenii* **23**, 139
84 Tumanov, I.I., Butenko, R.G. and Ogolevets, I.V. (1968). *Fiziologiya Rastenii* **15**, 749
85 Steponkus, P.L. and Bannier, L. (1971). *Cryobiology* **8**, 386
86 Sugawara, Y. and Sakai, A. (1978). In *Plant Cold Hardiness and Freezing Stress,* p. 197. P.H. Li and A. Sakai (eds.). New York: Academic Press
87 Sakai, A. (1965). *Plant Physiology* **40**, 882
88 Sakai, A. and Nishiyama, Y. (1979). *Horticultural Science* (in press)
89 Hatano, S., Sadakane, H., Tutumi, H. and Watanabe, T. (1976). *Plant Cell Physiology* **17**, 451
90 Morris, G.J. (1976). *Archives of Microbiology* **107**, 309
91 Morris, G.J. (1976). *Journal of General Microbiology* **94**, 395
92 Steponkus, P.L. and Weist, S.C. (1978). In *Plant Cold Hardiness and Freezing Stress,* p. 75. P.H. Li and A. Sakai (eds.). New York: Academic Press
93 Steponkus, P.L., Garber, M.P., Myers, S.P. and Lineberger, R.D. (1977). *Cryobiology* **14**, 305

94 Levitt, J. (1967). In *Cellular Injury and Resistance in Freezing Organisms,* p. 51. A. Asahina (ed.). Hokkaido University, Sapporo, Japan: Institute of Low Temperature Science
95 Polyansky, G.I. (1963). *Acta Protozoologica* **1**, 166
96 MacLeod, R.A. and Calcott, P.H. (1976). In *The Survival of Vegetative Microbes,* p. 81. T.R.G. Gray and J.R. Postgate (eds.). Cambridge: University Press
97 Jansz, E.R. and MacLean, F.I. (1973). *Canadian Journal of Microbiology* **19**, 381
98 Siva, V., Rao, K., Brand, J.J. and Myers, J. (1977). *Plant Physiology* **59**, 965
99 Tischner, R., Helse, K.P., Nelle, R. and Lorenzen, H. (1978). *Planta* **139**, 29
100 Dix, P.J. and Street, H.E. (1976). *Annals of Botany* **40**, 903
101 Brandts, J.F. (1967). In *Thermobiology,* p. 25. A.H. Rose (ed.). New York: Academic Press
102 Lyons, J.M. (1973). *Annual Review of Plant Physiology* **24**, 445
103 Raison, J.K. (1973). *Symposia of the Society for Experimental Biology* **27**, 285
104 Simon, E.W. (1974). *New Phytologist* **73**, 377
105 James, E.R. and Branton, D. (1973). *Biochimica et Biophysica Acta* **323**, 378
106 Kleeman, W. and McConnell, H.M. (1974). *Biochimica et Biophysica Acta* **345**, 220
107 Speth, V. and Wunderlich, F. (1973). *Biochimica et Biophysica Acta* **291**, 621
108 Verkley, A.J., Ververgaert, P.H.J., van Deenan, L.L.M. and Elbers, P.F. (1972). *Biochimica et Biophysica Acta* **288**, 326
109 Lyons, J.M. and Raison, J.K. (1970). *Plant Physiology* **45**, 386
110 Niki, T., Yoshida, S. and Sakai, A. (1978). *Plant Cell Physiology* **19**, 139
111 Wilson, J.M. and Crawford, R.H.M. (1974). *Journal of Experimental Botany* **25**, 121
112 Sheetz, M.P. and Singer, S.J. (1974). *Proceedings of the National Academy of Sciences* **71**, 4457
113 Waring, A.J., Breidenbach, R.W. and Lyons, J.M. (1976). *Biochimica et Biophysica Acta* **443**, 157
114 Ilker, R., Waring, A.J., Lyons, J.M. and Breidenbach, R.W. (1976). *Protoplasma* **90**, 229
115 Farrant, J. and Morris, G.J. (1973). *Cryobiology* **10**, 134
116 Green, F.A. and Jung, C.Y. (1977). *Journal of Membrane Biology* **33**, 249
117 Syrett, P.J. (1973). *New Phytologist* **72**, 37
118 Fineran, B.A. (1970). *Journal of Microscopy* **92**, 85
119 Robards, A.W. and Parish, G.R. (1970). *Journal of Microscopy* **93**, 61
120 Collander, R. (1954). *Physiologica Plantarum* **7**, 420
121 Ashwood-Smith, M.J. and Lough, P. (1975). *Cryobiology* **12**, 517
122 James, E.R. and Farrant, J. (1976). *Cryobiology* **13**, 625
123 Mazur, P., Miller, R.H. and Leibo, S.P. (1974). *Journal of Membrane Biology* **15**, 137
124 Daw, A., Farrant, J. and Morris, G.J. (1973). *Cryobiology* **10**, 126
125 Jackowski, S.C. and Leibo, S.P. (1976). *Cryobiology* **13**, 646
126 Heber, U., Tyankova, L. and Santarius, K.A. (1971). *Biochimica et Biophysica Acta* **241**, 578
127 Heber, U., Tyankova, L. and Santarius, K.A. (1973). *Biochimica et Biophysica Acta* **291**, 23
128 Santarius, K.A. (1971). *Plant Physiology* **48**, 156
129 Finkle, B.J. (1976). *Plant Physiology* **57**, 34
130 Finkle, B.J. and Ulrich, J.M. (1978). In *Plant Cold Hardiness and Freezing Stress,* p. 373. P.H. Li and A. Sakai (eds.). New York: Academic Press
131 Fishbein, W.N. and Winkert, J.W. (1977). *Cryobiology* **14**, 389
132 Duppel, W. and Dahl, G. (1976). *Biochimica et Biophysica Acta* **426**, 408
133 Asahina, E. (1978). In *Plant Cold Hardiness and Freezing Stress,* p. 17. P.H. Li and A. Sakai (eds.). New York: Academic Press
134 Withers, L.A. and Davy, M.R. (1978). *Protoplasma* **94**, 207
135 Luyet, B.J. and Gehenio, P.H. (1938). *Biodynamica* **42**, 1
136 Malek, L. and Bewley, J.D. (1978). *Plant Physiology* **61**, 334
137 Sun, C.N. (1959). *Botanical Gazette* **119**, 234
138 Kessler, E. (1974). *Archives of Microbiology* **100**, 51
139 Koehler, J.K. and Johnson, L.K. (1969). *Cryobiology* **5**, 375
140 Osborne, J.A., Morris, G.J. and Lee, D. (1973). *European Journal of Biochemistry* **35**, 445

141 Stout, D.G., Steponkus, P.L., Bustard, L.D. and Cotts, R.M. (1978). *Plant Physiology* **62**, 146
142 Mazur, P. (1963). *Journal of General Physiology* **47**, 347
143 Mansoori, G.A. (1975). *Cryobiology* **12**, 34
144 Lovelock, J.E. (1953). *Biochimica et Biophysica Acta* **11**, 28
145 Lovelock, J.E. (1954). *Biochemical Journal* **56**, 265
146 Elford, B.C. and Walter, C.A. (1972). *Cryobiology* **9**, 82
147 Taylor, M.J., Walter, C.A. and Elford, B.C. (1978). *Cryobiology* **15**, 452
148 Farrant, J. (1969). *Nature* **222**, 1175
149 Sakai, A. (1960). *Nature* **185**, 392
150 Sakai, A. (1966). *Plant Physiology* **41**, 1050
151 Sakai, A. and Otsuka, K. (1967). *Plant Physiology* **42**, 1680
152 Sakai, A. and Yoshida, S. (1967). *Plant Physiology* **42**, 1695
153 Moor, H. (1971). *Philosophical Transactions of the Royal Society, Series B* **261**, 121
154 Moor, H. and Riehle, U. (1968). *Proceedings of the 4th European Conference on Electron Microscopy* **2**, 33
155 Luyet, B.J. and Keane, J. (1955). *Biodynamica* **7**, 281
156 Farrant, J., Walter, C.A., Lee, H. and McGann, L.E. (1977). *Cryobiology* **14**, 273
157 McGann, L.E. and Farrant, J. (1976). *Cryobiology* **13**, 269
158 Morris, G.J., Coulson, G. and Clarke, A. (1979). *Cryobiology* (in press)
159 Walter, C.A., Knight, S.C. and Farrant, J. (1975). *Cryobiology* **12**, 103
160 Withers, L.A. (1978). *Protoplasma* **94**, 235
161 Cella, R., Sala, F., Nielsen, E., Rollo, F. and Parisi, B. (1978). *Abstract of the Meeting of the Federation of European Societites of Plant Physiology*, p. 127
162 Palta, J.P., Levitt, J. and Stadelmann, E.J. (1977). *Plant Physiology* **60**, 398
163 Levitt, J. (1962). *Journal of Theoretical Biology* **3**, 355
164 Chen, P.M., Gusta, L.V. and Stout, D.G. (1978). *Plant Physiology* **61**, 878
165 Dexter, S.T. (1956). *Advances in Agronomy* **8**, 203
166 Siminovitch, D., Therien, H., Wilner, J. and Gfeller, F. (1962). *Canadian Journal of Botany* **40**, 1267
167 Palta, J.P. and Li, P.H. (1978). In *Plant Cold Hardiness and Freezing Stress*, p. 93. P.H. Li and A. Sakai (eds.). New York: Academic Press
168 Shechter, E., Letellier, E. and Gulik-Krzywicki, T. (1974). *European Journal of Biochemistry* **49**, 61
169 Blok, M.C., van Deenan, L.L.M. and de Gier, J. (1976). *Biochimica et Biophysica Acta* **433**, 1
170 van Zoelen, E.J.J., van der Neut-Kok, E.C.M., de Gier, J. and van Deenan, L.L.M. (1975). *Biochimica et Biophysica Acta* **394**, 463
171 de la Roche, I.A., Pomeroy, M.K. and Andrew, C.J. (1975). *Cryobiology* **12**, 506
172 Singh, J., de la Roche, I.A. and Siminovitch, D. (1975). *Nature* **257**, 699
173 Siminovitch, D., Singh, J. and de la Roche, I.A. (1975). *Cryobiology* **12**, 144
174 Greenway, H. (1970). *Plant Physiology* **46**, 254
175 Greenway, H. (1974). *Australian Journal of Plant Physiology* **1**, 247
176 Matile, P.H. (1975). The lytic compartment of plant cells. *Cell Biology Monographs 1*. Wien: Springer Verlag
177 Yoshida, S. (1974). *Contributions from the Institute of Low Temperature Science, Series B* **18**, 1
178 Yoshida, S. (1978). In *Plant Cold Hardiness and Freezing Stress*. P.H. Li and A. Sakai (eds.). New York: Academic Press
179 Yoshida, S. and Sakai, A. (1974). *Plant Physiology* **53**, 509
180 Gusta, L.V. and Weiser, C.J. (1972). *Plant Physiology* **49**, 91
181 Pitt, D. (1978). *Planta* **138**, 79
182 Sukumaran, N.P. and Weiser, C.J. (1972). *Plant Physiology* **50**, 564
183 Siminovitch, D. and Chapman, D. (1971). *FEBS letters* **16**, 207

CHAPTER TWELVE

Practical Aspects

J Farrant and M J Ashwood-Smith

Introduction

In this chapter several general practical points are first considered. Subsequently, observations relevant to the practical preservation of the different systems described elsewhere in the book are put forward. It is not intended that this chapter will provide protocols and recipes that should be followed rigorously. Instead, it is hoped that the information will help individuals to design their own procedures.

General points

Cooling rate techniques

The initial step is to choose one or several cryoprotective agents and establish the tolerance of the cell type to these compounds at room temperature or 0°C. Chemical toxicity will be less at 0°C than at room temperature. Usually the addition of cryoprotective agents may be made in a single step. Since the total time of exposure of the cells to additive both before and after freezing may be kept to a minimum and often can be less than 10 minutes the time of exposure used in the toxicity trial should be of this order. The removal of the cryoprotective agent before testing the cells for function requires care; some of the factors are: slow dilution is better than rapid; dilution at higher temperatures (+20°C or + 37°C) is better than at 0°C; and serum in the diluting medium is beneficial.

Once a toxicity study has been done the cells can be frozen using as high a concentration as possible of the chosen protective agent that can be used without incurring toxic injury. As soon as possible after the addition of cryoprotective agent the cells may be cooled. If the agent penetrates the cells (e.g. dimethylsulphoxide or glycerol) the subsequent conditions of cooling and rewarming may differ depending upon whether time has been left for permeation.

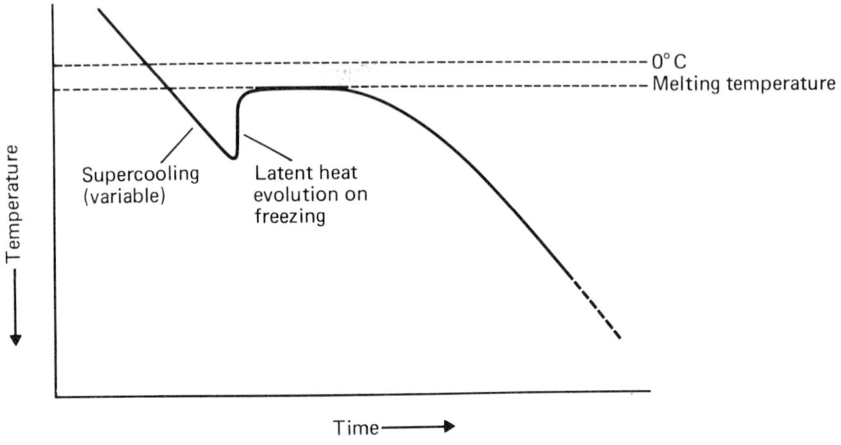

Figure 12.1 Diagram of cooling curve showing freezing plateau

Figure 12.1 illustrates a typical cooling curve showing variable supercooling and the rise in temperature on freezing. Most methods of controlling the rate of cooling whether homemade or commercial do not allow the samples to be nucleated simultaneously. There is little evidence to indicate that either seeding (ice nucleation) or 'manipulation' of the latent heat of ice phase has much influence on viability. If required, however, the cells can be cooled in a constant temperature bath at about 1°C below the melting point of the medium containing cryoprotective agent and nucleated with an ice crystal in the tip of a sterile Pasteur pipette. Alternatively, cooling cells in a bath set at about 3°C below the melting point allows freezing to be induced by tapping the vials. The frozen samples can then be transferred to the cooling equipment which has been arranged to be at the same temperature. Cooling over a range of cooling rates is then performed (see later for discussion on methods of cooling). Initially, rates of between 0.3 and 10°C/min can be chosen. Once cooling has reached −50 or −60°C samples can be plunged directly into liquid nitrogen. Usually, thawing may be done rapidly by agitation of the sample in a + 37°C water bath. Thawing rates are not only controlled by the temperature of the rewarming bath but also by the volume of the sample and nature of its container. In most instances survival improves with rapid rewarming, and as discussed in Chapter 1 rapidly cooled cells are more sensitive to injury if rewarming is slow. Using a single rate of thawing, this approach may indicate a maximal survival at a particular rate of cooling. A reduction in concentration of protective agent often leads to a lower survival than occurs at a more rapid rate of cooling. Sometimes thawing more slowly does improve survival (see Chapter 4). As well as interactions between the rates of cooling and rewarming, another variable of great importance, but often neglected, is the temperature of transfer of the sample from the slow cooling phase directly into liquid nitrogen. An interesting example of this phenomenon is also discussed in Chapter 4.

Two-step techniques

A very similar approach is needed for the two-step method. Both cooling rate and two-step methods require the creation of conditions, during freezing, that shrink the cells without harming them. As before, the tolerance of the cells should be established to a range of concentrations of the cryoprotective agents chosen. Replicate samples of cells with cryoprotective agent can then be held for 5 minutes at different subzero temperatures (perhaps initially at 5°C steps from −15°C to −40°C) before thawing. This can be done in dewars by using a convenient alcohol medium cooled by solid CO_2. Survival estimates of these samples after rapid thawing should indicate the lowest holding temperature that does not reduce survival under these conditions. In a subsequent study the effect of time of exposure at this temperature can be investigated. This can be done by cooling replicate samples in a constant temperature bath or refrigerator (see later). After different times. samples can be thawed directly and rapidly to monitor the onset of time dependent injury to the cells at this subzero holding temperature. At the same times replicate samples at the holding temperature can be plunged directly into liquid nitrogen. Rapid thawing of these samples should indicate initially an increased survival with time at the holding temperature. In ideal circumstances the survival of these plunged samples should approach or equal that of the samples thawed from the holding temperature. In general, if the survival of the plunged samples is always less than that of the other samples, protection is not complete and the holding temperature should be slightly lowered. If, however, injury is severe in both sets of samples, the holding temperature is too low. Increasing the concentration of protective agent usually demands a lower holding temperature, as does a reduction in the rate of re-warming.

Cryoprotective agents

The choice of these is determined primarily by their toxicity to the cell type under investigation and also, of course, to the ultimate use to which the thawed material is put. The traditional classification of cryo-protective agents is into those which can penetrate cells and those which do not. Of the penetrating type the following are the most common: dimethylsulphoxide, glycerol, ethylene glycol, methanol and dimethyl acetamide. Non-penetrating additives used include poly-vinylpyrollidone (PVP), hydroxyethyl starch (HES), dextrans, albumin and polyethylene glycols. Glycerol should be of BP, USP or analytical grade and dimethylsulphoxide preferably of spectroscopic grade. Dextrans of different molecular weights and of a consistently high quality are obtainable from the Pharmacia Company of Sweden. Dextran 70 (mol. wt. 70000 daltons) is recommended. PVP is often of variable quality. If used PVP (mol. wt. 30000) is probably the best choice. It should be dialysed for 24 hours against distilled water or pH 7.0 buffer before use. The best preparations are made by dialysing against distilled water and freeze-drying. The powder thus obtained is readily soluble and lacking in some of the toxic properties possessed by its commercial 'parent'.

There are two preferred ways for adding protective agents to the cells. First, a cell pellet can be resuspended in medium containing the required final concentration of cryoprotective agent. Second, equal volumes of cell suspension can be mixed with medium containing twice the desired final concentration of additive. It is hazardous to add undiluted cryoprotective agent directly to cell suspensions, particularly with DMSO, since there is a positive heat of solution, often severe enough to kill the cells, or cause cells to be lost in a 'gel' of precipitated proteins.

Double strength solutions can be sterilised by Millipore filtration (0.22 μ filter). In some instances addition at +4°C is preferable to addition at 37°C. Serum at a final concentration of 10–15% gives added cryoprotection.

The kinetics of the uptake of protective agents can be followed using conventional radioactive tracer techniques; in addition particularly with DMSO and glycerol, any concentration changes in the bulk medium may be followed by measuring the refractive index.

Cooling and rewarming rates

There are many excellent and sophisticated pieces of commercial equipment that can control the rates of cooling and rewarming of biological samples with great precision. Alternatively, or at least initially, temperature changes can be controlled using simple homemade equipment. Cooling and rewarming rates are best controlled by varying the lagging around the sample vials. One common method of doing this is to place the vials in different polystyrene boxes in a −70°C refrigerator and subsequently cooling the sample vials in liquid nitrogen. Another way is to place the sample vials in a stirred alcoholic solution in a convenient lagged vessel which is itself immersed into liquid nitrogen. The rate of cooling is then determined by a combination of the sample volume and container, the volume of alcoholic solution and the nature of the vessel. A very useful vessel to use for this purpose is an unsilvered dewar (either evacuated or not). Once the samples have reached a sufficiently low temperature, e.g. −40°C to −60°C, they can be removed from the vessel and placed into liquid nitrogen. Similarly, slow rewarming can be controlled simply by transferring the samples from liquid nitrogen to a lagged container on the bench at room temperature.

Another simple and commonly used method of controlling the rate of cooling is to place the sample vials in the neck of a liquid nitrogen storage container. Careful positioning of the sample in the severe temperature gradient present allows some control of the rate of temperature reduction.

Constant temperature environments for nucleating samples or for providing the holding temperature for the two-step method can be done in several ways. For very short periods, 5 minutes or less, an alcoholic solution in a dewar can be cooled with solid CO_2; excess CO_2 can be removed and the solution stirred just before use. For longer periods of time, equipment is required that controls the temperature at a preset value. As with cooling rate equipment, baths of this nature are commercially available. In the two-step method the usual requirements are for the maintenance for a short period (5–10 min) of the sample at the holding temperature. In a well stirred bath samples may reach the bath temperature rapidly and can therefore be plunged in liquid nitrogen 5–10 min after being placed in the bath. Alternatively, they can be placed in a refrigerator, set at the required temperature, and left there for sufficient time to allow them both to cool down to that temperature and then be held there for the required 5–10 min. This may take 30–40 min, but can be reduced if the sample vials are placed into a precooled liquid (e.g. an alcoholic solution or a mixture of equal parts of glycerol and water) in vessels in the refrigerator. It has been noted on many occasions that the protection given to cells at the holding temperature against damage on cooling in liquid nitrogen can be lost very readily if any rewarming is allowed at the time of transfer from the holding temperature into liquid nitrogen.

Measurement of rates

The most commonly used method for measuring rates of cooling and rewarming has been the use of copper–constantan thermocouples using a potentiometric recorder. The simplest reference temperature is provided by melting ice. Table 12.1 indicates the millivolt differences between the junctions induced for copper–constantan with the test junction at various subzero temperatures and the reference junction at 0°C.

Table 12.1 Potential differences (mV) between copper–constantan thermocouple junctions with the reference junction at 0°C

Temperature of test junction (°C)	*Potential difference (mV)*	*Temperature of test junction (°C)*	*Potential difference (mV)*
0	0.00	−100	3.35
−10	0.38	−110	3.62
−20	0.75	−120	3.89
−30	1.11	−130	4.14
−40	1.47	−140	4.38
−50	1.81	−150	4.60
−60	2.14	−160	4.82
−70	2.46	−170	5.02
−79	2.73	−180	5.20
−80	2.77	−190	5.38
−90	3.06	−196	5.48

When rates of temperature change are low, the thermocouple can be placed in the alcoholic medium around the sample vials. If the copper–constantan junction is within the sample, it must be remembered that other samples may behave differently, for example, with varying amounts of supercooling. Another factor is that of the temperature recorded by the thermocouple being influenced more by heat transfer to the junction along the thermocouple wire than by the temperature of the sample itself. Another major problem with the measurement of temperatures is that of temperature gradients within the sample itself. Different samples will behave differently in this respect and recent detailed studies of this effect has been published by Bald [1, 2]. All in all measurement of rates of cooling and rewarming is a subject fraught with difficulty. Although cooling and rewarming rates are often non-linear they are frequently calculated arbitrarily from the time taken for the recorded temperature to pass through two values, e.g. −15°C to −45°C.

Storage equipment

The choice for storage conditions is usually between −70°C to −80°C or −196°C. The higher temperature is provided either by a lagged cabinet containing solid CO_2 or, more frequently, by a −70°C refrigerator. The lower temperature is, of course, that of liquid nitrogen at atmospheric pressure. Although under some conditions, as with high concentrations of glycerol used to preserve red cells (see Chapter 5), a reasonable period of storage can be obtained at −70°C, the most popular and safest storage temperature is that of −196°C. Below the glass transition temperature (−139°C for pure water and at higher temperatures when solutes are present) molecules still vibrate but do not move from one position to another, thus preventing chemical reactions. The use of liquid nitrogen as the storage medium is thus to be preferred. There are excellent containers available for the storage of samples in liquid nitrogen.

Post-thaw handling

Two underlying principles are helpful when considering cell handling after thawing. The first is that this step of the preservation procedure should be considered to be an integral part of the whole protocol. One example given in Chapter 6 is that different cooling conditions affect the requirements for the post-thaw handling. Second, cells are more fragile after thawing. This is perhaps because of an increased sensitivity of partly injured cells together with a requirement for a period of cell repair after the traumas of freezing and thawing.

Assay of function after thawing

Several pitfalls occur in this area. Conditions worked out for the assay of control unfrozen material may no longer apply to the same material after thawing. As already mentioned an alteration to the freezing and thawing procedure may change the requirements for a meaningful assay technique. One logical way of considering the assay of thawed material is that it too forms an integral part of the whole procedure. An example is given for lymphocytes in Chapter 6.

Interaction of variables

A final general hint in the design of preservation procedures is that of the interaction of all variables. Optimisation of a factor such as the rate of cooling giving maximum survival may need to be reconsidered if any other condition in the total procedure is altered.

Practical hints relevant to specific systems

Enzymes

Some enzymes are inactivated by freezing and thawing, others are not. Catalase (IUB 1.11.1.6) is damaged by freezing and thawing and this damage is preventable by the addition of small amounts (1–3%) of dextran, glycerol, DMSO or PVP. A number of other 'frost sensitive' enzymes are protected by these additives.

The higher the initial enzyme concentration the less is the enzyme inactivated by freezing and thawing. The presence of a specific co-enzyme, or an analogue, will give added stability.

Molecular hybridisation caused by protein subunit disaggregation and reaggregation associated with the freezing and thawing of mixtures of different lactic acid dehydrogenases (IUB 1.1.1.27) such as beef muscle and heart enzymes can be prevented by ethylene glycol, propylene glycol, glycerol, DMSO and sucrose.

Tissues and organs

Although there is some controversy regarding the need for intra-cellular penetration by cryoprotective agents such as glycerol or DMSO there is no question that extracellular fluids must be replaced by these agents prior to cooling. Thus proper perfusion and equilibration is essential. Whether this is done at +4°C or 37°C must be established. Equilibration time is easily established by weighing, or in cellular systems, by measuring changes in cell volume. Both DMSO and glycerol can be measured at concentrations between 1 and 20% in balanced salt solutions by refractive index techniques. The use of microwave thawing, if carefully controlled, may be a useful procedure for some systems (Chapter 2). With single cell suspensions, however, it is no more or less effective than conventional warming methods. A series of viability assays appropriate to the cell or tissue must be used. If possible assays based on the characteristic functions of the cells should be used and examples of this are illustrated in Chapter 2 and in several of the other chapters concerned with the cryopreservation of specific cell types.

Spermatozoa

If faced with the prospect of trying to preserve spermatozoa of a species that had not previously been examined, the following approach might be adopted.

Effect of holding time: Since the spermatozoa of some species acquire increased resistance to freezing and thawing during incubation *in vitro*, the effect of varying the interval between collection of semen and freezing should be examined. This might be affected by temperature of incubation, presence of seminal plasma (i.e. semen of some species might be collected in separate fractions or spermatozoa separated by centrifugation), or, if the spermatozoa are diluted immediately, the composition of the medium might have an effect (i.e. inclusion of sugars, egg yolk components, etc.)

Dilution and composition of diluting media: There is no guide as to the best diluent for the spermatozoa of any particular species and one which supports optimal motility during storage *in vitro* may not necessarily be the best for their preservation during freezing and thawing. Osmolarity, ionic strength and pH are important and spermatozoa generally tolerate hypertonic media rather better than hypotonic ones. Media buffered to a pH of 6.5–7.2 have been commonly used, but seminal plasma and egg yolk may provide sufficient buffering capacity. Very simple media might be examined first, e.g. egg yolk–citrate, but in some species better results might be obtained in media of low ionic strength and in which osmolarity is increased by non-permeating solutes such as sugars, e.g. egg yolk-lactose. TRIS buffered diluents have also been widely used or ones in which TRIS forms an important osmotic component of the diluent in combination with sugars. The important protective action of components of egg yolk might also be provided by other substances such as milk or casein. Since there are complex interactions between diluent components, experiments relating to their effects on spermatozoa during freezing and thawing should be of a factorial design. Media containing sugars, when used as the basic diluent for spermatozoa, may provide considerable protection under some conditions of freezing and thawing, but other cryoprotective agents should be examined. Glycerol is an obvious candidate, but others might be better or equally effective. Nevertheless, the permeability of spermatozoa to different solutes and the concentrations that can be tolerated vary considerably between species and these factors are major considerations in diluent composition. The temperature and speed at which addition of various media is carried out may also affect viability of the spermatozoa as may the interval between dilution and freezing. The extent to which spermatozoa are diluted can in some circumstances affect survival.

Cooling rate: Very useful information can be obtained from the two-step freezing type of experiments. Survival after freezing to various temperatures in the first step and the effect of holding time at these temperatures should be checked. For uninterrupted freezing at various rates, liquid nitrogen vapour can be conveniently used either in vessels in which the rate of cooling is controlled automatically or samples may be suspended at different levels above the surface of liquid nitrogen. The plastic 'mini-straw' is a very good container for semen samples because variations in cooling rate within the thin column of semen are reduced to a minimum. 'Pelleting' is another simple method of freezing. The small drops of semen frozen in indentations on the surface of solid CO_2 can be stored in liquid nitrogen.

Thawing rate: The effects of different thawing rates should always be examined.

Interactions: Since important interactions between different factors are to be expected, for example between freezing rate and composition of medium, it is best to vary several factors within each experiment.

Tests for survival *in vitro*: The motility and morphology of the spermatozoa after thawing have been the characteristics most commonly examined. Maintenance of motility during incubation after thawing may give more information than just one examination immediately after thawing. If a thawing diluent is added, the composition of this medium may also affect survival. Various organelles within the spermatozoa may be affected in different ways during freezing and thawing. In some species the acrosome has been found to be particularly sensitive and morphological changes are often detectable by simple light microscopy. Other tests that might be useful are permeability to certain dyes such as eosin or measurements of enzyme leakage.

Fertilising capacity: There is no test or even combination of tests *in vitro* that can be considered as a reliable index to fertilising capacity. If possible, fertilising capacity should be examined in combination with other tests, before as well as after freezing and thawing. Factors which might affect fertility are the number and concentration of spermatozoa used and the time of insemination relative to the time of ovulation. Useful information may also be obtained by insemination into different regions of the female reproductive tract. Differences in the fertilising capacity of spermatozoa following two separate treatments can best be determined by competitive fertilisation experiments. One population of spermatozoa in the mixed semen sample used for insemination are marked in such a way that the eggs fertilised by these spermatozoa can be identified.

Embryos

Specific details for collecting and handling embryos *in vitro* are described adequately elsewhere ([3, 4] and see references 9, 10, 11 of Chapter 4). So, too, are details of culture and transferring embryos to foster mothers (see references 9, 10 and 11 of Chapter 4). The procedures described below are based on the methods currently used for mouse embryos but, with minor modifications, are similar to those used for other species.

Embryos are collected from superovulated or naturally mated females at specific times after ovulation depending upon the embryonic stage required for storage. They are collected in PB1 medium (for details of its composition see reference 14 of Chapter 4) contained in an embryological watchglass. After washing through several changes of PB1 (2 ml/wash) they are transferred either to 5 ml test tubes or 2 ml glass ampoules containing 0.15 ml of PB1 medium. Between 10 and 30 embryos are transferred to each ampoule or tube. The samples are placed in an icebath at 0°C for 15 minutes before adding an equal volume of PB1 medium contain 3 M DMSO also at 0°C. The final concentration of DMSO is 1.5 M. The glass ampoules are heat sealed. After a further 15 minutes at 0°C the embryos are transferred to a 'seeding' bath at −9°C and ice nucleation is induced 2–3 minutes later by sharply tapping each sample.

Immediately the samples are transferred to a cooling vessel at −6°C and slow cooling commenced at 0.3–0.8°C/min. One possible cooling apparatus consists simply of an inner evacuated unsilvered glass dewar (900 ml capacity, 85 mm o.d. × 285 mm long) containing 500 ml of constantly stirred ethanol in which the samples are placed and an outer 4 litre evacuated silvered dewar containing liquid nitrogen. By varying the depth to which the inner dewar is immersed, the cooling rate can be accurately controlled to temperatures as low as −100°C. The temperature of the cooling bath is monitored by means of a copper–constantan thermocouple attached to a temperature recorder. A thermocouple is also placed in a blank sample to give a more accurate recording of the cooling rate of the samples within the bath. At −40°C or −80°C the samples are transferred directly to liquid nitrogen for storage.

For recovery samples transferred at −40°C to liquid nitrogen are thawed rapidly by agitation in a +37 to +40°C water bath (450–500° C/min) and samples transferred at −80°C are thawed slowly by placing samples on the bench at room temperature (20°C/min). Once the ice has melted all samples are left at room temperature for 5 minutes before rapidly diluting the DMSO (fivefold). Finally, they are recovered from the ampoule or test tube in an embryological watchglass. The ampoule or test tube is washed several times with approximately 1 ml of medium. After recovery, the embryos are washed through several changes of PB1 and the number of normal, damaged and degenerate embryos recorded for each sample. The embryos are usually placed in culture for 24–48 hours to assess viability and then transferred to suitably prepared foster mother if live progeny are required.

Red blood cells

Many blood banks use frozen blood routinely with either a 'low glycerol' or a 'high glycerol' process. In the sections following the essential steps of these processes are summarised.

The 'low glycerol–rapid cooling' procedure: The low glycerol–rapid cooling procedure which The New York Blood Center uses has been described in the literature (Chapter 5, references 32, 95). A summary of the essential steps and recent modifications of the low glycerol-rapid cooling technique for preservation of red cells for transfusion is described below.

Collection: After collection of a unit of blood in ACD or CPD anti-coagulant, the packed red cells are prepared and the remainder of the unit is used for component fractionation of platelets, leucocytes, and various proteins such as cryoprecipitate (factor VIII), albumin, gamma-globulin and fibrinogen. In order to preserve optimal concentrations of essential metabolites, i.e. ATP, 2,3-DPG, Na/K ratio, etc. red cells should be frozen as soon after collection as possible, preferably before they are five days old.

Glycerolisation: The packed red cells are glycerolised by adding directly, using a transfer set, with manual agitation, an equal volume (by weight) of glycerol freezing solution to achieve a final red cell concentration of 14% (v/v). Each 100 ml of glycerol freezing solution contains 28 ml (v/v) (equivalent to 35 g (w/v)) glycerol, 3 g mannitol and 0.65 g NaCl.

Freezing: The glycerolised red cells are then transferred into a UCAR-Hemoflex cryogenic freezing bag, which has replaced the stainless steel container formerly used for blood freezing.

Before freezing, a label 'Red Blood Cells (Human) Frozen' is affixed to the blood freezing bag. The label contains the donor number, freezer location, ABO and Rh types, serology and hepatitis test results, and expiration date. (Note: conventional PVC plastic bags *cannot* be used, as they become extremely brittle and will crack when frozen in liquid nitrogen). After the plastic neck tube entrance port is heat sealed, the blood freezing bag is placed between two metal plates (holders) to maintain the bag in a flat configuration, similar to the geometrics of the stainless steel freezing container originally used in this process. The bag of blood between the two plates is then frozen by direct and complete immersion into liquid nitrogen (−196°C) without agitation. A metal retaining device (which is clamped around the holder) should be used to prevent excessive bulging of the blood container during freezing. Complete freezing is achieved in 2.5–3 minutes when the liquid nitrogen ceases to boil.

Storage: The frozen unit within its metal holder is stored either in the vapour (−165°C) or liquid phase (−196°C) in a liquid nitrogen refrigerator. Blood stored in the frozen state should be organised for ready retrieval based on blood group information, e.g., rare blood, 'fully typed' blood, partially typed (ABO, Rh) blood, autotransfusion, 'hold' status (for particular patient or hospital), etc. Frozen blood has been stored successfully in liquid nitrogen for more than ten years with full retention of viability.

In the United States, 'Red Blood Cells (Human) Frozen' can be stored in liquid nitrogen for no longer than three years from the date the blood was collected and frozen. However, the FDA-imposed three-year storage limitation may be extended in the case of rare units or those types of blood which are difficult to obtain.

Thawing and deglycerolisation: Rapid thawing is accomplished by gentle agitation of the frozen unit in a warm water bath (42–45°C) for 2.5 min. The post-thaw washing protocol for deglycerolisation employs volumes of 300–500 ml of a 3.5% buffered NaCl solution or a 15% mannitol–saline solution (Wash 1), followed by 1 or 2 litres of physiological saline preferably containing glucose (e.g. 0.9% NaCl or 0.9% NaCl–0.2% Dextrose) (Wash 2 and 3). Wash volume is dependent on the method of wash. The use of concentrated (3.5%) sodium chloride solution for Wash 1 has been introduced into the technique to replace the 15% mannitol solution originally reported by Rowe *et al.* (Chapter 5, references 11, 23, 151).

Glycerol removal from the thawed cells can readily be accomplished by any of the following manual or automated methods of washing:

Manual serial centrifugation: Batch washing of thawed blood, although somewhat time consuming and laborious, is still the most simple, reliable and inexpensive way of removing glycerol from the cells. After the thawed blood is transferred into a dry quadruple plastic bag, it is centrifuged to remove the supernatant-containing free haemoglobin, glycerol, and cell debris. The three separate washes are then introduced aseptically via the access ports of the satellite packs using volumes equivalent to or slightly greater than the amount of supernatant removed.

Automatic serial centrifugation (IBM): An automated Blood Cell Processor (IBM-Model 1991) which utilises the principle of 'manual' batch washing has recently been introduced for use in red cell deglycerolisation. The thawed blood is drained directly into a special cell washing pack. The Blood Cell Processor is then programmed to centrifuge, decant the supernatant, and introduce the washing solutions.

Continuous flow centrifugation (Elutramatic): A continuous flow washing apparatus, the Elutramatic, has also been used for deglycerolisation and is programmed to remove glycerol, free haemoglobin, and cell debris. The Elutramatic allows for the washing of two units of thawed blood at the *same time.* This apparatus, however, no longer is manufactured and thus is of historical interest only.

Continuous flow centrifugation (Haemonetics–Latham bowl): Two types of continuous flow cell washing units, a gravity-flow-feed (Haemonetics Models 15 and 115) and a programmed unit (Model 10) also can be used for washing low glycerol frozen blood. Solutions are introduced through a plastic harness. Predilution of the thawed unit with 500 ml of Wash 1 (3.5% sodium chloride, 15% mannitol–saline) is preferred to a direct wash with this solution. The diluted thawed blood is then fed into the Haemonetics wash bowl and deglycerolised by washing with 2 litres of 0.9% saline with or without 0.2% dextrose.

Post-thaw storage and outdating: The thawed-washed unit of blood should be stored at 4°C, and transported on wet ice in the same manner as fresh blood. However, in the United States, a thawed-washed unit of blood *must* be used within 24 hours after thawing. This limited outdating period is regulated by the FDA Bureau of Biologics. The label on the thawed-washed cells, i.e., 'Red Blood Cells (Human) Deglycerolised' must indicate the date and time of outdating, which is 24 hours from the time of thawing. Although refreezing of thawed blood is not recommended, it is possible to do so if indicated (e.g. in the case of rare blood etc.).

The 'high glycerol–slow cooling' process: The high glycerol–slow cooling process by which the American Red Cross prepares red cells for freezing and deglycerolises them prior to transfusion has been described in detail (Chapter 5, references 82, 84). The centrifugal method of Tullis (Chapter 5, references 19, 72) forms the basis of the process developed by the American Red Cross. A summary of (1) the standard protocol for glycerolising and deglycerolising using the Haemonetics or IBM cell processors; (2) the modified protocol using these two processors; and (3) the manual procedure using a clinical centrifuge is presented below.

Collection: Approximately 250 ml of packed cells are obtained from a unit of ACD or CPD whole blood. The packed cells are allowed to reach a temperature between ambient and 37°C prior to glycerolisation, since glycerol permeation across the red cell membrane is markedly temperature dependent.

Glycerolisation:

Standard protocol. (Chapter 5, reference 82): The collection bag is suspended from a shaking device and connected through plastic tubing with the storage bag and the glycerol solution. To the cells is added 400 ml of a solution containing 6.2 M glycerol, 0.14 M sodium lactate,

5 mM potassium chloride, and 5 mM sodium phosphate (pH 6.8) in the following way. Of the 400 ml of solution, 100 ml are added with simultaneous shaking and the cell suspension is then allowed to equilibrate for at least 5 minutes, after which the remainder of the glycerol solution is added without shaking.

Modified Protocol. (Chapter 5, reference 84): The red cells are glycerolised as in the standard protocol but are transferred to an 850 ml freezing bag instead of the larger 2 litre bag. The glycerolised cell suspension is subsequently centrifuged with the collection bag still attached. After sedimentation the supernatant solution is transferred back into the collection bag and the collection bag then discarded. The glycerolised cells are then frozen in the form of packed cells.

Freezing: The glycerolised cells are transferred to a 2 litre freezing bag (UCAR–Hemoflex). The transfer tube containing glycerolised cells is sectioned at intervals to provide samples from which the blood may be cross-matched prior to the thawing of the entire unit. These segments are stored in a separate compartment of the aluminium storage canister in which the UCAR bag is maintained. The freezing bag (in its canister) is placed in a −80°C (mechanical) freezer where the cells will cool to this temperature over a period of 6–12 hours.

Storage: Storage of the frozen units is accomplished under mechanical refrigeration which should not allow the storage temperature to rise above −80°C.

Thawing and deglycerolisation:

Standard protocol. (Chapter 5, reference 82): The frozen cells are thawed by direct immersion of the storage canister into a stirred water bath at 37°C. Following thawing, the storage bag is removed from its aluminium holder and the thawed cell suspension diluted with 150 ml of 12% NaCl solution. The cells are deglycerolised further using either the Haemonetics Cell Washing System or the IBM Cell Processor.

(1) *Continuous flow centrifugation (Haemonetics–Latham bowl)*– After a 3 minute equilibration with the 12% NaCl solution, the cell suspension is pumped into a rotating centrifugal Cohn–ADL–Latham washing bowl (Haemonetics Models 10, 15, 102 or 115). Simultaneously, a 1.6% NaCl solution is pumped into the bowl. The cells are retained in the bowl by the centrifugal force, and the

excess supernatant leaves the bowl and is collected in a waste bag. When all the cells have been introduced into the bowl, the 1.6% NaCl solution continues to be pumped through the cells, progressively washing away the glycerol solution. After a wash with 1–2 litres of 1.6% NaCl, the wash is continued with 1 litre of 0.9% NaCl–0.2% dextrose solution.

At the end of the wash, the cells are suspended in a sufficient volume of saline-glucose solution to obtain a haematocrit of 35–55%, depending on the size of the original unit. The cell suspension is siphoned from the bowl into a conventional blood bag which has been connected to the disposable washing set. The transfer tubing to this blood pack is segmented to provide samples for cross-match.

(2) *Serial Centrifugation (IBM)*–After the 3-minute equilibration with the 12% NaCl solution, approximately 600–650 ml of 1.6% NaCl is added to the cell suspension in the freezing bag. The mixture of cells and concentrated NaCl solution is introduced into the washing bowl and concentrated in two sedimentation cycles. The cells are then washed in one cycle with 1.6% NaCl and two cycles with 0.9% NaCl–0.2% dextrose solution.

Modified Protocol (Chapter 5, reference 84). After thawing, the freezing bag is removed from the metal canister and the thawed cell suspension diluted with 100 ml of 8.5% NaCl solution with vigorous mixing. After 3 minutes for equilibration, 0.9% NaCl–0.2% glucose solution is added from a 2 litre container with vigorous mixing until the bag is filled.

Following dilution, the contents of the bag can be deglycerolised using either the Haemonetics Cell Washing System or the IBM Blood Cell Processor or by means of a manual procedure which requires two sedimentation cycles in a clinical centrifuge. These procedures are summarised below and described in more detail elsewhere (Chapter 5, reference 84).

(1) *Haemonetics (e.g. Model 102 or 115)*–The diluted cell suspension is transferred into the centrifugal bowl while the centrifuge is in operation. When all the cells have drained into the bowl, they are washed with at least 1 litre of the remaining 0.9% NaCl–0.2% dextrose solution.

(2) *IBM (e.g. Model 2991)*–The diluted cell suspension is introduced into the cell washer in two cycles followed by two additional wash cycles using 0.9% NaCl–0.2% dextrose solution.

(3) *Manual deglycerolising*–The diluted cell suspension is sedimented

in the freezing bag at 4100 *g* for 6 minutes at room temperature, the supernatant solution is expelled with a plasma extractor, and the bag refilled with 0.9% NaCl–0.2% dextrose solution. The sedimentation and removal of supernatant is repeated once.

The 'Cytoagglomeration' Process. The 'cytoagglomeration' process by which Huggins of the Massachusetts General Hospital prepares red cells for freezing and subsequently removes the glycerol cryoprotectant is based on the natural phenomenon of reversible agglomeration, a process by which red cells may be made to clump and settle in the presence of sugars and in the absence of electrolytes, and eliminates the need for centrifugal equipment to add or remove the cryoprotectant from the cells.

Collection and Glycerolisation. Packed red cells prepared from blood collected in ACD or CPD solution are placed in a special elongated disposable Blood Freezing Unit. After determining the weight of the packed cells, a measured amount of glycerol solution is added and mixed with a magnetic stirrer. The glycerol freezing solution for ACD-collected blood contains 8.6 M (79% w/v) glycerol, 8.0% w/v dextrose, 1.0% w/v fructose and 0.3% w/v Na_2 EDTA. The glycerol freezing solution for CPD-collected blood contains 8.6 M glycerol, 6.0% w/v dextrose, 1.0% w/v fructose, 0.3% w/v Na_2 EDTA and 0.0006% w/v citric acid.

Freezing: After heat sealing the inlet tube, the Blood Freezing Unit is laid flat for even distribution of the glycerolised cells, after which the unit is folded and placed in a box which is then laid horizontally in a −85°C (mechanical) freezer for 24 hours.

Storage: The boxes containing the frozen units of blood are stacked vertically for easy retrieval in a mechanical freezer at −85°C. Frozen blood cells may also be packed on solid CO_2 for shipping purposes.

Deglycerolisation:

Cytoglomerator method: To prepare the blood for transfusion, the blood freezing unit is removed from its box and thawed in a warm water bath for 7 minutes. After stripping the bag to get the thawed red cells to one end, the blood freezing unit is placed on the operating station of a Huggins cytoglomerator.

Washing consists of the introduction to the thawed cells (in the presence of a magnetic stirrer), of a diluent solution of 50% aqueous glucose bottled in 2 litre quantities.

In an aqueous non-electrolyte solution such as this, with a pH between 5.2 and 6.1, clumping (or agglomeration) takes place when the magnetic stirrer is stopped. The cells then rapidly settle to the bottom of the container and the supernatant may be decanted. This phenomenon is based on the fact that in the natural state, gamma-globulins and red blood cells exist together in the plasma as separate entities. When the pH range is lowered to between 5.2 and 6.1, however, it is believed that a reversible complex is formed between the gamma-globulins of the plasma and the lipoproteins of the red cell membrane. If the ionic strength of the slightly acidic medium is lowered further, the globulins precipitate, causing coprecipitation (or agglomeration) of the adherent red blood cells. It is a fortuitous coincidence that the ACD solution in which the blood was originally collected combines with the autoclaved sugar solution and results in a solution which has the optimum pH and ionic strength for agglomeration (Chapter 5, references 66, 86). In this way the concentration of glycerol in the thawed red cells is gradually reduced in a manner which minimises osmotic lysis of the cells.

After repeating the agglomeration and decantation steps several times, the red cells are resuspended either by the addition of an electrolyte (e.g. 0.9% NaCl) which breaks the globulin–globulin bond, or by raising the pH which breaks the globulin–red cell bond, rendering the washed cells suitable for transfusion.

Continuous flow centrifugal method: Hornblower and Meryman (Chapter 5, reference 85) have reported that cells glycerolised in a low salt medium (Huggins) and deglycerolised by agglomeration could also be deglycerolised by a centrifugal procedure using a high salt procedure and a Haemonetics disposable wash bowl and apparatus.

Leucocytes

Lymphocytes: Lymphocytes can be stored successfully using either permeating or non-permeating, cryoprotective compounds but the most widely used has been dimethylsulphoxide (DMSO). Adding DMSO to the cells at 0°C rather than at 37°C before freezing will reduce the toxic effects of this compound although little toxicity for human lymphocytes can be seen using 10% DMSO at room temperature for short periods (e.g. 15 min). A variety of storage methods ranging from continuous reduction in temperature (cooling rate) to prefreezing (two-step) have been used successfully. Examples of the former approach would be to take samples (1 ml) of lymphocytes in DMSO (10% v/v) with 20% serum and freeze them at 1°C/min to −60°C before plunging them into liquid nitrogen or alternatively to cool them at 1°C/min to between −30°C and −60°C followed by a faster cooling rate (5–10°C/min) to between −80°C and −150°C before plunging them into liquid nitrogen. For a two-step technique samples with 5% DMSO can be allowed to reach an intermediate holding temperature between −25 and −35°C and then be held for 5–10 min before plunging the samples into liquid nitrogen. Thus 0.25 ml samples of lymphocytes in DMSO (5% v/v) in polypropylene tubes (Sterilin 12 × 35 mm) require to be put in a constant temperature bath at −30°C (e.g. a bath of 50% glycerol in a deep freeze) for 30 min before *rapid* transfer to liquid nitrogen.

For most techniques used fast thawing of the sample in a 37°C water bath is reported to be preferable and for this reason it may be better to have samples of small volume. Dilution of the DMSO from the sample by slow addition to a five- to tenfold volume of warm medium (+20 to +37°C) containing some protein is recommended. Thawed lymphocytes may be more susceptible to mechanical trauma (e.g. high '*g* values' in centrifugation) than normal cells. The interrelationships of these variables and methods of improving poor recovery are discussed more fully in Chapter 6.

Granulocytes: There is no accepted method for successful preservation of functional human granulocytes. Evidence suggests that the best recovery using conventional cooling rate techniques with cells in DMSO (10% v/v) would be with very slow cooling rates, probably less than 0.3°C/min. Alternatively, using a two-step technique 30% of the cells can be recovered in a functional state when a small sample of cells (0.2 ml in 0.5 dram glass vials, Johnson and Jorgenson) are placed at −35°C for 2 hours before being plunged directly into liquid nitrogen. Granulocytes are extremely sensitive to osmotic stresses particularly during dilution out of DMSO but damage is minimised by diluting slowly with warm medium (+20 to +37°C) containing some protein. As mentioned in Chapter 6, granulocytes from some animal species are much easier to preserve.

Bone marrow stem cells

One possible storage protocol is described. Heparinised bone marrow was suspended in Hanks' buffered saline solution (HBSS). Cryoprotective agents were added to make final concentrations of 10% glycerol, 10% PVP, 10% PVP + 10% glycerol, 20% HES, always in combination with 20% serum. With DMSO a 20% concentration should be prepared from a stock solution. The final cell concentration of marrow separated from red cells should be 10×10^6 to 200×10^6 per ml before transfer to the freezing container (glass ampoules, polypropylene tubes, polyolefine bags). Samples should be cooled at a controlled rate of 1–2°C/min to −40°C and then rapidly to −196°C. The samples were stored in liquid nitrogen. Rapid thawing was done using a water bath (+40 to +45°C). The protective agent was removed by dilution in HBSS at room temperature. With glycerol or DMSO the dilution should be slowly stepwise. First a very small amount of HBSS was added which equalled 2% of the thawed suspension volume. Then increasing amounts of HBSS were added to the suspension. Roughly doubling quantities were added each time. After approximately 45 minutes a tenfold dilution should have been obtained. Samples were shaken carefully between each step for 3–5 minutes. During the procedure cells were kept at room temperature.

Cell suspensions were treated with DNAase to avoid gel formation. After centrifugation for 10 minutes at 500 *g* the supernatant was poured off and cells were resuspended in HBSS.

Protozoa and helminth parasites

There are several general points which should be made for anyone attempting to cryopreserve parasites.

(1) Different species, and even different stages or strains of the same species may have very different cooling rate requirements for optimum survival.
(2) The preferred cooling schedule can also be altered by the addition of, and the concentration and type of cryoprotectant present in the suspending medium.
(3) With stepped cooling schedules, the intermediate temperature, and the time for which samples are held at this temperature, are critical, as too are the cooling rates employed either side of this temperature.
(4) The final storage temperature may well be critical.
(5) The warming rate can be as important for survival as the cooling rate.
(6) The specific cryoprotectant used is important, as are the temperature of addition, the time allowed for equilibration and the method by which the compound is removed.

Previous published reports are useful as guidelines, but it may be necessary to re-evaluate all these parameters independently as well as in conjunction, in order to obtain the optimum survival.

Insects and their cells

Cold tolerant insects are readily collected in temperature zones, at high altitudes, or at high latitudes. The Canadian Arctic, for instance, has a rich and varied insect fauna. An appropriate acclimation regimen is an important factor in eliciting maximum cold tolerance, and under laboratory conditions this usually requires a relatively long period (at least 2–6 weeks) at low temperature for completion of the biochemical and physiological adjustments. There is a number of estimates of how long an insect may take to reach its new equilibrium state following step-function transfer to low temperatures, but this probably varies according to the species and the thermal history of the individual. In experimental studies the cooling rate may be critical in ensuring maximal frost tolerance, and should be kept to less than 0.2°C/min if possible. Two-step freezing methods allow some frost tolerant insects to survive relatively long periods of storage under deep-freeze conditions, and the following procedure may serve as a useful general guide:

(1) an acclimation period at temperatures around 0°C or slightly above;
(2) further slow cooling to relatively high subzero temperatures (*c.* −30°C);
(3) rapid freezing to low subzero temperature (*c.* −80°C).

Thawing rate does not appear to be critical for the survival of frozen insects, although it has to be admitted that few detailed studies have been carried out. Rapid thawing to room temperature or in a warming bath at +30 to +40°C is employed in most studies. Unlike acclimation, the process of deacclimation is much more rapid and the benefits of low temperature acclimation are usually lost within 2–6 days of the insect being placed at a higher temperature. Recent evidence indicates that frost tolerant insects may synthesise nucleating agents to ensure freezing at high subzero temperatures.

The cold hardiness and frost tolerance of isolated cells and tissues reflect the degree of tolerance which has been acquired by the donor insect. Frost tolerant cells, therefore, are usually obtained from diapause, overwintering stages of insects that have developed cold tolerance either naturally or by acclimation at low temperatures under laboratory conditions. Further development of cryobiological studies on insect cell systems is limited by the lack of available completely defined growth media which can support insect cell lines indefinitely. Such an advance would have an important impetus in the study of low temperature preservation of insect cells and would also allow closer cooperation among scientists in different parts of the world.

The laboratory techniques in insect cryobiology are not complicated and are similar to those employed in other areas of low temperature biology. Some standard equipment and supplies of major importance would be:

(1) a series of acclimation chambers (10 to −10°C);
(2) a series of low temperature cabinets or rooms (0 to −30°C);
(3) availability of liquid gases for testing viability under deep-freeze conditions.

Cryoprotectants such as DMSO and glycerol can be used for cryopreservation, and cells are usually processed according to the standard techniques used for mammalian tissue culture.

Microorganisms

Most microorganisms and viruses can be preserved for long periods of time without great difficulty and in many instances the simple and well tried methods involving desiccation or freeze-drying are adequate. Often, however, the initial survival levels after the completion of drying by either process are very low, although sufficient to serve as inocula. The titres become increasingly lower with increasing storage time; the lower the storage temperature the longer the 'shelf-life'.

For freeze-drying or desiccation, protective and supportive media are necessary and one that is recommended is a solution containing 5 g of dextran (110000 mol. wt.), 7.5 g of saccharose and 1 g of monosodium glutamate. Survival in this medium will usually be of the order of 5–20% depending on the drying process and residual moisture levels as measured by the Karl Fischer technique (usually between 1 and 3%). The higher the initial cell density the greater is the survival. Bacterial suspensions (0.1 ml) are shell frozen in ampoules (2.5 ml) and freeze-dried over a period of 24 hours. The initial sublimation of ice should take place from about −30°C over a period of several hours and the secondary drying can be conducted at room temperature. A good vacuum is required and a low condensor temperature is essential. Samples should be sealed under vacuum or nitrogen and stored in the dark, at the lowest convenient temperature. Rehydration in distilled water followed by immediate plating or inoculation is recommended. Often damaged cells can recover if given a suitable metabolic milieu.

If survival is low or if selection or mutation is considered a problem then cryopreservation is recommended rather than desiccation or freeze-drying. Bacteria or viruses should be suspended in buffer, broth or distilled water containing either 5–10% DMSO or glycerol. Freezing rates of 1–20°C/min to −40 or −50°C prior to plunging in liquid nitrogen (−196°C) will normally give survival values of nearly 100% when combined with rapid thawing (samples thawed in water at 37°C). This cryopreservation technique has the advantage of being reliable and possessed of no mutagenic hazards. It is also without the problems of decreasing survival with increasing storage time (at −196°C). At −80°C survival will decrease slowly over the years but is quite adequate for routine use.

Whichever preservation method is used some problems may still occur with some very sensitive organisms. If survival values are low or lower than expected, it is advisable to investigate one of more of the following points:

(1) the medium in which the cells are suspended;
(2) cooling and thawing rates;
(3) cell density;
(4) phase of cell growth when harvesting took place;
(5) plating medium after preservation process;
(6) growth conditions (see reference 29 of Chapter 10, Mycoplasma).

Plant cells

Estimation of cellular multiplicity: The frequency of cellular aggregation is determined in at least three samples, 500 cells being counted for each treatment. The multiplicity of the sample is defined as:

$$\text{Multiplicity} = \frac{\text{Number of cells}}{\text{Number of groups}}$$

The total cell number is then:

Number of colonies (as determined by assay in agar) × Multiplicity

Following freezing and thawing there is a breakdown of cell aggregates; typical results are illustrated in Figure 12.2. In this example there is a reduction in multiplicity from 1.21 to 1.09. The recovery following freezing and thawing is 86% if cellular aggregation is taken into consideration or 95% if it is ignored. The extent of the reduction in multiplicity is a factor of cell type, rates of cooling and warming, concentration and type of additive, therefore multiplicity should be determined for each frozen and thawed sample. If it is not

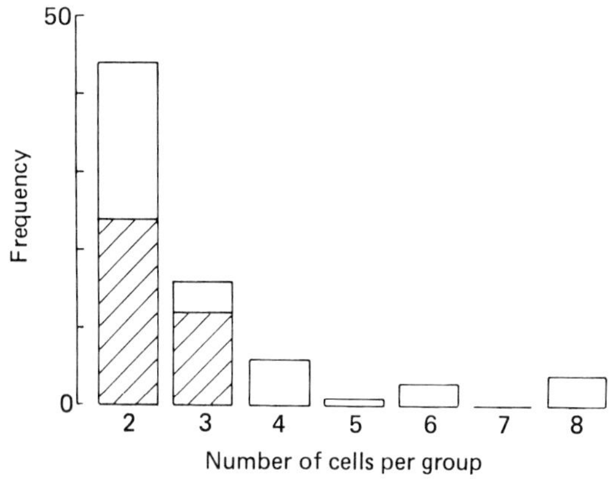

Figure 12.2 Distribution of cellular aggregation in a culture of *Chlorella protothecoides*. Open figures unfrozen, hatched figures following freezing and thawing. The number of single cells was 480 unfrozen, 505 frozen and thawed

taken into consideration then results should be expressed as colony forming units (CFU).

Fluorescein diacetate (FDA) staining: To a dilute solution (0.01% w/v) of freshly prepared FDA (Sigma) is added an equal volume of cells. After a 5 minute incubation at room temperature they are then observed microscopically under tungsten light and the number of cells counted in one field. The light source is then changed to ultra-violet and the percentage of cell survival is estimated by the number of cells showing fluorescence (Figure 12.3). Prolonged exposure leading to non-specific uptake and breakdown should be avoided.

Evans' blue staining: To permit cytological examination of surviving cells, frozen and thawed preparations are first stained with Evans' blue (Searle Diagnostic UK) to mark lethally damaged cells. The stain, in aqueous solution (0.025% w/v) is applied to the cells for 5 minutes and then removed by washing twice with distilled water. A good correlation is observed between Evans' blue exclusion and a fluorescein diacetate positive reaction (Figure 12.3).

Triphenyl tetrazolium chloride (TTC) reduction: After thawing, cells are left at room temperature for 2 hours to minimise non-specific TTC reduction before the assay is started. The cells are then incubated, without shaking, in a 1% (w/v) TTC solution buffered at pH 7.5 for 20 hours at room temperature in the dark. After incubation the cells are pelleted by centrifugation and washed once with distilled water; the formazan is extracted with ethanol (95%) for 30 minutes. Heating

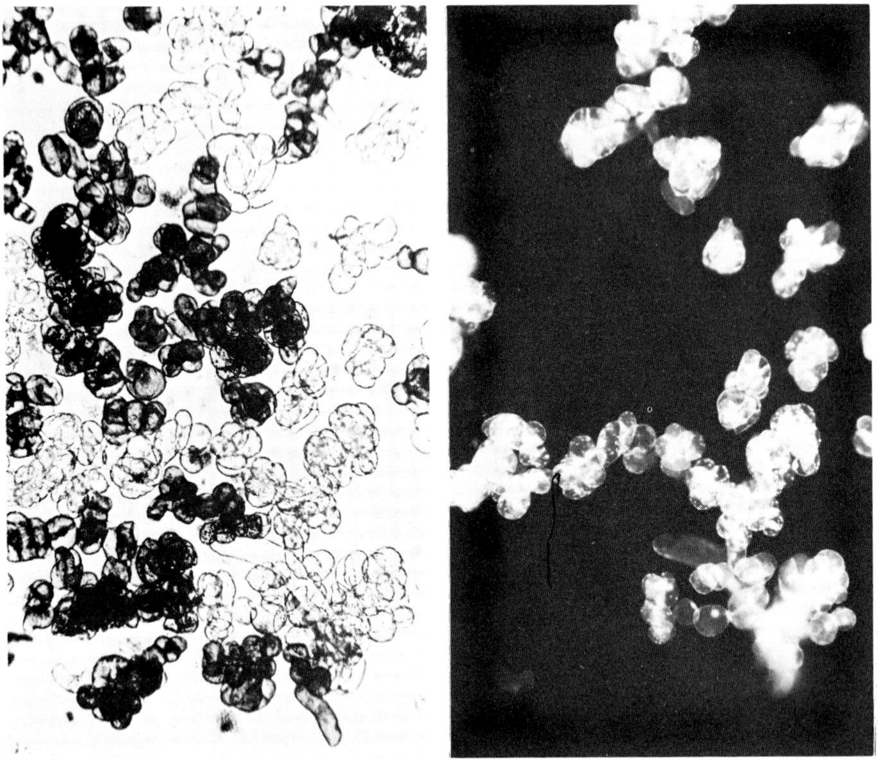
(a) (b)

Figure 12.3 Specific staining for cell viability. A mixture of live and dead suspension culture cells of *Acer pseudoplatanus* was stained with fluorescein diacetate and Evans' blue. Observation (a) under ultra-violet illumination revealed fluorescing live cells, and (b) under bright field tungsten illumination revealed Evans' blue stained dead cells. (From Withers, L.A. (1978). *Cryobiology* **15**, 87, by permission of the author and publishers)

is not necessary to remove formazan from single cells; brief heating (60°C for 5–15 min) extracts formazan from cell clumps. Absorbance of the extract is measured at 485 nm. The assay is calibrated using artificial mixtures of live and heat-killed cells.

References

1 Bald, W.B. (1975). *Journal of Experimental Botany* **26**, 90 and 103
2 Bald, W.B. and Robards, A.W. (1978). *Journal of Microscopy* **112**, 3
3 Whittingham, D.G. (1971). *Journal of Reproduction and Fertility* **14**, 7
4 Rafferty, K.A., Jr. (1970). *Methods in Experimental Embryology of the Mouse.* Baltimore and London: The Johns Hopkins Press

Index

Compiled by William Hill